AF846956

Cyberphysical Systems for Epilepsy and Related Brain Disorders

Nikolaos S. Voros • Christos P. Antonopoulos
Editors

Cyberphysical Systems for Epilepsy and Related Brain Disorders

Multi-parametric Monitoring and Analysis for Diagnosis and Optimal Disease Management

Springer

Editors
Nikolaos S. Voros
Computer & Informatics Engineering Department
Technological Educational Institute of Western Greece
Antirio, Greece

Christos P. Antonopoulos
Computer & Informatics Engineering Department
Technological Educational Institute of Western Greece
Antirio, Greece

ISBN 978-3-319-20048-4 ISBN 978-3-319-20049-1 (eBook)
DOI 10.1007/978-3-319-20049-1

Library of Congress Control Number: 2015942893

Springer Cham Heidelberg New York Dordrecht London

Printed on acid-free paper

Springer International Publishing AG Switzerland is part of Springer Science+Business Media (www.springer.com)

Foreword

Writing books on epilepsy for the past three millennia is well justified since epilepsy is the most common serious chronic neurological disease affecting almost 65 million worldwide. It affects people of all ages, with maximal rates of occurrence in children and in old age. Unlike many other serious diseases, epilepsy is readily treatable with inexpensive medications in the majority of cases. At least 70 % of people with epilepsy can have their seizures controlled and can enjoy a normal life if they are diagnosed and treated appropriately. However, in certain cases, definitive diagnosis of a paroxysmal event cannot be made even in hospital setting despite thorough neurological and EEG assessment. These cases often demand multimodal, personalized, and long-term monitoring, ideally at home and especially during sleep, for proper seizure diagnosis, differential diagnosis, and classification.

At the "International Bureau for Epilepsy," we look forward to efforts aiming to exploit modern technology for the benefit of patients with epilepsy avoiding misdiagnosis, delayed or inappropriate diagnosis.

This book stems from an effort to contribute in facing this problem by exploiting the latest advancements in Information and Communication Technology (ICT).

In the relevant chapters of this book, we read how the above challenge can be approached and in some cases how novel approaches can provide new solutions that nevertheless require clinical validation. The approaches include: (a) developing a holistic, personalized, medically efficient, convenient to the patient, and economical home-monitoring system; (b) managing and analysis of a large number of data from brain and body activities of patients with epilepsy; and (c) allowing to detect, monitor, predict, and make diagnosis remotely. Arguments are given that such a system will facilitate and increase the yield of diagnosis, treatment, and practice, reduce epilepsy-related management costs, and increase understanding of epileptic seizures and their relationship to other non-epileptic paroxysmal events. In parallel, the advances made in ICT are described, i.e., developing novel multi-parametric data processing, management, and analysis tools, both for real-time and off-line monitoring, adopting and adapting of communication platforms providing robust and flexible end-to-end communication, and assuring security and privacy on sensitive medical data.

I consider it a virtue of this book that it reflects a rarely successful but increasingly demanded common understanding and effective close collaboration by blending theory and practice, between academic research laboratories and epilepsy clinics as well as accomplished industrial partners from five European countries. I believe that this book can be easily read by scientists of both epileptology and ICT background and by people generally interested in epilepsy and/or long-term monitoring. The effort will advance their capacity for mutual understanding and effective collaboration.

The book conveys an optimism that advanced home monitoring systems like ARMOR can provide reliable detection of life-threatening seizures and improve the level of care and the quality of life of the patient, while avoiding hospital stays and thus drastically reducing costs. With one more push, this effort can lead to a much-needed tool.

Dr. Athanasios Covanis
President of International Bureau for Epilepsy

Acknowledgments

The research work that provided the material for this book was carried out during 2011–2015 mainly in the ARMOR STREP Project (**A**dvanced multi-pa**R**ametric **M**onitoring and analysis for diagnosis and **O**ptimal management of epilepsy and **R**elated brain disorders) partially supported by the European Commission under the Grant Agreement number 287720.

The guidance and the comments of EU Project Officer *Dr. Amalia Irina Vlad* and EU Reviewers *Mr. Jean Delbeke*, *Mr. Henry Kanoui*, and *Mr. Paulo Mendes* on research direction are highly appreciated.

In addition to the authors, the management teams of the partners participating in the ARMOR consortium are gratefully acknowledged for their valuable support.

Last but not least, the editors express their special thanks to *Mr. Athanasios Covanis, M.D., Epileptologist* and *President of International Bureau of Epilepsy* for his valuable remarks and his willingness to contribute to the final review of the material presented in the book.

Contents

Chapter 1
Introduction to ARMOR Project

Nikolaos S. Voros, Christos P. Antonopoulos, Michalis Koutroumanidis, George K. Kostopoulos, and Andreas A. Ioannides

Abstract Epilepsies affect 1–2 % of general population, especially in childhood and adolescence. Epileptic seizures, manifest with a wide range of paroxysmally recurring motor, cognitive, affective, and autonomic symptoms and EEG changes. Their recognition and full understanding is the basis of their optimal management. The yield of epilepsy diagnosis is considered unsatisfactory, as seizures occur unpredictably and typically outside hospital, other paroxysmal disorders are often misdiagnosed as epilepsy, and hospital evaluation costs of patients with uncertain clinical features or possibly mixed disorders are quite substantial. Reliable diagnosis requires state of the art monitoring and communication technologies providing real-time, accurate and continuous brain and body multi-parametric data measurements, suited to the patient's medical condition and normal environment and facing issues of patient and data security, integrity and privacy.

In this context, a cornerstone objective of the ARMOR project was to manage and analyze a large number of already acquired and new multimodal and advanced technology data from brain and body activities of epileptic patients and controls (MEG, multichannel EEG, ECG, GSR, EMG, etc.) aiming to design a more holistic, personalized, medically efficient and economical monitoring system. New methods and tools have been developed for multimodal data pre-processing and fusion, real-time and offline data mining of multi-parametric streaming and archived data to discover

N.S. Voros (✉) • C.P. Antonopoulos
Embedded System Design and Application Laboratory, Department of Computer and Informatics Engineering, Technological Educational Institute of Western Greece, Patras, Greece
e-mail: voros@teiwest.gr

M. Koutroumanidis
Department of Clinical Neurophysiology and Epilepsy, Guy's and St Thomas' NHS Foundation Trust, London, UK

Department of Academic Neurosciences, Kings College London, London, UK

G.K. Kostopoulos
Neurophysiology Unit, Department of Physiology, Medical School, University of Patras, Patras, Greece

A.A. Ioannides
Laboratory for Human Brain Dynamics, AAI Scientific Cultural Services Ltd., Nicosia, Cyprus

N.S. Voros, C.P. Antonopoulos (eds.), *Cyberphysical Systems for Epilepsy and Related Brain Disorders*, DOI 10.1007/978-3-319-20049-1_1

patterns and associations between external indicators and mental states, lag correlation detection, identification of motifs or outliers (vital signs changing significantly), automatic summarization of results and efficient medical context data management. In addition to the technical advances, work within research produced significant clinical results and important new insights on the nature of sleep and its putative reciprocal relationship with sleep.

1.1 Introduction

ARMOR project addresses the needs of the epileptic patient and healthcare professionals, aiming at the design and development of a non-intrusive Personal Health System (PHS) for the monitoring and analysis of epilepsy-relevant multi-parametric data (i.e. EEG, EOG, EMG, EKG, skin conductance data), and the documentation of the epilepsy related symptoms.

Epilepsy is a common, devastating and still incurable disorder (a detailed introduction in epilepsy is presented in Chap. 2) [1, 2]. Although in most cases its symptoms can be ameliorated by life-long pharmaceutical treatment, still this treatment needs continuous adjustment and change to retain its efficacy. Due to its multifactorial causes and paroxysmal nature, epilepsy needs multi-parametric monitoring for purposes of accurate diagnosis, prediction, alerting and prevention, treatment follow-up and presurgical evaluation. The incidence of epilepsy is age-related, higher in children; epileptic seizures occur in 1–2 % of the general population and in 4 % of children. During the periods of childhood and adolescence non-epileptic paroxysmal events (NEPE) also occur more frequently than in adult life with similar clinical features. It is important to note that 30 % of people with epilepsies have also NEPE. Furthermore, epileptic seizures differ with respect to motor, cognitive, affective and autonomic and EEG manifestations. Their recognition and full understanding is the basis for the optimal management (including additional diagnostic tests and genetics) and treatment. The total cost of epilepsy in the European Union is counted upwards of 15 billion euros per year, without counting the severe psychological impact of having a seizure in public and having the feeling that such an event may occur any time [1–7].

Current diagnostic methodologies to accurately diagnose epilepsy have limitations. The paroxysmal and multifaceted nature of the disease demand long-term multimodal monitoring which is a great burden to hospitals' financial and human resources, while it is not always diagnostically effective. To face this challenge and observe seizures day or night at the patient's own environment, ambulatory EEG is in practice for some time. However video-EEG at home which would further capture important behavioral features before, during and after the electrographic seizures are only recently being evaluated [5]. Sensors able to detect crucial autonomic, motor or other changes that cannot be appreciated by the video and the scalp EEG electrodes are seldom used in the EEG departments and when used a limited coverage is applied, while ambulatory EEG does not include such sensors. Therefore, there is a need for more accurate diagnosis of integrated seizure phenotype in individual patients, which will allow better understanding of underlying mechanisms, prediction (and alert) of time and type of seizure (and alert) and availability of medical assistance and advice.

ARMOR is visualized as an ambulatory monitoring system for diagnosis and management, limited, but optimally selected for each patient, scalp EEG covering and custom-designed multi-polygraphy (textile based EMG, body activity sensors, autonomic and other biological data such as blood pressure, temperature, sugar blood levels and O_2 and CO_2 saturation continuous monitoring). Diagnosis of a disease as multifactorial and unpredictable as epilepsy demands continuous observation and correlation analysis of as many parameters as possible of the patient's brain, body and the environment. Such a system enables successful stepwise diagnosis i.e. *Step-1*: Decision on whether the seizure belongs to Epilepsy or NEPE, 1a: What type of Epilepsy- classification, 1b: Which NEPE—appropriate further diagnostic workout and treatment. *Step 2*: Delineation of clinical and EEG expression of the epilepsy. 2a: Identification of different seizure types and mechanisms (for example a fall may be due to atonic, myoclonic, negative myoclonic or tonic muscle activity that can be monitored with chronically attached electromyography (EMG) electrodes, and therefore accurately diagnosed). 2b: Full clinical semeiology (identification of all clinical symptoms and signs); what are the constituents of a particular seizure (behavioral, muscle tone, autonomic, and other changes) and their timing.

We recognized in the above a major medical problem which can be solved with current advanced ICT technology and further advancement in data analysis, combined to benefit both the patient and the economy of the health care system. From the very beginning to the end of the project the theoretical background was elaborated and the effectiveness of the ARMOR sensors improved by targeted research work that proceeded in parallel with the steps described above. This research involves sophisticated analysis of existing data from expensive devices (that are not routinely available in clinics, e.g. multichannel EEG and/or MEG). As the project matured and new data were collected, the detailed analysis of selected subset of the ARMOR data were analyzed in detail, partly as prototypical examples or critical cases for diagnosis and classification and partly as part of the final evaluation of the ARMOR project.

Privacy and security [7] also constitute major concerns within ARMOR. Different levels of security and privacy are considered depending on the type of data being managed and the type of function being executed/used. ARMOR particular privacy and security requirements are identified and enriched with regulatory constraints and societal requirements.

The next sections provide a brief overview of ARMOR project perspectives, and challenges as well as its main contributions both at medical and technological domain.

1.2 ARMOR Perspectives

The technology developed in ARMOR project offers on-line and off-line analysis of data with the help of medical databases and the patient's medical file for the purpose of prediction and description/classification of seizures, prior to delivering to the patient appropriate alerts and treatment advises.

The user scenario escalation of the ARMOR system indicating the importance and usefulness of such a system are the following:

1. Differential Diagnosis of seizures—Is it epilepsy or non-epileptic paroxysmal event?
2. Delineation of the clinical EEG expression of different types of epilepsy.
3. Localized signs in childhood idiopathic generalized epilepsy.
4. Seizures dependence on the level of arousal.
5. Nocturnal seizures and their relationship to sleep macro- and microstructure.
6. Presurgical evaluation of intractable seizures.

1.2.1 Medical Perspective

ARMOR project aims to deliver considerable medical benefits adding to both clinical praxis and the background theoretical foundation that supports it. From clinical perspective, ARMOR offers a full phenotypical (clinical/EEG) description of the seizure disorder that can guide further diagnostic workout. It includes use of imaging results and detailed anatomical and electrophysiological measurements and information about genetic factors and clinical history. As part of the workout of ARMOR, additional examinations may be called for that could include further studies, specific screening and genetic testing, delineation of baseline state before treatment and accurate assessment of the treatment response, as well as guidance of further diagnostic workout with a view towards epilepsy surgery if medical treatment fails. ARMOR technology can be deployed at home avoiding hospital stays and reducing health expenses. The application of ARMOR technology at the clinic or at home can be done for diagnostic purposes (as it was tested at the last stages of ARMOR in Kings College London hospital). Long term monitoring that cannot be provided routinely in hospitals becomes accessible at home with ARMOR, including detection of possible life threatening ictal autonomic changes, seizure detection. Crucially the consideration of the effect of environmental factors in the home environment influencing the occurrence of seizures are included and monitored for the first time on a continuous basis. Since these are factors to which the patient is constantly exposed new possibilities for the prevention of seizure occurrence by intervening to the person's immediate surroundings are possible thanks to the research and development conducted in ARMOR.

ARMOR is poised to increase our understanding of the prevalence of different types of epilepsy and other NEPE. The ARMOR project also advances the theoretical perspective of epilepsy in two parallel directions. The first, is simply through the huge increase of the yield of events that home monitoring allows and it will become apparent soon after the system is successfully deployed. The second is through research undertaken within the project. We simply mention two important research strands here, that will be described in more detail in later chapters. The first is through tomographic analysis of high density EEG and MEG data that within

ARMOR are employed to enhance the details of each individual case and define better the optimal ARMOR design for each subject (described in more detail in Chaps. 4 and 13). These methods can also be used, once the ARMOR is deployed, to specifically improve our understanding of epilepsy mechanisms. The second strand is original new research on sleep. Within ARMOR this research has focused on elucidating the changes in background activity from awake state to the first two sleep stages constituting light sleep and the emergence of the large graphoelements of the second sleep stage, the spindles and K-complexes and their dynamic relationships. This research is described in more detail in Chap. 3 where its relationship to epilepsy is also highlighted. Beyond furthering our understanding of this relationship, healthy whole night sleep multimodal recordings were analyzed for the practical purposes of deriving recommendations and metrics for long term monitoring, developing novel EEG analysis tools and establishing the normative database of those quantitative features of sleep which are of importance in seizure diagnosis.

1.2.2 ICT Perspective

From Information and Communication Technologies (ICT) perspective ARMOR advances a novel holistic monitoring and analysis approach combining feasibility (individualized choice of limited number of sensors, non-contact wearable sensors, wireless recording etc.) with advanced data analysis, telecommunication and medical management tools. The main research issues tackled and main contributions to ICT are detailed in the following paragraphs.

The proposed system provides novel functionality for both real-time (online) and offline analysis of data (Chaps. 12 and 14 provide insight in the algorithms developed for online and offline data analysis). New techniques have been developed for multi-parametric sensor data mining using data fusion and correlation analysis, integration of information from various data sources/modalities, similarity analysis of signals, clustering, classification and prediction. Novel real time (online) analysis methods for multi-parametric stream data are also available through ARMOR technology aiming in detecting signals beyond the limits, identify seizure premonitory signs, discover typical patterns of activity followed by seizures and detect any typical patterns of activity/behavior based on models that will be created. Trade-offs for automated analysis taking place at the local site of each patient (instead of at the Health Center) aiming to reduce processing time, storage requirements and communication cost, facilitating the reduction of raw data to secondary and tertiary parameters (that will be correlated), have also been considered. All analysis and emergency alert mechanisms are based on a personalized model according to the patient's health profile. New decision support tools for advising the patient, triggering an alarm and detecting emergency situations are also available through ARMOR platform.

New informatics tools have also been developed for offline analysis of multi-parametric data correlation with other stored data about the patient (EEG, PET, SPECT, fMRI, genetic data) and the disease. In addition, offline data fusion for certain combinations of modalities (e.g. MEG, MRI) have also been developed, which are taking place at the Health Center with the participation of medical experts as well as new functionality providing feedback to the online analysis model. ARMOR has also made novel contributions in:

- The analysis of multidimensional time series.
- Similarity analysis of signals.
- Detection of patterns and associations between external indicators and mental states.
- Analysis of associations among signals and symptoms, discovery of lag correlation among different signals, detection of vital signs of a person changing in a significant manner.
- Identification of motifs (in spatio-temporal signals) and frequently repeated patterns or outliers (corresponding to seizure signs).
- Automatic summarization of results for each patient.

In addition, new techniques have been investigated for the detailed offline tomographic analysis of multichannel EEG and MEG data recorded simultaneously with measurements of heart activity (EKG), Galvanic skin Response (GSR) and other measurements that can be easily incorporated in the online monitoring for normal and epileptic patients in awake state and at different sleep stages obtaining the most direct insight of what is happening in the brain.

For sensor data acquisition and pre-processing (cleaning, integration, transformation, reduction) existing techniques have been considered, while new contributions have been made to data reduction and summarization techniques to deal with streaming data (details are provided in Chaps. 8 and 9). Existing database technology has been extended to support the organization of multi-parametric data including the support of efficient storage and retrieval capabilities such as multidimensional indexing. Data compression issues were also investigated. The ARMOR databases store logs of all events, recorded values from sensors and other metrics that are monitored, personalized patient health profiles, medical information including guidelines for diseases, symptoms, medication, potential side effects of medication, etc. Chapters 11 and 14 provide extensive descriptions of the database related technology employed and developed in the context of ARMOR.

1.3 What ARMOR Project Offers?

In addition to the theoretical advances the main contributions of ARMOR project are:

- Requirements concerning all key aspects such as real-time data acquisition, middleware, sensors etc.
- The design and development of multi-parametric data analysis models and personalized patient health profile.
- On/off line data analysis of the data collected

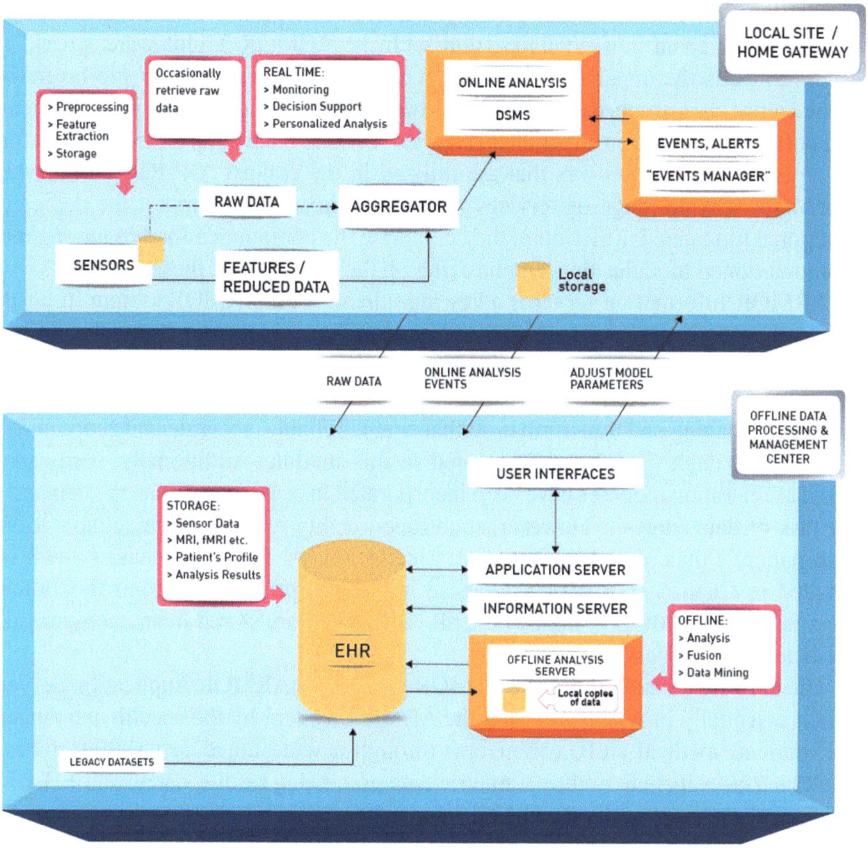

Fig. 1.1 An overview of ARMOR approach

- ARMOR test cases implementation and evaluation
- The set-up and configuration of the ARMOR end-to-end service delivery platform emphasizing in security provision level.

Aiming towards a holistic, medically efficient and economical monitoring system ARMOR platform, depicted in Fig. 1.1, tackles and addresses all functional aspects described in Sect. 1.2. Thus, following a bottom up description the first sub-layer comprises the actual sensors enabling the data acquisition. Based on extensive experience as well as equipment provided by the ARMOR consortium partners, multi-parametric data acquisition has been made possible through a wide range of possible sensors gathering a wide range of medical data continuously and in real-time. At the same time, attention has been paid in issues related to minimizing discomfort caused by the acquisition equipment thus allowing the patient to perform his/her normal activities.

The first level of data aggregation is the patient's equipment data acquisition sub-layer, which in close collaboration with the ARMOR Middleware, provides access towards the upper modules and vice versa. The role of these sub-layers is crucial since they comprise the gateway point between sensor hardware equipment and software application functional modules. On one hand support is provided for all types of data and sensors that are utilized in the context ARMOR and on the other hand a wide range of services is provided in order to support the depicted functional modules. Furthermore, they comprise the communication bus among the aforementioned modules thus emphasizing on the criticality of these sub layers.

ARMOR Information server is a key ingredient of the ARMOR system. It hosts the models derived from extended research effort, which intend for the multi-parametric data analysis. As indicated in Fig. 1.1, the development of new methods and tools regarding online/offline analysis, data fusion, real time processing, data profiling, cleaning and transformation that enable reliable, accurate and economical diagnosis through models are integrated in this module. Additionally, hardware-based acceleration modules have been incorporated in order to facilitate the demanding task of data analysis. However, such functionality requires close collaboration with patient's data stored in Electronic Health Record (EHR) database (which is detailed in Chap. 11), a secure database which comprises the second functional module of the ARMOR system where all patient data are stored in an anonymized, yet efficient to process way.

The third functional module of ARMOR platform is ARMOR Application Server which essentially provides access to the ARMOR system by the specific personnel (i.e. patient, medical stuff, caregivers) through a wide range of user interfaces. Such interfaces include multi-parametric data processing results visualization, EHR access and personal tracking and vital signs information.

The ARMOR system provides also novel functionalities for managing and analyzing both new and already acquired multimodal data of epileptic patients (e.g. EEG, EMG, ECG, PET, sensor data, genetic data). As it will be presented in Chap. 14, as a first step some of the existing well-established techniques for data cleaning, integration and transformation were reused and enhanced for the pre-processing of multi-parametric data. In order to deal with the large amount of ARMOR's data, new contributions have been made in data and dimensionality reduction, involving feature selection and extraction methodologies together with new representation techniques for time series data. ARMOR's data, such as logs of all events, recorded values from sensors and other metrics that are monitored, together with personalized patient health profiles, medical information for diseases, medication, etc., have been organized in a data management system.

In order to achieve more efficient and accurate inferences, data fusion techniques have also been explored. Existing fusion techniques have been extended and new ones have been developed for the integration and fusion of information from sensors, already stored data about the patients that the clinical collaborators of ARMOR project collected or had already available, along with the personalized patient health profiles. Novel real time (online) analysis methods for multi-parametric streaming data that have also been developed, which aim at detecting signals beyond the limits, identify seizure premonitory signs, and discover typical patterns of activity followed by seizures and atypical patterns of activity, based on the models created. All analysis and emergency alert mechanisms offered by ARMOR platform are based on a personalized model according to the patient's health profile. Moreover, new decision support tools for advising the medical professional, and the patient, triggering an alarm, and detecting emergency situations have been developed. Chapters 10 and 11 provide an in depth analysis of the corresponding aspects.

Last but not at all least, a vertical layer concerning all security issues is also included as part of ARMOR platform. ARMOR pays special attention on security issues regarding sensitive medical data, therefore the respective security layer does not concern just a single or a sub-set of the ARMOR layers but all of them, thus the vertical orientation. Again following a bottom up approach the first layer required tackling security issues are sensors ensuring secure data acquisition and transferring towards the aggregation point/s. All storage sites also employ security techniques ensuring data integrity and privacy. Data communication and data transferring were very challenging issues that have been successfully addressed especially when communication is performed over the air. Both sensor communication sections and backhaul communication sections are susceptible to a wide range of dangers and possible attacks requiring specially attention. Moving on the upper layers, access rights, user authentication and authorization are also significant issues and capabilities to be provided by an efficient and robust Personal Health System (PHS) as ARMOR intends to be. Details on the above issues are thoroughly addressed in Chaps. 6, 9 and 11.

1.4 Conclusions

In this introductory chapter we presented a brief overview of ARMOR project, where its main challenges and perspectives have been analyzed. Although the major contributions of ARMOR have been described, we have intentionally avoided providing too many technical details. The rationale is that the specific book chapter intents to introduce the main concepts behind ARMOR project and not presenting them in detail.

The book chapters that follow provide all the necessary information both for medical and ICT experts. For that purpose the rest of the book is dived in three main parts. **Part I**: *EPILEPSY MEDICAL BACKGROUND* where the medical experts of ARMOR provide insight to epilepsy, **Part II**: *CYBERPHYSICAL SYSTEMS FOR NEUROPHYSIOLOGICAL MONITORING*, where the ARMOR technology is presented in detail and **Part III**: *EVALUATION* where the evaluation results of ARMOR technology are presented.

References

1. Engel J Jr, Pedley TA (2005) What is epilepsy? In: Engel J (ed) Epilepsy: a comprehensive textbook. Lippincott Williams & Wilkins\Raven Press, Philadelphia, pp 1–11
2. Panayiotopoulos CP (2010) A guide to epileptic syndromes and their treatment. Springer, London
3. Brunnhuber F, Amin D, Nguyen Y, Goyal S, Richardson MP (2014) Development, evaluation and implementation of video-EEG telemetry at home. Seizure 23(5):338–343
4. Korff CM, Scheffer IE (2013) Epilepsy classification: a cycle of evolution and revolution. Curr Opin Neurol 26(2):163–167. doi:10.1097/WCO.0b013e32835ee58e
5. Berg AT, Cross JH (2012) Classification of epilepsies and seizures: historical perspective and future directions. In: Stefan H, Theodore WH (eds) Handbook of clinical neurology, Epilepsy, part I. Elsevier, Amsterdam
6. Newton CR, Garcia HH (2012) Epilepsy in poor regions of the world. Lancet 380(9848): 1193–1201. doi:10.1016/S0140-6736(12)61381-6
7. Al Ameen M, Liu J, Kwak K (2012) Security and privacy issues in wireless sensor networks for healthcare applications. J Med Syst 36(1):93–101

Chapter 2
Introduction to Epilepsy and Related Brain Disorders

Evangelia Giourou, Alkistis Stavropoulou-Deli, Aspasia Giannakopoulou, George K. Kostopoulos, and Michalis Koutroumanidis

Abstract Epilepsy affecting 1 % of the world's population and is the most common serious disorder of the brain, greatly impacting on the quality of life of affected individuals, particularly those whose seizures are not fully controlled. Epilepsy has a multifactorial origin and a multifaceted expression. It is caused by clusters of nerve cells in the brain which sometimes signal abnormally, causing seizures. Anything that disturbs the normal pattern of neuronal activity—from illness to brain damage to abnormal brain development—can lead to seizures. Epilepsy may develop because of an abnormality in brain wiring, an imbalance of nerve signaling chemicals called neurotransmitters, changes in important features of brain cells called membrane receptors and channels, or some combination of these and other factors.

Depending on the brain area affected and its physiological role, these disturbances of neuronal activity that occur during seizures may cause strange sensations, emotions, and behaviors. They also sometimes cause convulsions, abnormal movements, and loss of consciousness. In some people, seizures happen only occasionally. Other people may experience hundreds of seizures a day. There are many different forms of epilepsy, and symptoms vary greatly from one person to another. About three-quarters of the individuals diagnosed with the epilepsies can control their seizures with medicine or surgery. However, about 30 % will continue to experience seizures even with the best available treatment. In some cases, people experience a type of seizure that last so long that they can damage the brain and may be life-threatening. Having a single seizure as the result of a high fever (called febrile seizure) or head injury does not necessarily mean that a person has epilepsy. Only when a person has had two or more seizures is he or she considered to have epilepsy.

E. Giourou • A. Stavropoulou-Deli • A. Giannakopoulou • G.K. Kostopoulos (✉)
Neurophysiology Unit, Department of Physiology,
Medical School, University of Patras, Patras, Greece
e-mail: gkkostop@upatras.gr

M. Koutroumanidis
Department of Clinical Neurophysiology and Epilepsy, Guy's and St Thomas'
NHS Foundation Trust, London, UK

Department of Academic Neurosciences, Kings College London, London, UK

N.S. Voros, C.P. Antonopoulos (eds.), *Cyberphysical Systems for Epilepsy and Related Brain Disorders*, DOI 10.1007/978-3-319-20049-1_2

A measurement of electrical activity in the brain and brain scans such as magnetic resonance imaging or computed tomography are common diagnostic tests for epilepsy.

Research efforts need to be stepped up to better understand pathophysiologic mechanisms and to develop more effective therapies. Current understanding is on the mechanisms underlying seizure expression and thus allow us only symptomatic treatment (and still some seizures are or become with time drug resistant). To develop a cure of epilepsy we have to understand epileptogenesis, the long process which makes brain neurons vulnerable to hyperexcitability and abnormal synchronization. Equally important is to raise awareness on the nature of epilepsy through public education so that the lives of people with epilepsy are not adversely affected by stigma, prejudice, and discrimination, neither face unjust restrictions in their human rights, employment, marriage, and daily activities such as driving.

In this chapter we introduce the historical and current efforts to define and categorize epilepsy and we briefly describe the variety of its causes and our current ideas on the mechanisms underlying the expression of its main types.

2.1 Introduction

This chapter aims to present a concise view of the major clinical and neuroscience aspects of epilepsy for an academic but not necessarily medical audience. More comprehensive accounts can be found in recent excellent reviews and books covering both clinical [1–5] and pathophysiological aspects [6, 7].

One can find descriptions of seizures since the beginning of recorded history, but they were usually attributed to demonic possessions or other supernatural influence [8, 9]. Babylonians astutely described many of the seizure types we recognize today as tonic clonic seizures, absences, drop attacks, simple and complex partial seizures and even focal motor (Jacksonian) or gelastic attacks. A writing from China's seventh to second century BC describes symptoms resembling both partial and generalized convulsions. All three Indian medical systems recognize epilepsy and a compendium of the sixth century BC mentions loss of consciousness during particular seizures. The Babylonians and the other advanced Asiatic civilizations, although apparently kin observers, were not motivated to search for underlying mechanisms, being content with associating each seizure type with the invasion of the body by a particular evil spirit. The concept of pathophysiology had to wait till diseases were dissociated from the supernatural. Greeks imported all this medical tradition by coining the verb ΕΠΙΛΑΜΒΑΝΕΙΝ (to seize someone) from which the word epilepsy, while however questioning the nature of its causes and finally recognized epilepsy as a disorder of the body rather than a sacred disease. This was boldly asserted by Hippocrates in the fifth century BC—"ΑΛΛΑ ΓΑΡ ΑΙΤΙΟΣ Ο ΕΓΚΕΦΑΛΟΣ ΤΟΥΤΟΥ ΤΟΥ ΠΑΘΕΟΣ … ΟΤΩ ΔΕ ΤΡΟΠΩ ΚΑΙ ΕΞ ΟΙΗΣ ΠΡΟΦΑΣΙΟΣ ΓΙΝΕΤΑΙ, ΕΓΩ ΦΡΑΣΩ ΣΑΦΑ" (Hippocrates Corpus 90, On the sacred disease §§ 1–6, 21). Hippocrates was absolutely right on his first statement that "the cause of epilepsy is in the brain". But the second, that "he will proceed to explain clearly the mechanism and cause of seizures" is far from being true even today. Of historical interest to medicine is also his proposal to search for abnormal humors in the head of the epileptic goats—which seems to be the first ever proposal to experiment with animal models of any disease.

The idea of the brain as a cause of epilepsy was well accepted in specific circles even up to the second century AD when we see Galen and Aretaeus treating and theorizing about epilepsy in the Latin era. Unfortunately it had little influence on the public's view of supernatural causes, which remained up to the middle ages (at least, given the still lingering progress of our society to abolish the stigma of epilepsy). The main holdup to progress in this field appears to have been the prohibition of anatomical studies, which ended with the European Renaissance of the fourteenth to seventeenth centuries, with Thomas Willis writing of convulsive disorders and the advance of pathology in the nineteenth century. Two fundamental developments, the concepts of animal electricity (from Galvani to Todd) and of functional localization in the brain (i.e. motor cortex) lead to the study of "epileptiform" or "partial" and "generalized" seizures. About this time Caton discovered EEG in animals and 52 years later Berger discovered human EEG. This led to confirmation of Todd's hypothesis that seizures are the result of electrical discharges by Lennox in 1935. In the beginning of the twentieth century the road to basic mechanisms of epilepsy is paved by the introduction of the neuron doctrine (Cajal), the role of synapses (Sherrington) and their transmitters (Levy, Dale). In the second half of the twentieth century research is accelerated by enormous methodological progress in both directions: reductionistic i.e. discovery of ionic channels (Hodgkin and Huxley) and molecular genetics and integrating i.e. structural and functional brain imaging, video-telemetry, MEG, computer assisted analysis and multimodal data fusion. Every advance seems to add to the enormous complexity of the nervous system and the probability that multiple elusive genetic–molecular–metabolic mechanisms contribute to the wide range of epilepsies. We seem to know enough at the micro- and macroscopic level but not much at the mesoscopic one. This is probably why we know ictogenesis well enough to have relatively efficient drugs to suppress seizures, but we know almost nothing about epileptogenesis and how to prevent it.

Epilepsy, a chronic condition that is characterized by recurrent seizures, affects people of any gender, age and geographic region. Approximately 1 % of the general population in western societies suffers from some sort of epilepsy while up to 10 % will have at least one seizure during their lifetimes [10].

The incidence and prevalence of epilepsy in the EU is estimated to be respectively 3.3–7.8 per 1000 and 44/100.000 [11]. The total number of people affected with epilepsy in EU is estimated to 2,64 million. The financial burden for EU expressed in disability adjusted life years is 245,475 and in cost-of-illness 5221 € per person and 13,800 in total [12]. However, besides the calculated financial burden, there is also an enormous and uncalculated cost to the patients and their families. Even today, epilepsy still remains a stigmatizing disease with many social consequences. The term itself, which means that the control over one's behavior is lost and remains unchanged over three millennia [8], shows the awe experienced by both the patient and the bystanders to a seizure. Epilepsy can alter patients' everyday life causing impairments in quality of life [13] and psychological distress to caregivers. One of the most stressful aspects of the disease is the unpredictability of seizure occurrence [14]. A large number of epilepsy patients fail to remain seizure free despite adequate treatment; there is no such thing as "cured" epilepsy since seizure threshold remains reduced compared to that of an unaffected person [15].

Moreover, epileptic seizures can be lethal. Common causes of death include accidents while Sudden Unexpected Death in Epilepsy Patients (SUDEP) is often attributed to ictal cardiac arrhythmias [16].

Epilepsy is a chronic condition of multifactorial causes (from genetics to brain trauma and from inflammation to tumor) and with a multifaceted expression (from a mere brief loss of consciousness to focal or generalized convulsions) depending on which brain area has developed a tendency for neuronal hyperexcitability and hypersynchronization and what is the physiological role of this area.

In newly diagnosed patients with epilepsy, the initial treatment option is usually choosing one of the available antiepileptic drugs (i.e. monotherapy) based on seizure type, age, co-morbidities and other factors. Should this fail to control seizures, an alternative drug can be selected or be additionally prescribed.

Antiepileptic drugs accomplish seizure reduction by suppressing neuronal intrinsic or synaptic excitation (usually mediated by the neurotransmitter glutamate) and promoting synaptic inhibition (usually mediated by the neurotransmitter gamma-amino-butyric acid or GABA). Mechanisms of action include blockage of voltage-gated Na^+ (e.g. carbamazepine) and Ca^{2+} channels, enhancement of GABA-mediated inhibition (e.g. benzodiazepines) and interference with glutamatergic excitation (e.g. felbamate). Some antiepileptics have multiple target systems (e.g. divalproate influences Na^+, Ca^{2+} and GABA), while some others' mechanism of action remains unknown (such as that of gabapentin) [17]. Unfortunately all available drugs treat ictogenesis (the expression of seizures) rather than epileptogenesis (the long term development of conditions leading to seizures). In that sense all treatment is symptomatic rather than causal, i.e. we do not yet have a cure for epilepsy. This symptomatic treatment is usually effective in about 70 % of the cases.

Surgery can be employed in selected cases of drug-resistant epilepsy or in the presence of structural lesions such as a brain tumor or hippocampal sclerosis. In certain cases of uncontrolled seizures vagal nerve stimulation might also be considered [18].

Undoubtedly epilepsy and its multiple often long term neurobiological, cognitive, psychological and social consequences require tailor-made and multitargeted treatment.

2.2 Definitions

Epilepsy is defined as a neurological condition which is characterized by a predisposition to generate recurrent epileptic seizures; it is not a single disease entity but points toward multiple underlying neurological defects and structural or functional changes in the brain. This is fundamentally considered to be independent of readily identified, transient factors that can induce seizures in the normal brain.

The current definition of epilepsy requires at least two unprovoked seizures occurring 24 h apart [15].

The word *seizure* may refer to many sudden and severe events often including psychogenic seizures, dissociative seizures, conversion seizures many of which can often resemble epileptic seizures without being due to epilepsy [15].

Epileptic seizures are defined as events (behavioral expressions) of a paroxysmal nature often accompanied by transient alteration in consciousness level and with signs and symptoms due to abnormal, excessive or synchronized neuronal discharges of the brain which can be widespread or localized [19].

Non epileptic seizures on the other hand are not caused by abnormal neuronal discharges. They can be divided in two large groups; the organic non epileptic seizures (i.e. atypical syncope and parasomnias) and the psychogenic non epileptic seizures—NEPS (i.e. conversion symptoms and dissociative states). Typically, during a NEPS paroxysmal event sustained forceful eye closure can be present while eye closure in any form is uncommon during epileptic seizures [20].

There are currently several definitions employed to define the actual lesion or zone which is responsible for or correlated to the origin of epileptic seizures, mostly in terms of presurgical evaluation. The (actual and the larger 'potential') Epileptogenic Zone is described as the total cortical area which is necessary and sufficient to generate the seizures. The Epileptogenic Lesion refers to an abnormal structural brain area which is presumed to be causal of epileptic seizures in the symptomatic epilepsies and can be usually indentified by MRI. The Irritative Zone is the cortical area which is capable of generating interictal discharges while the Ictal Onset Zone is the cortical area from which ictal discharges arise. Both the Irritative Zone and the Ictal Onset Zone can be identified by scalp or intracranial EEG recordings. The Functional Deficit Zone is the cortical area with focal nonepileptic dysfunction which is responsible for functional deficits identified by neurological medical examination. The Symptomatogenic Zone refers to the cortical area which when activated by a seizure is responsible for producing the first clinical ictal signs and symptoms [21, 22].

Diagnostics of epilepsy still rely on clinical features. EEG, video and brain imagining such as MRI and CT are used for differential diagnostic purposes once a suspicion on a possible epileptic syndrome is made. Clinical manifestations depend on several factors such as the type of epilepsy or the particular epilepsy syndrome, the patient's age, the area of the brain that generates seizures, and whether ictal discharges remain localized or propagate to other brain areas.

Every person can suffer an epileptic seizure given the right trigger and circumstances [15]. Acute symptomatic seizures occur at the time of a systemic insult or in close temporal association with a documented brain insult. Such precipitating events include metabolic disturbances, trauma, fever, infection, intoxication and substance withdrawal [23]. Alcohol withdrawal may precipitate seizures. Alcohol decreases CNS excitability by facilitating GABA action and acting as an NMDA receptor antagonist. Chronic alcohol abuse results in a down regulation of GABA and an up regulation of NMDA receptors and glutamate production. When withdrawal occurs, this increased potential for excitation can result in seizures [24]. Metabolic disturbances and trauma can further contribute to seizures in patients with alcohol abuse.

There has been a long discussion on classification of the epilepsies which has mainly been triggered by the recent technological advances both in imaging and genetics. Diagnosing a specific electroclinical syndrome is not always possible while the underlying cause might be of equal if not more value. Terminologies such as idiopathic, symptomatic, cryptogenic and also partial, complex and simple are not anymore used. A more descriptive approach has been recommended taking into account the underlying aetiology while retaining electrophysical syndromes when this is possible [25–28]. In parallel there is a growing appreciation of the fact that classification schemes should be based on current pathophysiological explanations but also serve to best distinguish conditions demanding different treatment and often these two goals may be conflicting, while mutually dependent.

2.3 Classification

Epilepsy is not a single disease entity but it is rather consisted of a range of underlying neurological disorders. The International League against Epilepsy—ILAE [25] in response to concerns about the existing classification systems, proposed a multiaxial diagnostic scheme which is summarized in Table 2.1. This diagnostic scheme aimed in categorizing individuals according to a standardized terminology that could be used by the vast majority of physicians in any relevant specialty while it would be flexible enough to include the dynamic aspects of the disease.

Yet, a syndromic diagnosis is not always possible while presumed seizures types and syndromes might alter as new information becomes available. Furthermore, a classification system is used for a variety of purposes; epidemiological investigations, basic research, clinical aspects (e.g. screening patients before surgery) and clinical trials. Thus a classification system should be able to address several different needs of diverse areas. Therefore, in the light of new basic and clinical science advances and led by the need to further simplify terminology used, a new classification has

Table 2.1 A proposed diagnostic scheme for people with epileptic seizures and epilepsy

<u>Axis 1:</u> Ictal phenomenology. Description of Ictal events using the Glossary of Descriptive Ictal Terminology (Blume, 1991)
<u>Axis 2:</u> Seizure type. Specifying localization within the brain and precipitating stimuli for reflex seizures using the List of Epileptic Seizures
<u>Axis 3:</u> Syndrome. Specifying possible syndrome from the List of Epilepsy Syndromes. A syndromic diagnosis may not always be possible though
<u>Axis 4:</u> Aetiology. Defining when is possible genetic defects and pathologic underlying causes for symptomatic focal epilepsies from the Classification of Diseases Frequently Associated with Epileptic Seizures or Epilepsy Syndromes
<u>Axis 5:</u> Impairment. Classification of impairment. Optional additional diagnostic parameter

Modified from: Engel J. A proposed diagnostic scheme for people with epileptic seizures and with epilepsy: report of the ILAE task force on classification and terminology, 2001 [25]

Table 2.2 Classification of seizures (2010)

Generalized seizures: Tonic-clonic; Absence (Typical, Atypical, Absence with special features); Myoclonic (Myoclonic, Myoclonic atonic, Myoclonic tonic); Clonic; Tonic; Atonic
Focal seizures: according to severity (consciousness impairment, motor or autonomic components, subjective sensory or psychic phenomena, evolving to bilateral, convulsive seizures); according to site of origin according to sequence of clinical features
Unknown

Modified from: Berg AT, Berkovic SF, Brodie MJ, Buchhalter J, Cross JH, van Emde Bas W, Engel J, French J, Glauser TA, Mathern GW, Moshe SL, Nordli D, Plouin P, Scheffer I. (2010) Revised terminology and concepts for organisation of seizures and epilepsies: Report of the ILAE Commission on Classification and Terminology 2005-2009. Epilepsia;51:676-685 [26]

Table 2.3 Electroclinical syndromes and other epilepsies (2010)

By age at onset
Neonatal period (Benign familiar neonatal epilepsy; Early myoclonic encephalopathy; Ohtahara syndrome
Infancy (Epilepsy of infancy with migrating focal seizures; West syndrome; Myoclonic epilepsy in infancy; Benign infantile epilepsy; Benign familiar infantile epilepsy; Dravet syndrome; Myoclonic encephalopathy in nonprogressive disorders
Childhood (Febrile seizures plus; Panayiotopoulos syndrome; Epilepsy with myoclonic atonic seizures; Benign epilepsy with centrotemporal spikes; Autosomal-dominant nocturnal frontal lobe epilepsy; Late onset childhood occipital epilepsy; Epilepsy with myoclonic absences; Lennox-Gastaut syndrome; Epileptic encephalopathy with continuous spike and wave during sleep; Landau-Kleffner syndrome; Childhood absence epilepsy
Adolescence-Adult (Juvenile absence epilepsy; Juvenile myoclonic epilepsy; Epilepsy with generalized tonic–clonic seizures alone; progressive myoclonus epilepsies; Autosomal dominant epilepsy with auditory features; other familial temporal lobe epilepsies
Less specific age relationship (Familial focal epilepsy with variable foci; Reflex epilepsies)
Distinctive (MTLE with HS; Rasmussen syndrome; Gelastic seizures with hypothalamic hamartoma; Hemiconvulsion-hemiplegia-epilepsy)
Epilepsies not fitting into any of the above are separated by the presence or not of a known presumed cause (structural or metabolic) and by the primary mode of seizure onset (generalized or focal)
By structural or metabolic causes
(Malformations of cortical development; Neurocutaneous syndromes; Tumor; Infection; Trauma; Angioma; Perinatal insults; Stroke etc.
Of unknown cause
Conditions with epileptic seizures not diagnosed as epilepsy per se

Modified from: Berg AT, Berkovic SF, Brodie MJ, Buchhalter J, Cross JH, van Emde Boas W, Engel J, French J, Glauser TA, Mathern GW, Moshe SL, Nordli D, Plouin P, Scheffer I. (2010) Revised terminology and concepts for organization of seizures and epilepsies: Report of the ILAE Commission on Classification and Terminology 2005-2009. Epilepsia;51:676-685 [26]

arisen concerning mainly seizures classification but also introducing a simplification of terminology used on epilepsies (Tables 2.2 and 2.3) [26].

2.4 Etiology

The possible identification of specific brain abnormalities associated with epilepsy has crucial implications in the treatment and prognosis of the disease. A broad category of brain abnormalities stems from a strongly genetic/developmental component, while other cases mainly originate from acquired insults (such as infection, trauma and hypoxia). While still classified together with other acquired pathologies, hippocampal sclerosis is considered to originate from a combination of genetic risk factors and initiating insults (such as febrile seizures) [29].

Lesions such as tumors and severe trauma are in no way transient and reversible. Seizures occurring in patients with a potential cause of epilepsy (such as a highly epileptogenic oligodendroglioma) are not classified as acute symptomatic seizures [30]. Moreover, if the lesion generates an enduring predisposition for unprovoked seizures, with a risk comparable to those who have had two unprovoked seizures, then the person should be considered to have epilepsy despite having only one seizure [15].

Although there is an apparent variance in the power of potential seizure triggers, a recognized epilepsy syndrome comes to blur the lines. Seizures in reflex epilepsies are triggered by sensory, motor or cognitive stimuli, such as bright lights, eating and music. The pathophysiology of this syndrome involves the activation of hyperexcitable diffuse cortical pathways, with different triggering points according to the precipitating stimulus (f. ex. occipital triggering and propagation through cortico-cortical pathways in photosensitivity [31].

2.4.1 Febrile Seizures

Febrile seizures (FS) encountered in children are a well-known example of provoked seizures. Factors involved in their pathogenesis include genetic susceptibility (reflected in a positive family history), inflammatory mechanisms (with a particular significance of IL-1β) [32], mutations in the GABA-A receptors and participation of sodium channels [33]. FS are benign and not to be confused with the distinct epileptic syndrome of Generalizes Epilepsy with FS (GEFS+).

2.4.2 Developmental Brain Abnormalities

Abnormalities in neuronal cell migration and dysplasias of the neural ectoderm occur in embryonic life and result in malformations of cortical development (such as hemimesencephaly, heterotopias and lissencephaly) and epileptogenic neurocutaneous syndromes (Tuberous Sclerosis Complex and Sturge-Weber syndrome), respectively. A strong genetic component is involved in Tuberous Sclerosis (autosomal dominant inheritance) as well as in vascular malformations such as cavernomas. Such lesions mainly result in intractable seizures that require surgical intervention.

A variety of aggressors can disrupt the normal migration of neurons from the periventricular germinal matrix to their final destination and disrupt cortical lamination, thus resulting in a failure of normal circuit formation. These include infectious agents (such as rubella and the TORCH complex), toxins (e.g. alcohol) [34], while genetic factors may also play a role. Histologically, a scattering of large anomalous neurons is observed, associated with thickening and white matter "balloon cells" (focal cortical dysplasia), a blurring of the cortical/white matter border (hemimegalencephaly), while heterotopic gray matter islands in the white matter are observed in heterotopias. Lissencephaly and pachygyria are notable for disrupted gyri architecture [35].

In Tuberous Sclerosis, blurred cortical lamination results from masses of astrocytic cells and calcifications, while in Sturge-Weber cortical atrophy is the result of overlying angiomas. This finding is also present in vascular malformations. Other epileptogenic factors include the presence of hemosiderin from recurrent hemorrhages, especially in cavernomas [36].

2.4.3 *Acquired Lesions*

Epilepsies resulting from non-developmental abnormalities include post-traumatic epilepsy (PTE), seizures originating after hypoxic brain injury (as in stroke and perinatal insults) and tumor-associated epilepsy (TAE). Hippocampal sclerosis (HS) and medial temporal lobe epilepsy (MTLE), classified as a distinctive constellation, will also be mentioned here.

Although infectious and immunologic aetiologies are important in acquired epilepsy, with the former playing an important role in the developing world (mainly in the form of tuberculomas and cystic brain lesions due to neurocysticercosis) and the latter in surgical constellations such as Rasmussen's Syndrome [35], a systematic review is beyond the scope of this chapter.

2.4.3.1 Brain Injury (Traumatic, Hypoxic/Ischemic)

Post traumatic epilepsy (PTE) is the most common cause of new-onset epilepsy in young adults [37] and accounts for 20 % of structural epilepsy (5 % of all epilepsy cases) [38]. Nevertheless, only a sub-group of brain trauma patients will develop epilepsy. The latency from the time of the injury to the onset of epilepsy is extremely variable (weeks to years) [39] while scalp EEG may be unable to detect initial epileptiform activity [40]. Higher injury severity and the presence of an intracranial hematoma are important risk factors for both early (in the first week after injury) and late seizures [41]. Other risk factors for PTE include advanced patient age, multiple concussions and seizures within 24 h post-injury [42].

Histopathological consequences of penetrating injuries include the formation of an epileptogenic cortical scar, while non-penetrating trauma results in axonal shearing, edema and ischemia in the gray–white matter junction [43]. Haemoglobin breakdown products, resulting from haemorrhage, have been implicated in epileptogenesis [44].

Changes in molecular signaling involving gene induction and modifications of neurotransmitter receptors and ion channels occur early after an injury. Axonal sprouting and dendritic modifications (such as mossy fiber sprouting) happen later on. A variety of mechanisms, including blood-brain barrier disturbances, inflammatory responses and release of related cytokines have also been implied [43]. All of the above changes result in an increased excitability that lowers seizure threshold.

Neuronal damage extends past an acute hypoxic insult and into the reperfusion phase. Pathophysiological mechanisms include apoptosis, activation of inflammatory mediators [45] as well as excessive extracellular glutamate excitotoxicity and intracellular accumulation of calcium [46]. Astrocytes post-insult release signals (such as thrombospondins) that increase excitatory synapse formation [47].

Apart from mechanisms leading to increased neuronal excitation, tissue necrosis and liquefaction can result in the formation of isolated cortical foci. Animal models of cortical isolation ("undercut" models) have shown a selective loss of GABAergic interneuron [48], resulting in a limitation of inhibitory mechanisms, which also contributes to epileptogenesis. It is worth noting that cortical isolation is present in both acquired (PTE, hypoxic injury) and developmental structural epilepsies.

2.4.3.2 Hippocampal Sclerosis (HS)

HS is present in at least 30 % of all epilepsy cases, according to both surgical [49] and post-mortem [50] series. HS is mainly associated with mesial temporal lobe epilepsy (MTLE) and reported in 50–70 % of patients with TLE and MTLE [51].

Whether HS is the cause or a consequence of seizures remains a highly controversial subject. Although *status epilepticus* has been proved to cause neuronal death, whether brief seizures amount to a similarly deleterious effect is still debated. Studies supporting the claim that the type of precipitating injury, seizure frequency and severity influence HS abound [52]. On the other hand, it has been proposed that hippocampal abnormalities in TLE play a pathogenetic role. Further support to this claim comes from imaging studies, showing that the degree of hippocampal atrophy is not correlated with the duration and severity of seizures [53]. Also, patients with multiple or poorly controlled seizures, as well as patients in a distinct subgroup of TLE, do not always display hippocampal sclerosis and neuronal loss. Finally, it has also been suggested that the pathologic alterations observed in sclerotic hippocampi can result from an abnormal cell migration during brain development [54].

Histopathologically, two areas of the hippocampus are mainly affected in HS: Ammon's horn and the dentate gyrus. In Ammon's horn, there is a marked neuronal depletion and astrocytic proliferation. These astrocytes display increased expression of glutamatergic receptors and are more capable of generating action potential-like responses *in vitro* compared to normal astrocytes [55]. In the dentate gyrus, loss and dispersion of granule cells, growth of new fiber systems (mossy fiber sprouting), loss of inhibitory interneurons and upregulation of inhibitory neurotransmission (perhaps in an attempt to curb increased excitation) are observed. In contrast to these changes, the subiculum (which is the main output region of the hippocampus) remains intact [56]. Nevertheless, it has been reported to contribute to epileptogenesis by initiating spontaneous interictal discharges [57].

ILAE classifies HS in three distinct categories, based on the regions (CA1 or CA4) involved in neural cell loss and gliosis. These categories differ in epilepsy history and prognosis of postsurgical seizure control [54].

2.4.3.3 Tumor-Associated Epilepsy (TAE)

Epilepsy is far more common in brain tumor patients than in the general population. Compared to high-grade tumors and cerebral metastases, low-grade gliomas are linked to a higher risk of seizures. Although the longer survival rate of patients with low-grade tumors, as well as molecular differences of each tumor category, could contribute to these findings, slow growing tumors might deafferentate cortical areas and thus create highly epileptogenic foci, as opposed to high-grade tumors that could induce epilepsy mainly through rapid tissue disruption and necrosis [58]. The most common seizure types in TAE are generalized tonic-clonic and complex partial seizures.

Evidence from both animal models [59, 60] and patients with TAE [61, 62] point towards the fact that crucial changes occur in the peritumoural region and not the immediate tumor invasion zone. These include glial cell swelling and damage resulting from hypoxia, acidosis [63] and altered expression of glutamate receptors in reactive astrocytes.

Numerous studies have shown that a frontal location of the tumor (and thus close to the premotor and motor cortex) is associated with a higher probability of associated epilepsy [64, 65]. Propagation of abnormal signals and/or a disruption of electrical activity transmitted to the primary motor cortex could be a logical explanation for the higher incidence of epilepsy when a tumor is located closer to the premotor cortex. In addition, the limbic and temporal lobe, the primary somatosensory cortex (S-I) and the opercula and insula regions of the secondary somatosensory area (S-II) also have a low threshold for producing seizures [65]. Notable exceptions include medial sphenoid wing meningiomas (where incidence of seizures is low) and seizure-like phenomena associated with posterior temporal-inferior occipital lesions [66].

2.4.4 Seizure Precipitants and Modulators

2.4.4.1 Sleep Deprivation

There is a controversy regarding seizures occurring after sleep deprivation. EEG recordings after sleep deprivation in epilepsy patients have been widely used, while sleep deprivation is known to activate epileptiform discharges independently of the activating effects of sleep [67]. However, seizures occurring after sleep deprivation are not classified as provoked by ILAE [15] and epilepsy specialists [68], since recurrence is far more likely than for patients with a provoked seizure. Further confounding to this issue, sleep deprivation is causing severe stress, a known seizure promoting factor, so it is hard to say whether epilepsy aggravation upon sleep deprivation is a direct effect (Chap. 3 will elaborate on the complex relationship between sleep and epilepsy).

2.4.4.2 Stress and Epilepsy

Seizures, particularly infantile spasms, have been linked to a stress-related elevation of CRH, which is responsible for the cascade leading to the production of adrenal steroids (such as cortisol). CRH has been found to act proconvulsively on seizure-prone brain regions such as the limbic system [69], by enhancing the actions of glutamate and suppressing afterhyperpolarisation [70]. Glucocorticoids limit the production of CRH as part of a negative feedback loop. Nevertheless, they activate CRH gene expression in the amygdala (a potential proconvulsive action) while adrenal-derived neuropeptides have both pro- and anticonvulsive properties [71]. Adrenocorticotropic hormone (ACTH) is released from the pituitary in response to hypothalamic CRH and has been proved to be anticonvulsive. ACTH also suppresses CRH, while inducing steroid synthesis, and thus is superior to oral steroids in the treatment of infantile spasms [72]. Under the influence of CRH, β-endorphin is released in an attempt of stress-level reduction. It acts as an endogenous opioid (μ-receptors), inhibiting the release of GABA and increasing dopamine levels [73] hence its excitatory effect.

2.4.4.3 Epilepsy and Reproductive Hormones

While men are reported to be more susceptible to epilepsy [74], sex-specific seizure propensity differs in various epilepsy syndromes. Sex-related differences are mainly thought to be the consequence of estrogen, progestin and androgens, which are also responsible for sexual brain dimorphism and behaviour patterns [75].

Ten to seventy percent of female patients with epilepsy has a greater propensity to seize in specific phases of their menstrual cycle, a "clustering" termed catamenial epilepsy, and observed in premenstrual and preovulatory periods, where the estrogen/progesterone ratio is high. Progesterone has antiepileptic properties, mediated by the GABAergic neurosteroid allopregnanolone, while it also has receptors in the limbic system. Estrogen, on the other hand, is primarily proconvulsive, by directly affecting glutamate receptor subtypes and increasing the density of hippocampal dendritic spines and excitatory synapses in animal models. Nevertheless, it has also been reported to increase neuropeptide Y concentration, with a possible antiepileptic effect [76].

2.5 Pathophysiology of Seizures and Epilepsy

Medicine is both art and science. The Hippocratic art of healing becomes a medical science to the extent that we succeed to identify which physiological function is changed in each disease and by what mechanism. Decades of neurological and neurosurgical observations have made numerous such links between the expression of seizures (loss of motor control, consciousness etc.) and the normal function of the particular brain areas which appear to initiate epileptic discharges and/or show lesions in each patient. In the process epileptology has become the greatest school for neuroscience, i.e. the description of the motor homunculus (for motor control), correlates of consciousness etc. More recently basic research in epileptology has elucidated many mechanisms underlying the expression of seizures at the macroscopic (brain imaging) and microscopic levels, especially in the ionic channels and neurotransmitter systems involved in the expression of seizures. Knowledge at the mesoscopic level (brain circuits) is seriously lagging behind. The most crucial question of mechanisms underlying epileptogenesis is still unanswered, although strides are currently made in this direction by molecular genetics and neuroplasticity research.

2.5.1 *Basic Neurophysiology*

GABA is the major inhibitory neurotransmitter in the brain. $GABA_A$ receptors coupled to chloride are involved to the hyperpolarizing of membranes which enables the inhibition of neuronal activity. The chloride mediated hyperpolarizing current counterbalances the excitatory input's depolarizing currents and needs to be overcome for an action potential to fire. As data of animal models suggest, the acute $GABA_A$ receptor blockade will produce epileptic activity while in addition the $GABA_A$ receptors are the targets of many anticonvulsant medications which are currently in use. Benzodiazepines and topiramate increase the rate of chloride channels opening while barbiturates increase the duration that these channels remain open thus decreasing excitability. $GABA_B$ receptors show a similar action through a different mechanism. $GABA_B$ receptors couple to potassium channels and form a current with longer duration of action compared to the $GABA_A$ receptor activation [77].

Glutamate on the other hand, is a major excitatory neurotransmitter in the brain, mediated by three main receptors: *N*-methyl-D-aspartic acid (NMDA), alpha-amino-3-hydroxy-5-methyl-4-isoxazole propionic acid (AMPA)/kainite and metabotropic. Selective NMDA receptor agonists show proconvulsant action while antagonists of the AMP/kainite and NMDA receptors have antiepileptic properties [78].

Glia cells and gap junctions between neuron cells facilitate voltage spread. GABA regulates the release and reuptake of neurotransmitters and transporters located both on glia and neurons. Furthermore, glia regulates the extracellular potassium concentration, which is correlated to neuronal excitability and epilepsy [79].

Two main types of ion channels exist. The ligand-gated channels which are activated by GABA, glutamate and acetylcholine, mediate cell communication at the synapse while the voltage-gated channels mediate the action potential and the axonal conduction of electrical signals. During resting state, the latter are closed and they open with the local membrane potential changes. Abnormalities of these channels lead to a disruption of membrane's depolarization and repolarization that is necessary for the action potential. Several studies have suggested that channelopathies can be the underlying cause of certain epileptic syndromes. Generalized Epilepsy with Febrile Seizures Plus (GEFS+) is linked to missense mutations in sodium (NaV1.1) channels, while complete loss-of-function mutations in NaV1.1 cause Severe Myoclonic Epilepsy of Infancy. Benign Neonatal Infantile Seizures is also caused by mutations in sodium (NaV1.2) channels [80]. Potassium channelopathies have been implicated in Sudden Unexplained Death in Epilepsy (SUDEP, Kv1.1) [81], while monogenic idiopathic generalized epilepsies (IGE) have been associated with mutations in $GABA_A$ receptor genes [82]. Absence epilepsies have also been linked to $GABA_A$ [83], $GABA_B$ [84] as well as voltage-gated calcium channel dysfunction [85].

Neuromodulators (such as neuropeptides and neurosteroids) are endogenous factors that influence the balance between excitation and inhibition in the brain. Neuropeptides regulate GABAergic and glutamatergic neurotransmission, as well as the monoaminergic system. Neurosteroids are locally synthesized from cholesterol and circulating steroid hormones, and mostly target membrane receptors.

Many neuropeptides have been identified as endogenous antiepileptics. However, only one is currently in use as a treatment (ACTH). Other anticonvulsant neuropeptides include neuropeptide Y, which increases neocortical GABAergic neurotransmission and hippocampal dopamine, and somatostatin (expressed, most importantly, in GABA-ergic, inhibitory hippocampal hilar interneurons) [86].

The list of proconvulsant neuropeptides includes, among others, corticotropin-releasing hormone (CRH), β-endorphin and arginine-vasopressin peptide (AVP). While the former two are implicated in stress-related mechanisms, high levels of AVP have been associated with febrile seizures, given that this peptide in lower levels also acts as an endogenous antipyretic [87].

2.5.2 Hypersynchrony, Hyperexcitation and Epileptogenesis

Current research converges on the notion that seizures erupt when two fundamental measures of neuronal activity, excitability and synchronization, surpass certain normal level and that epileptogenesis is based on an aberrant exploitation of the most fundamental property of our brain, neuronal plasticity, the very one that enables us to develop, adopt to the environment and learn.

The early Penfield and Jasper's hypothesis on generalized seizures introduced the idea of seizures correlating to synchronous brain activity characterized by decreased inhibition and enhanced excitation leading to the transient hypersynchronous activity

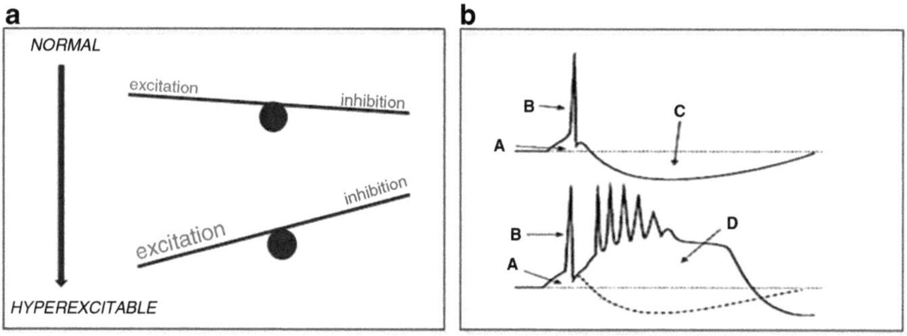

Fig. 2.1 (**a**) Simplified view of excitability changes towards an epileptiform activity. (**b**) Comparison of a normal neuronal discharge (*upper*) to an epileptiform one (*lower*). A = excitatory postsynaptic potential, B = sodium channel mediated action potential, C = inhibitory postsynaptic potential. Blockage of C leads to delayed and persistent expression of NMDA receptor channels mediated Ca^{++} spikes and paroxysmal depolarization shift (D)

of electrical epileptiform discharges [88]. The hallmarks of epilepsy which refer to hypersynchrony and hyperexcitation, still remain. Yet, synchronization in epilepsy is complex and it strongly depends on how 'synchrony' is defined, on the signals being measured (i.e. neuronal spikes or field potentials) and the spatial scale [89].

There is a variety of possible causes of hyperexitability. They can be *intrinsic* to neurons, like changes of types, numbers and opening times of voltage gated ion channels for Na^+ or Ca^{2+}, which may occur post-translationally, through 2nd messengers or through modulation of gene expression. They can also be *extrinsic*, like changes of local ion concentrations i.e. in $[K^+]$, and in extracellular space or modulation of transmitter metabolism or uptake by glial cells. Finally they can be *synaptic*, like increased excitatory post synaptic potentials (usually mediated by glutamate) and/or decreased inhibitory potentials (usually mediated by GABA), alterations in expression of transmembrane gated ionotropic channels, remodeling of synapse location or configuration, changes in gap-junction synaptic function etc. Simply put, it is a matter of balance tipped in favor of excitation in specific neurons (Fig. 2.1).

Synchronization can be increased in an equally large number of different ways like: recurrent excitation of neighbouring neurons, rhythms offering a narrow time window of higher opportunity to fire, gap junctions and rebound from synchronous inhibition.

Neuronal populations synchrony caused by an excitation/inhibition imbalance along with alterations in neural communication are key elements of epileptogenesis, while glia cells, through their extracellular modulation of environment, also play a role to it.

Hyperexcitability, i.e. burst of action potentials and paroxysmal depolarization shifts coupled with hypersynchronization of neighboring neurons may lead to seizure local initiation. Activation of connected neurons with concomitant loss of surround inhibition can lead to seizure propagation to other brain areas.

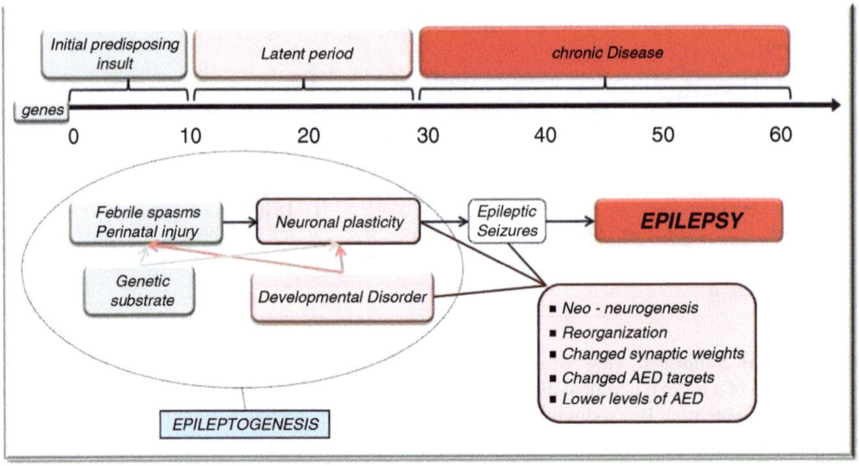

Fig. 2.2 Putative mechanisms of epileptogenesis

Epileptogenesis is a dynamic process which progresses until the manifestation of the first clinical seizure. Changes on neuronal interconnectivity and excitability are present [5], either in relation to a genetic susceptibility to seizures [90] or to an identifiable structural brain lesion [25].

One feature that facilitates seizures in the cortex is the existence of positive feedback loops between the pyramidal neurons which enable them to excite each other. Therefore, the connection structure of the neuronal network (rather than the density of existing neurons) seems to be of high importance to epileptogenesis [91, 92].

The fact that there is a wide diversity of seizure types, apparent causes and the variety of epileptic syndromes phenomenology [93] does not make it possible for a single pathophysiologic mechanism to underlie it all.

Epileptogenesis consists of a variety of intracellular, intrinsic membrane and extracellular mechanisms with neuronal plasticity and glia offering adaptation to environmental changes [94]. Therefore, the existence of a small group of neurons being responsible for epileptic discharges which then spread throughout the rest of the brain is also unlikely.

Epileptogenesis is considered to take several years in three stages (Fig. 2.2) and mainly unexplored remains the 2nd one: the latent period. If the initial insult is not countered several modification in the brain are escalated leading to epileptogenesis and expression of the first spontaneous seizures. Things may stabilize there or worsen in which case chronic epilepsy is established. Pharmacoresistance may develop either because antiepileptic drugs cannot cope with the new hypersynchronous and complex networks which newly emerge or because their membrane targets have changed or because mechanisms develop which decrease their brain levels.

The putative mechanisms of epileptogenesis as they emerge from research in several animal models have been recently reviewed [85, 95–101]. In retrospect, probably the most influential animal model has been that of kindling induced epilepsy in rodents [102]. An example of significant relevance between animal and human studies is the observation that hippocampus is the brain area most vulnerable to epileptogenesis and specifically its anterior part (in humans—ventral in rodents a distinction based on still unresolved underlying mechanisms (see [103, 104].

Epilepsy syndromes can be broadly separated in two categories: generalized and focal epilepsies. Focal onset seizures and generalized onset seizures differ in terms of the main mechanisms involved. While in both types of seizures cellular excitability is increased, the mechanisms of synchronization differ.

2.5.3 Pathophysiology of Focal Epilepsy

Pyramidal neurons receive a large amount of excitatory input in contrast to the relatively less inhibitory one. Yet, with inhibitory synapses being on the soma and close to axon hillock, inhibition is able to counterbalance the large amount of excitation received through 'gating it out', an effect known as the 'inhibition veto' [105].

A paradoxical action of inhibitory interneurons, being active during seizure onset, suggests this as the basis of a cellular correlate to the large scale discharges that are recorded in EEG. A period of intense activation of interneurons is followed by recurrent discharges of principal excitatory cells which had been 'silenced' during the inhibitory activation. These fast oscillations are important markers of the seizure onset zone. Possible mechanisms of this action is firstly the extracellular increase of potassium concentration which will cause depolarization of membranes and as a result the increase of excitatory neurons discharges and secondly the intracellular increase of chloride within the principal excitatory neuron which results to the impairment of inhibition and thus an excitation/inhibition imbalance. It has been shown that the reversed chloride gradient is correlated to GABA having an excitatory effect on neurons [106].

There are suggestions that the seizure onset zone is organized in microdomains which present synchronous high frequency firing in the form of 'microseizures' [107]. The merging of these cortical microdomains is proposed as the mechanism underlying the ictal transmission and the emergence of the macroseizure [89].

At the time that a seizure initiates, an early ictal desynchronization is noted. As the seizure progresses, this is followed by an increasing large scale synchronization [108, 109]. A possible explanation of this phenomenon is the 'inhibition veto' induced by the area ahead of the ictal waveform [105].

Long lasting repetitive seizures, as seen in status epilepticus, lead to a progressive and permanent modification of cortical neuronal networks thus repetitive seizures will lead to spontaneous recurrent seizures. This might also be true in the cases of acquired epileptogenesis for example after brain damage. The degeneration of GABAergic interneurons and the spouting of new glutamatergic circuits both leading to increased excitation is a consistent finding in both animal models and human tissue studies. This diminished inhibition permits the recurrent excitation and multisynaptic network activation while the remaining GABAergic inhibition prevents it from happening continually [110]. Repetitive seizures cause neuronal damage which strongly depends on the seizure severity as this is recorded by electrographic recordings. Neuronal death and synaptic regeneration appear to be important mechanisms to epileptogenesis.

Beyond the cell level though, epileptic seizures are also "multi- scale network phenomena". Networks can be represented as a graph made of nodes and links between them. Small worlds as they have been described by Watts and Strogatz [111], have many local links and a few of long-range connections. This characteristic allows them to locally process information and at the same time coordinate this local activity by long range connections, therefore balancing 'segregation' and 'integration' which are crucial for information processing. A network's topography (i.e. how the nodes are connected or in regard to epileptic seizures in the brain, cortex topography) will influence the dynamics that take place on the network.

Spatial scale's impact of synchronization can be seen in the high frequency interictal discharges. Populations of neurons synchronize their action potential firing which manifests as high-frequency oscillations with spectral peaks correlating to the inter-spike frequency of individual neurons at the 200–300 Hz. Also, in chronic epilepsy, the presence of interictal high frequency oscillations (i.e. fast ripples) which have a much higher firing frequency than the maximum firing rate of the pyramidal neuron, suggests the emergence of a network phenomenon generated by out-of-phase neuronal populations [89].

Finally, synchronization reaches its maximum close to seizure termination. It has been suggested that synchronization's enhancement or disruption promotes the seizure termination, proposing that once all of the available neural correlates are integrated within the synchronous paroxysmal activity, the seizure terminates because of the emergence of an extended 'hypoexcitable area'. A synchronized neural bursting can be terminated either by the decreased excitatory transmission caused by the massively increased membrane conductance [112] or by the synchronized inhibition; inhibition can be synchronized by synchronized excitation due to the interconnectivity between inhibitory interneurons and their connections to the principal excitatory cells [113]. On top of that, extracellular environmental changes can further contribute to seizure termination [114].

2.5.4 Pathophysiology of Generalized Epilepsy

Generalized seizures can be classified as atonic, tonic, clonic, tonic-clonic, myoclonic, or absence seizures on the basis of clinical symptoms and EEG abnormalities. Typical absence seizures and the interactions of the thalamocortical circuitry have been used as examples to understand the pathophysiological mechanisms underlying generalized seizures.

Especially the mechanisms underlying the conspicuous EEG rhythm of 3/s generalized spike and wave discharges (GSWD) and the transient "loss of consciousness" characterizing these seizures have been investigated for over eight

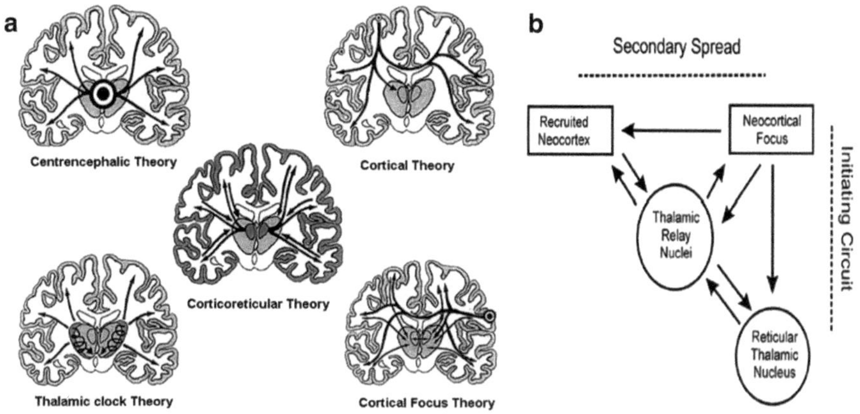

Fig. 2.3 (**a**) The historical development of ideas regarding the generation of synchronous activity in generalized absence epilepsy (modified from [120] see text therein). (**b**) The prevalent today prototype of epilepsy circuits (from [121])

decades because—although relatively benign—they are thought to constitute a unique electrographic and behavioral marker of the genetic predisposition to most types of epilepsy [115–117]. Interestingly, the subject is still controversial since both its classification terms, idiopathic and primary generalized, are recently proposed to be abandoned. The question of the neuronal mechanisms underlying the generation of GSWD and the concomitant "loss of consciousness" will be more properly dealt with in the next chapter (Chap. 3), since it relates to some mechanisms of sleep [118, 119]. Regarding the second term, all seizures, even those associated with what have historically been thought of as 'primary generalized' epilepsies, are considered now to originate in local microcircuits and then propagate from that initial ictogenic zone [7]. The matter goes beyond semantics and is of crucial practical importance to the clinic where the pharmacology of seizures known as generalized is quite distinct from that of focal epilepsies. The long history of conceptual developments as for the mechanisms by which rhythmical GSWD appear so very synchronous all over the brain (in titles in Fig. 2.3a) can be found in [6, 115, 117, 118, 120].

According to the prototype epileptic circuit (Fig. 2.3b) there is always a focus from where generalized seizures start. This neocortical focus develops in mutual influence with the thalamocortical neurons of its sector and initiates activity in GABAergic neurons of reticular thalamic nucleus. The interaction of thalamocortical and reticular neurons sets the pace of the oscillatory neuronal activity, while seizures spread through cerebral networks and subsequent involvement of the newly recruited cortex in cortico-thalamo-cortical reverberations, resulting in a globally synchronous EEG rhythm. According to this general prototype, locally-lead (mostly frontally) GSWD and absences have been referred to as "frontal absences" and thought to represent fast secondary generalization by a frontal focus [122]. They correspond to those with apparently localized onset of the ILAE guidance [26].

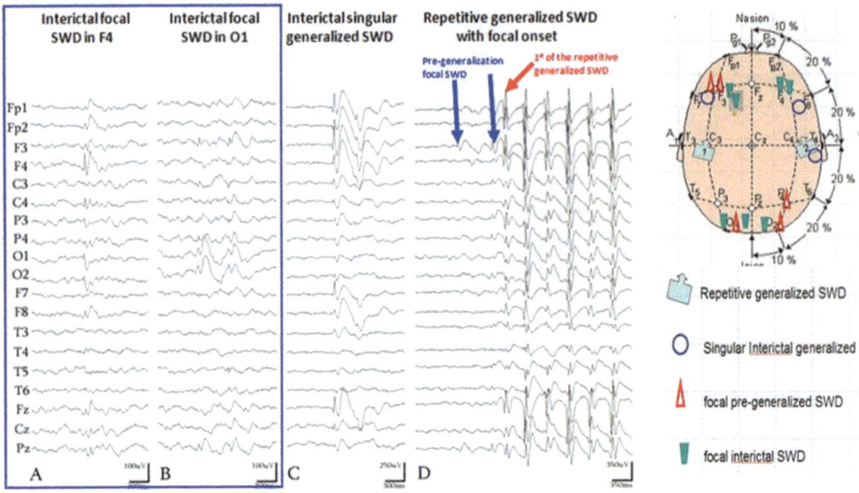

Fig. 2.4 Comparison of topographic features of the first SWD in a sustained 3/s GSWD (*red arrows*), singular interictal focal SWD (in *blue frame*) and "lead in" or pre-generalization focal SWD (modified from [124])

To better understand the nature of focal SWDs (FSWDs) in relation to GSWD in idiopathic generalized epilepsy Koutroumanidis and colleagues [123, 124] studied video-EEG recordings of children with typical childhood absence epilepsy. In particular they studied their behavior during the different phases of sleep microstructure and the topographic relation of the FSWDs to the EEG leading areas of the absences in each child. They found significant concordance (80 %) between the interictal FSWD (in blue frame in Fig. 2.4) and the "lead-in" or pre-generalization-FSWD (indicated by blue arrows in Fig. 2.4) in waveform, topography, in onset and propagation patterns as well in their affiliation to CAP-B periods. On the contrary the first generalized spike of the sustained 3/s GSWD (indicated by red arrows in Fig. 2.4) showed limited variability and only 8.2 % concordance with the focal SWD and the focal (pre-generalisation) SWD, while appearing in proximity to CAP-A and sleep transition periods. In general, focal SWDs were frontal or occipital, while the generalized mostly fronto-temporal or temporal (Fig. 2.4 right). These studies concluded that focal "lead in" or pre-generalization SWD are not likely the determinants of GSWD. They may reflect a system of multifocal non-localizing electrically unstable cortical areas that under the facilitatory influence of exogenous or endogenous factors like sleep instability can foster a corticothalamic response strong enough to generate 3 Hz GSWD in long preset networks—of autonomous onset location—that are conditionally sustainable and potentially ictal. FSWD can be viewed as incomplete forms of the GSWD; together they define the EEG identity of idiopathic "generalized" epileptogenesis. It is worth exploring the relationship between the characteristics of the focal as well as the generalized EEG signs and the variation in behavioral deficits in different absence seizures. Such electroclinical investigation may reveal what is lost in particular absence seizures [123, 124].

The recognition of focal onset generalized epilepsies does not take the magic away from the spectacular synchronization of GSWD and the associated behavioral changes of IGE. On the contrary it kindles interest on the brain mechanisms which have prepared (probably through long epileptogenesis) specific brain-wide circuits to be ready to respond to cortical hyperexcitability in this so well organized way. These considerations are in line with increasing acceptance of epilepsy as a network disease. Epileptic activity is increasingly considered as the dysfunction of a neuronal networks, a multi-entrance circuits rather than a single pinpoint source [125]. Gloor's 1968 paper on corticoreticular epilepsy (see Fig. 2.3a) had a pioneering influence in this direction [126]. Brain networks are neither orderly nor random; they are very complex and non-linear; but both of the latter are quantifiable. Recent observations suggest that networks acquire larger path lengths and clustering coefficients near the beginning of the seizure and that become more small-world during seizure propagation and more random at seizure termination, when there is also increased coupling. Contradiction in data exist, but the total evidence converges to the suggestion of a refinement of the traditional idea that seizures are hypersynchronous events [127]. In parallel or perhaps in consequence to the network aspects of epilepsy there is a very interesting fermentation about the concept of "system epilepsies" [128–130] to describe some types of epilepsy that depend on the dysfunction of specific functional neural systems.

Typical absence seizures and the interactions of the thalamocortical circuitry are examples used to understand the pathophysiological mechanisms underlying generalized seizures. The thalamocortical circuitry consists of the pyramidal neurons of the neocortex, the thalamic relay nucleus and the thalamic nucleus reticularis and it generates the sleep spindles observed during NREM sleep. The thalamocortical systems thus appears as the primary hub of neural processes common to sleep and generalized epilepsy and possibly underlying their mutual relationships.

References

1. Staley K (2015) Molecular mechanisms of epilepsy. Nat Neurosci 18(3):367–372. doi:10.1038/nn.3947
2. Engel J Jr (2013) Progress in the field of epilepsy. Curr Opin Neurol 26(2):160–162. doi:10.1097/WCO.0b013e32835ee5a3
3. Panayiotopoulos CP (2005) Idiopathic generalized epilepsies: a review and modern approach. Epilepsia 46(Suppl 9):1–6. doi:10.1111/j.1528-1167.2005.00330.x
4. Fisher RS, van Emde BW, Blume W, Elger C, Genton P, Lee P, Engel J Jr (2005) Epileptic seizures and epilepsy: definitions proposed by the International League Against Epilepsy (ILAE) and the International Bureau for Epilepsy (IBE). Epilepsia 46(4):470–472. doi:10.1111/j.0013-9580.2005.66104.x
5. Engel J Jr, Pedley TA (2005) What is epilepsy? In: Engel J (ed) Epilepsy: a comprehensive textbook. Lippincott Williams & Wilkins\Raven Press, Philadelphia, pp 1–11
6. Avoli M (2012) A brief history on the oscillating roles of thalamus and cortex in absence seizures. Epilepsia 53(5):779–789. doi:10.1111/j.1528-1167.2012.03421.x

7. Paz JT, Huguenard JR (2015) Optogenetics and epilepsy: past, present and future. Epilepsy Curr 15(1):34–38. doi:10.5698/1535-7597-15.1.34
8. Temkin O (1945) The falling sickness. The John Hopkins Press, Baltimore, p 380
9. Saraceno BA, G, Lee P (2005) Atlas: epilepsy care in the World
10. Thurman DJ, Beghi E, Begley CE, Berg AT, Buchhalter JR, Ding D, Hesdorffer DC, Hauser WA, Kazis L, Kobau R, Kroner B, Labiner D, Liow K, Logroscino G, Medina MT, Newton CR, Parko K, Paschal A, Preux PM, Sander JW, Selassie A, Theodore W, Tomson T, Wiebe S, ILAE Commission on Epidemiology (2011) Standards for epidemiologic studies and surveillance of epilepsy. Epilepsia 52(Suppl 7):2–26. doi:10.1111/j.1528-1167.2011.03121.x
11. Forsgren L, Beghi E, Oun A, Sillanpaa M (2005) The epidemiology of epilepsy in Europe—a systematic review. Eur J Neurol 12(4):245–253. doi:10.1111/j.1468-1331.2004.00992.x
12. Effertz T, Mann K (2013) The burden and cost of disorders of the brain in Europe with the inclusion of harmful alcohol use and nicotine addiction. Eur Neuropsychopharmacol 23(7):742–748. doi:10.1016/j.euroneuro.2012.07.010
13. Guekht AB, Mitrokhina TV, Lebedeva AV, Dzugaeva FK, Milchakova LE, Lokshina OB, Feygina AA, Gusev EI (2007) Factors influencing on quality of life in people with epilepsy. Seizure 16(2):128–133. doi:10.1016/j.seizure.2006.10.011
14. Isaacs KL, Philbeck JW, Barr WB, Devinsky O, Alper K (2004) Obsessive-compulsive symptoms in patients with temporal lobe epilepsy. Epilepsy Behav 5(4):569–574. doi:10.1016/j.yebeh.2004.04.009
15. Fisher RS, Acevedo C, Arzimanoglou A, Bogacz A, Cross JH, Elger CE, Engel J Jr, Forsgren L, French JA, Glynn M, Hesdorffer DC, Lee BI, Mathern GW, Moshe SL, Perucca E, Scheffer IE, Tomson T, Watanabe M, Wiebe S (2014) ILAE official report: a practical clinical definition of epilepsy. Epilepsia 55(4):475–482. doi:10.1111/epi.12550
16. Surges R, Thijs RD, Tan HL, Sander JW (2009) Sudden unexpected death in epilepsy: risk factors and potential pathomechanisms. Nat Rev Neurol 5(9):492–504. doi:10.1038/nrneurol.2009.118
17. Simonato M, Brooks-Kayal AR, Engel J Jr, Galanopoulou AS, Jensen FE, Moshe SL, O'Brien TJ, Pitkanen A, Wilcox KS, French JA (2014) The challenge and promise of antiepileptic therapy development in animal models. Lancet Neurol 13(9):949–960. doi:10.1016/S1474-4422(14)70076-6
18. Morris GL, Gloss D, Buchhalter J, Mack KJ, Nickels K, Harden C (2013) Evidence-based guideline update: vagus nerve stimulation for the treatment of epilepsy: report of the Guideline Development Subcommittee of the American Academy of Neurology. Epilepsy Curr 13(6):297–303. doi:10.5698/1535-7597-13.6.297
19. Krumholz A, Wiebe S, Gronseth G, Shinnar S, Levisohn P, Ting T, Hopp J, Shafer P, Morris H, Seiden L, Barkley G, French J, Quality Standards Subcommittee of the American Academy of Neurology, American Epilepsy Society (2007) Practice parameter: evaluating an apparent unprovoked first seizure in adults (an evidence-based review): report of the Quality Standards Subcommittee of the American Academy of Neurology and the American Epilepsy Society. Neurology 69(21):1996–2007
20. DeToledo JC, Ramsay RE (1996) Patterns of involvement of facial muscles during epileptic and nonepileptic events: review of 654 events. Neurology 47(3):621–625
21. Rosenow F, Luders H (2001) Presurgical evaluation of epilepsy. Brain 124(Pt 9):1683–1700
22. Martinkovic L, Hecimovic H, Sulc V, Marecek R, Marusic P (2014) Modern techniques of epileptic focus localization. Int Rev Neurobiol 114:245–278. doi:10.1016/B978-0-12-418693-4.00010-8
23. Pohlmann-Eden B, Beghi E, Camfield C, Camfield P (2006) The first seizure and its management in adults and children. BMJ 332(7537):339–342. doi:10.1136/bmj.332.7537.339
24. Kattimani S, Bharadwaj B (2013) Clinical management of alcohol withdrawal: a systematic review. Ind Psychiatry J 22(2):100–108. doi:10.4103/0972-6748.132914
25. Engel J, Jr., International League Against E (2001) A proposed diagnostic scheme for people with epileptic seizures and with epilepsy: report of the ILAE Task Force on Classification and Terminology. Epilepsia 42(6):796–803

26. Berg AT, Berkovic SF, Brodie MJ, Buchhalter J, Cross JH, van Emde BW, Engel J, French J, Glauser TA, Mathern GW, Moshe SL, Nordli D, Plouin P, Scheffer IE (2010) Revised terminology and concepts for organization of seizures and epilepsies: report of the ILAE Commission on Classification and Terminology, 2005-2009. Epilepsia 51(4):676–685. doi:10.1111/j.1528-1167.2010.02522.x
27. Berg AT, Cross JH (2012) Classification of epilepsies and seizures: historical perspective and future directions. Handb Clin Neurol 107:99–111. doi:10.1016/B978-0-444-52898-8.00005-7
28. Berg AT (2012) Introduction: changing terms and concepts for epilepsy. Epilepsia 53(Suppl 2):1–2. doi:10.1111/j.1528-1167.2012.03549.x
29. Thom M (2014) Review: hippocampal sclerosis in epilepsy: a neuropathology review. Neuropathol Appl Neurobiol 40(5):520–543. doi:10.1111/nan.12150
30. Verhaert K, Scott RC (2010) Acute symptomatic epileptic seizures. In: Panayiotopoulos CP (ed) Atlas of epilepsies. Springer, London
31. Italiano D, Ferlazzo E, Gasparini S, Spina E, Mondello S, Labate A, Gambardella A, Aguglia U (2014) Generalized versus partial reflex seizures: a review. Seizure 23(7):512–520. doi:10.1016/j.seizure.2014.03.014
32. Reid AY, Galic MA, Teskey GC, Pittman QJ (2009) Febrile seizures: current views and investigations. Can J Neurol Sci 36(6):679–686
33. Pavlidou E, Hagel C, Panteliadis C (2013) Febrile seizures: recent developments and unanswered questions. Childs Nerv Syst 29(11):2011–2017. doi:10.1007/s00381-013-2224-3
34. Dahlgren L, Wilson RD (2001) Prenatally diagnosed microcephaly: a review of etiologies. Fetal Diagn Ther 16(6):323–326
35. Kim JH (2001) Pathology of epilepsy. Exp Mol Pathol 345–367
36. Wong JH, Awad IA, Kim JH (2000) Ultrastructural pathological features of cerebrovascular malformations: a preliminary report. Neurosurgery 46(6):1454–1459
37. Annegers JF, Coan SP (2000) The risks of epilepsy after traumatic brain injury. Seizure 9(7):453–457. doi:10.1053/seiz.2000.0458
38. Agrawal A, Timothy J, Pandit L, Manju M (2006) Post-traumatic epilepsy: an overview. Clin Neurol Neurosurg 108(5):433–439. doi:10.1016/j.clineuro.2005.09.001
39. Raymont V, Salazar AM, Lipsky R, Goldman D, Tasick G, Grafman J (2010) Correlates of posttraumatic epilepsy 35 years following combat brain injury. Neurology 75(3):224–229. doi:10.1212/WNL.0b013e3181e8e6d0
40. Spencer S (2010) Intracranial electrodes. In: Engel J (ed) Epilepsy, a comprehensive textbook. Lippincott, Williams & Wilkins, Philadelphia, pp 1791–1815
41. Frey LC (2003) Epidemiology of posttraumatic epilepsy: a critical review. Epilepsia 44(Suppl 10):11–17
42. Bratton SL, Chestnut RM, Ghajar J, McConnell Hammond FF, Harris OA, Hartl R, Manley GT, Nemecek A, Newell DW, Rosenthal G, Schouten J, Shutter L, Timmons SD, Ullman JS, Videtta W, Wilberger JE, Wright DW (2007) XIII. Antiseizure prophylaxis. J Neurotrauma 24(supplement 1):S-83–S-86. doi:10.1089/neu.2007.9983
43. Lamar CD, Hurley RA, Rowland JA, Taber KH (2014) Post-traumatic epilepsy: review of risks, pathophysiology, and potential biomarkers. J Neuropsychiatry Clin Neurosci 26(2):iv–113. doi:10.1176/appi.neuropsych.260201
44. Willmore LJ, Ueda Y (2009) Posttraumatic epilepsy: hemorrhage, free radicals and the molecular regulation of glutamate. Neurochem Res 34(4):688–697. doi:10.1007/s11064-008-9841-3
45. Chalak LF, Rollins N, Morriss MC, Brion LP, Heyne R, Sanchez PJ (2012) Perinatal acidosis and hypoxic-ischemic encephalopathy in preterm infants of 33 to 35 weeks' gestation. J Pediatr 160(3):388–394. doi:10.1016/j.jpeds.2011.09.001
46. Szydlowska K, Tymianski M (2010) Calcium, ischemia and excitotoxicity. Cell Calcium 47(2):122–129. doi:10.1016/j.ceca.2010.01.003
47. Lin TN, Kim GM, Chen JJ, Cheung WM, He YY, Hsu CY (2003) Differential regulation of thrombospondin-1 and thrombospondin-2 after focal cerebral ischemia/reperfusion. Stroke 34(1):177–186

48. Avramescu S, Nita DA, Timofeev I (2009) Neocortical post-traumatic epileptogenesis is associated with loss of GABAergic neurons. J Neurotrauma 26(5):799–812. doi:10.1089/neu.2008.0739
49. Blumcke I, Coras R, Miyata H, Ozkara C (2012) Defining clinico-neuropathological subtypes of mesial temporal lobe epilepsy with hippocampal sclerosis. Brain Pathol 22(3):402–411. doi:10.1111/j.1750-3639.2012.00583.x
50. Novy J, Belluzzo M, Caboclo LO, Catarino CB, Yogarajah M, Martinian L, Peacock JL, Bell GS, Koepp MJ, Thom M, Sander JW, Sisodiya SM (2013) The lifelong course of chronic epilepsy: the Chalfont experience. Brain 136(Pt 10):3187–3199. doi:10.1093/brain/awt117
51. de Tisi J, Bell GS, Peacock JL, McEvoy AW, Harkness WF, Sander JW, Duncan JS (2011) The long-term outcome of adult epilepsy surgery, patterns of seizure remission, and relapse: a cohort study. Lancet 378(9800):1388–1395. doi:10.1016/S0140-6736(11)60890-8
52. von Lehe M, Lutz M, Kral T, Schramm J, Elger CE, Clusmann H (2006) Correlation of health-related quality of life after surgery for mesial temporal lobe epilepsy with two seizure outcome scales. Epilepsy Behav 9(1):73–82. doi:10.1016/j.yebeh.2006.03.014
53. Bower SP, Kilpatrick CJ, Vogrin SJ, Morris K, Cook MJ (2000) Degree of hippocampal atrophy is not related to a history of febrile seizures in patients with proved hippocampal sclerosis. J Neurol Neurosurg Psychiatry 69(6):733–738
54. Blumcke I, Thom M, Aronica E, Armstrong DD, Bartolomei F, Bernasconi A, Bernasconi N, Bien CG, Cendes F, Coras R, Cross JH, Jacques TS, Kahane P, Mathern GW, Miyata H, Moshe SL, Oz B, Ozkara C, Perucca E, Sisodiya S, Wiebe S, Spreafico R (2013) International consensus classification of hippocampal sclerosis in temporal lobe epilepsy: a Task Force report from the ILAE Commission on Diagnostic Methods. Epilepsia 54(7):1315–1329. doi:10.1111/epi.12220
55. O'Connor ER, Sontheimer H, Spencer DD, de Lanerolle NC (1998) Astrocytes from human hippocampal epileptogenic foci exhibit action potential-like responses. Epilepsia 39(4): 347–354
56. de Lanerolle NC, Lee TS, Spencer DD (2012) Histopathology of human epilepsy. In: Noebels JL, Avoli M, Rogawski MA, Olsen RW, Delgado-Escueta AV (eds) Jasper's basic mechanisms of the epilepsies, 4th edn. National Center for Biotechnology Information, Bethesda
57. Cohen I, Navarro V, Clemenceau S, Baulac M, Miles R (2002) On the origin of interictal activity in human temporal lobe epilepsy in vitro. Science 298(5597):1418–1421. doi:10.1126/science.1076510
58. Riva M (2005) Brain tumoral epilepsy: a review. Neurol Sci 26(Suppl 1):S40–S42. doi:10.1007/s10072-005-0404-y
59. Senner V, Kohling R, Puttmann-Cyrus S, Straub H, Paulus W, Speckmann EJ (2004) A new neurophysiological/neuropathological ex vivo model localizes the origin of glioma-associated epileptogenesis in the invasion area. Acta Neuropathol 107(1):1–7. doi:10.1007/s00401-003-0771-5
60. Buckingham SC, Campbell SL, Haas BR, Montana V, Robel S, Ogunrinu T, Sontheimer H (2011) Glutamate release by primary brain tumors induces epileptic activity. Nat Med 17(10):1269–1274. doi:10.1038/nm.2453
61. Patt S, Steenbeck J, Hochstetter A, Kraft R, Huonker R, Haueisen J, Haberland N, Ebmeier K, Hliscs R, Fiehler J, Nowak H, Kalff R (2000) Source localization and possible causes of interictal epileptic activity in tumor-associated epilepsy. Neurobiol Dis 7(4):260–269. doi:10.1006/nbdi.2000.0288
62. Aronica E, Yankaya B, Jansen GH, Leenstra S, van Veelen CW, Gorter JA, Troost D (2001) Ionotropic and metabotropic glutamate receptor protein expression in glioneuronal tumours from patients with intractable epilepsy. Neuropathol Appl Neurobiol 27(3):223–237
63. Kempski O, Staub F, von Rosen F, Zimmer M, Neu A, Baethmann A (1988) Molecular mechanisms of glial swelling in vitro. Neurochem Pathol 9:109–125
64. Chang EF, Potts MB, Keles GE, Lamborn KR, Chang SM, Barbaro NM, Berger MS (2008) Seizure characteristics and control following resection in 332 patients with low-grade gliomas. J Neurosurg 108(2):227–235. doi:10.3171/JNS/2008/108/2/0227

65. Hamasaki T, Yamada K, Yano S, Nakamura H, Makino K, Hide T, Hasegawa Y, Kuroda J, Hirai T, Kuratsu J (2012) Higher incidence of epilepsy in meningiomas located on the premotor cortex: a voxel-wise statistical analysis. Acta Neurochir 154(12):2241–2249. doi:10.1007/s00701-012-1511-1
66. Beaumont A, Whittle IR (2000) The pathogenesis of tumour associated epilepsy. Acta Neurochir 142(1):1–15
67. Fountain NB, Kim JS, Lee SI (1998) Sleep deprivation activates epileptiform discharges independent of the activating effects of sleep. J Clin Neurophysiol 15(1):69–75
68. Lawn N, Lieblich S, Lee J, Dunne J (2014) Are seizures in the setting of sleep deprivation provoked? Epilepsy Behav 33:122–125. doi:10.1016/j.yebeh.2014.02.008
69. Baram TZ (2007) Models for infantile spasms: an arduous journey to the Holy Grail. Ann Neurol 61(2):89–91. doi:10.1002/ana.21075
70. Aldenhoff JB, Gruol DL, Rivier J, Vale W, Siggins GR (1983) Corticotropin releasing factor decreases postburst hyperpolarizations and excites hippocampal neurons. Science 221(4613):875–877
71. Reddy DS (2009) The role of neurosteroids in the pathophysiology and treatment of catamenial epilepsy. Epilepsy Res 85(1):1–30. doi:10.1016/j.eplepsyres.2009.02.017
72. Jaseja H, Jaseja B, Badaya S, Tonpay P (2012) Superior therapeutic efficacy of adrenocorticotrophic hormone (ACTH) in infantile spasms: emerging evidence. Epilepsy Behav 25(2):250. doi:10.1016/j.yebeh.2012.08.003
73. Koneru A, Satyanarayana SR, Rizwan S (2009) Endogenous opioids: their physiological role and receptors. Global J Pharmacol 3:149–153
74. McHugh JC, Delanty N (2008) Epidemiology and classification of epilepsy: gender comparisons. Int Rev Neurobiol 83:11–26. doi:10.1016/S0074-7742(08)00002-0
75. Scharfman HE, MacLusky NJ (2014) Sex differences in the neurobiology of epilepsy: a preclinical perspective. Neurobiol Dis 72(Pt B):180–192. doi:10.1016/j.nbd.2014.07.004
76. Reddy DS (2014) Neurosteroids and their role in sex-specific epilepsies. Neurobiol Dis 72(Pt B):198–209. doi:10.1016/j.nbd.2014.06.010
77. Kaila K, Ruusuvuori E, Seja P, Voipio J, Puskarjov M (2014) GABA actions and ionic plasticity in epilepsy. Curr Opin Neurobiol 26:34–41. doi:10.1016/j.conb.2013.11.004
78. Ure J, Baudry M, Perassolo M (2006) Metabotropic glutamate receptors and epilepsy. J Neurol Sci 247(1):1–9. doi:10.1016/j.jns.2006.03.018
79. Foresti ML, Arisi GM, Shapiro LA (2011) Role of glia in epilepsy-associated neuropathology, neuroinflammation and neurogenesis. Brain Res Rev 66(1–2):115–122. doi:10.1016/j.brainresrev.2010.09.002
80. Catterall WA (2012) Sodium channel mutations and epilepsy. In: Noebels JL, Avoli M, Rogawski MA, Olsen RW, Delgado-Escueta AV (eds) Jasper's basic mechanisms of the epilepsies, 4th edn. National Center for Biotechnology Information, Bethesda
81. Glasscock E, Yoo JW, Chen TT, Klassen TL, Noebels JL (2010) Kv1.1 potassium channel deficiency reveals brain-driven cardiac dysfunction as a candidate mechanism for sudden unexplained death in epilepsy. J Neurosci 30(15):5167–5175. doi:10.1523/JNEUROSCI.5591-09.2010
82. Cossette P, Lachance-Touchette P, Rouleau GA (2012) Mutated GABAA receptor subunits in idiopathic generalized epilepsy. In: Noebels JL, Avoli M, Rogawski MA, Olsen RW, Delgado-Escueta AV (eds) Jasper's basic mechanisms of the epilepsies, 4th edn. National Center for Biotechnology Information, Bethesda
83. Crunelli V, Leresche N, Cope DW (2012) GABAA receptor function in typical absence seizures. In: Noebels JL, Avoli M, Rogawski MA, Olsen RW, Delgado-Escueta AV (eds) Jasper's basic mechanisms of the epilepsies, 4th edn. National Center for Biotechnology Information, Bethesda
84. Han HA, Cortez MA, Snead OC (2012) GABAB receptor and absence epilepsy. In: Noebels JL, Avoli M, Rogawski MA, Olsen RW, Delgado-Escueta AV (eds) Jasper's basic mechanisms of the epilepsies, 4th edn. National Center for Biotechnology Information, Bethesda

85. Noebels JL (2012) The voltage-gated calcium channel and absence epilepsy. In: Noebels JL, Avoli M, Rogawski MA, Olsen RW, Delgado-Escueta AV (eds) Jasper's basic mechanisms of the epilepsies, 4th edn. National Center for Biotechnology Information, Bethesda
86. Clynen E, Swijsen A, Raijmakers M, Hoogland G, Rigo JM (2014) Neuropeptides as targets for the development of anticonvulsant drugs. Mol Neurobiol 50(2):626–646. doi:10.1007/s12035-014-8669-x
87. Richmond CA (2003) The role of arginine vasopressin in thermoregulation during fever. J Neurosci Nurs 35(5):281–286
88. Penfield W, Jasper H (1954) Epilepsy and the functional anatomy of the human brain. Little, Brown and Company, Boston
89. Jiruska P, de Curtis M, Jefferys JG, Schevon CA, Schiff SJ, Schindler K (2013) Synchronization and desynchronization in epilepsy: controversies and hypotheses. J Physiol 591(Pt 4):787–797. doi:10.1113/jphysiol.2012.239590
90. Zara F, Bianchi A (2009) The impact of genetics on the classification of epilepsy syndromes. Epilepsia 50(Suppl 5):11–14. doi:10.1111/j.1528-1167.2009.02113.x
91. Lothman EW, Bertram EH 3rd, Stringer JL (1991) Functional anatomy of hippocampal seizures. Prog Neurobiol 37(1):1–82
92. Schevon CA, Cappell J, Emerson R, Isler J, Grieve P, Goodman R, McKhann G Jr, Weiner H, Doyle W, Kuzniecky R, Devinsky O, Gilliam F (2007) Cortical abnormalities in epilepsy revealed by local EEG synchrony. Neuroimage 35(1):140–148. doi:10.1016/j.neuroimage.2006.11.009
93. Benbadis SR (2001) Epileptic seizures and syndromes. Neurol Clin 19(2):251–270
94. Lukasiuk K, Pitkanen A (2009) Seizure induced gene expression. In: Schwartzkroin PA (ed) Encyclopedia of basic epilepsy research. Academic, Oxford, pp 1302–1309
95. McNamara JO, Huang YZ, Leonard AS (2006) Molecular signaling mechanisms underlying epileptogenesis. Sci STKE 356:re12. doi:10.1126/stke.3562006re12
96. Armijo JA, Valdizan EM, De Las CI, Cuadrado A (2002) Advances in the physiopathology of epileptogenesis: molecular aspects. Rev Neurol 34(5):409–429
97. Scharfman HE, Brooks-Kayal AR (2014) Is plasticity of GABAergic mechanisms relevant to epileptogenesis? Adv Exp Med Biol 813:133–150. doi:10.1007/978-94-017-8914-1_11
98. Kandratavicius L, Balista PA, Lopes-Aguiar C, Ruggiero RN, Umeoka EH, Garcia-Cairasco N, Bueno-Junior LS, Leite JP (2014) Animal models of epilepsy: use and limitations. Neuropsychiatr Dis Treat 10:1693–1705. doi:10.2147/NDT.S50371
99. Ritter LM, Golshani P, Takahashi K, Dufour S, Valiante T, Kokaia M (2014) WONOEP appraisal: optogenetic tools to suppress seizures and explore the mechanisms of epileptogenesis. Epilepsia 55(11):1693–1702. doi:10.1111/epi.12804
100. Coppola A, Moshe SL (2012) Animal models. Handb Clin Neurol 107:63–98. doi:10.1016/B978-0-444-52898-8.00004-5
101. Scharfman HE (2007) The neurobiology of epilepsy. Curr Neurol Neurosci Rep 7(4):348–354
102. McNamara JO (1986) Kindling model of epilepsy. Adv Neurol 44:303–318
103. Kostopoulos G, Moschovos C, Papatheodoropoulos C (2005) Functional differentiation along the longitudinal axis of hippocampus and its possible relevance to epileptogenesis. Epileptic Disord 7(2):167–170
104. Bekiari C, Grivas I, Giannakopoulou A, Michaloudi-Pavlou H, Kostopoulos GK, Papadopoulos GC (2015) Dentate gyrus variation along its septo-temporal axis: structure and function in health and disease. In: Lowes Z (ed) Dentate gyrus: structure, role in disease, and potential health implications. Nova, New York
105. Trevelyan AJ, Sussillo D, Watson BO, Yuste R (2006) Modular propagation of epileptiform activity: evidence for an inhibitory veto in neocortex. J Neurosci 26(48):12447–12455. doi:10.1523/JNEUROSCI.2787-06.2006
106. Avoli M, de Curtis M (2011) GABAergic synchronization in the limbic system and its role in the generation of epileptiform activity. Prog Neurobiol 95(2):104–132. doi:10.1016/j.pneurobio.2011.07.003

107. Stead M, Bower M, Brinkmann BH, Lee K, Marsh WR, Meyer FB, Litt B, Van Gompel J, Worrell GA (2010) Microseizures and the spatiotemporal scales of human partial epilepsy. Brain 133(9):2789–2797. doi:10.1093/brain/awq190
108. Schiff SJ, Sauer T, Kumar R, Weinstein SL (2005) Neuronal spatiotemporal pattern discrimination: the dynamical evolution of seizures. Neuroimage 28(4):1043–1055. doi:10.1016/j.neuroimage.2005.06.059
109. Cymerblit-Sabba A, Schiller Y (2012) Development of hypersynchrony in the cortical network during chemoconvulsant-induced epileptic seizures in vivo. J Neurophysiol 107(6):1718–1730. doi:10.1152/jn.00327.2011
110. Ben-Ari Y, Dudek FE (2010) Primary and secondary mechanisms of epileptogenesis in the temporal lobe: there is a before and an after. Epilepsy Curr 10(5):118–125. doi:10.1111/j.1535-7511.2010.01376.x
111. Watts DJ, Strogatz SH (1998) Collective dynamics of "small-world" networks. Nature 393(6684):440–442. doi:10.1038/30918
112. Gutkin BS, Laing CR, Colby CL, Chow CC, Ermentrout GB (2001) Turning on and off with excitation: the role of spike-timing asynchrony and synchrony in sustained neural activity. J Comput Neurosci 11(2):121–134
113. Pavlov I, Kaila K, Kullmann DM, Miles R (2013) Cortical inhibition, pH and cell excitability in epilepsy: what are optimal targets for antiepileptic interventions? J Physiol 591(Pt 4):765–774. doi:10.1113/jphysiol.2012.237958
114. de Curtis M, Manfridi A, Biella G (1998) Activity-dependent pH shifts and periodic recurrence of spontaneous interictal spikes in a model of focal epileptogenesis. J Neurosci 18(18):7543–7551
115. Avoli M, Gloor P, Kostopoulos G, Naquet R (1990) Generalized epilepsy: neurobiological approaches. Birkhauser, Boston
116. Kostopoulos G (1997) Neuronal mechanisms underlying loss of consciousness during absence seizures. In: Kostopoulos G, Rakovic D, Koruga D, Rakic L (eds) Brain and consciousness. ECPD of the UN, Belgrade, pp 185–193
117. Kostopoulos GK (2001) Involvement of the thalamocortical system in epileptic loss of consciousness. Epilepsia 42(Suppl 3):13–19
118. Kostopoulos GK (2000) Spike-and-wave discharges of absence seizures as a transformation of sleep spindles: the continuing development of a hypothesis. Clin Neurophysiol 111(Suppl 2):S27–S38
119. Kostopoulos GK (2009) Brain mechanisms linking epilepsy to sleep. In: Schwartzkroin PA (ed) Encyclopedia of basic epilepsy research, vol 3. Academic, Oxford, pp 1327–1336
120. Meeren H, van Luijtelaar G, Lopes da Silva F, Coenen A (2005) Evolving concepts on the pathophysiology of absence seizures: the cortical focus theory. Arch Neurol 62(3):371–376. doi:10.1001/archneur.62.3.371
121. Bertram EH (2013) Neuronal circuits in epilepsy: do they matter? Exp Neurol 244:67–74. doi:10.1016/j.expneurol.2012.01.028
122. Holmes MD, Brown M, Tucker DM (2004) Are "generalized" seizures truly generalized? Evidence of localized mesial frontal and frontopolar discharges in absence. Epilepsia 45(12):1568–1579. doi:10.1111/j.0013-9580.2004.23204.x
123. Koutroumanidis M, Tsiptsios D, Kokkinos V, Kostopoulos GK (2012) Focal and generalized EEG paroxysms in childhood absence epilepsy: topographic associations and distinctive behaviors during the first cycle of non-REM sleep. Epilepsia 53(5):840–849. doi:10.1111/j.1528-1167.2012.03424.x
124. Kokkinos V, Koupparis AM, Kostopoulos GK (2013) An intra-K-complex oscillation with independent and labile frequency and topography in NREM sleep. Front Hum Neurosci 7:163. doi:10.3389/fnhum.2013.00163
125. Spencer SS (2002) Neural networks in human epilepsy: evidence of and implications for treatment. Epilepsia 43(3):219–227
126. Kostopoulos GK, Gloor P (2013) Epilepsy is a network disease. In: 30th International Epilepsy Congress, Montreal

127. Kramer MA, Cash SS (2012) Epilepsy as a disorder of cortical network organization. Neuroscientist 18(4):360–372
128. Koutroumanidis M (2007) Panayiotopoulos syndrome: an important electroclinical example of benign childhood system epilepsy. Epilepsia 48(6):1044–1053. doi:10.1111/j.1528-1167.2007.01096.x
129. Avanzini G, Manganotti P, Meletti S, Moshe SL, Panzica F, Wolf P, Capovilla G (2012) The system epilepsies: a pathophysiological hypothesis. Epilepsia 53(5):771–778. doi:10.1111/j.1528-1167.2012.03462.x
130. Halasz P (2013) How sleep activates epileptic networks? Epilepsy Res Treat 2013:425697. doi:10.1155/2013/425697

Chapter 3
Sleep Features and Underlying Mechanisms Related to Epilepsy and Its Long Term Monitoring

Evangelia Giourou, Alkistis Stavropoulou-Deli, Konstantinos Theofilatos, George K. Kostopoulos, Andreas A. Ioannides, and Michalis Koutroumanidis

Abstract The reciprocal relationship between sleep and epilepsy has been recognized since antiquity. The exact mechanisms underlying the precise nature of this relation though, remain unclear even today. The scope of this chapter is to describe the basic neurophysiologic mechanisms underlying sleep and its relation to the interictal epileptiform discharges and epileptic seizures. Furthermore, the theoretical background of the mechanisms involved in the interaction of sleep and epilepsy are discussed especially the effects that sleep mechanisms have on altering brain synchrony and excitability, which consist the hallmark of epileptiform activity in the brain. Finally, aspects of the open problems in polysomnographic long term monitoring of epilepsy are examined, which the ARMOR approached aimed to address.

3.1 Introduction

Epilepsy and sleep are strongly interrelated at many levels [1–3]. Several epilepsy syndromes present seizures only during sleep, like the nocturnal frontal lobe epilepsy (NFLE), while others like the Landau-Kleffner syndrome, electrical status epilepticus during slow wave sleep, childhood epilepsy with occipital paroxysms

E. Giourou • A. Stavropoulou-Deli • K. Theofilatos • G.K. Kostopoulos (✉)
Neurophysiology Unit, Department of Physiology, Medical School, University of Patras, Patras, Greece
e-mail: gkkostop@upatras.gr

A.A. Ioannides
Laboratory for Human Brain Dynamics, AAI Scientific Cultural Services Ltd., Nicosia, Cyprus

M. Koutroumanidis
Department of Clinical Neurophysiology and Epilepsy, Guy's and St Thomas' NHS Foundation Trust, London, UK

Department of Academic Neurosciences, Kings College London, London, UK

N.S. Voros, C.P. Antonopoulos (eds.), *Cyberphysical Systems for Epilepsy and Related Brain Disorders*, DOI 10.1007/978-3-319-20049-1_3

and benign epilepsy with centrotemporal spikes, are correlated with seizures occurring during sleep or around awakening [4, 5].

This close relationship between sleep and epilepsy has been recognized since the nineteenth century. William Gowers subdivided epilepsies with regards to their periodicity and highlighted the effects of sleep on epilepsy recognizing the different seizure distribution and semiology of epileptic syndromes during the sleep wake cycle [6].

Seizures or epileptiform activity during sleep, lead to sleep fragmentation and increased arousals, which has serious implications not only to the subjective sleep quality leading to morning tiredness and increased daytime sleepiness or even insomnia but also to the general patient's quality of life [7]. Results of epileptiform activity during sleep often extend to deterioration of cognitive functions as well, such as memory consolidation [8].

Furthermore, parasomnias, a common comorbidity, seem to infer to sleep quality but to also pose an extra diagnostic challenge in their distinction from nocturnal epilepsies [9, 10].

Antiepileptic medications are also major contributors to sleep disturbances, even though their exact role remains unclear. In studies in patients with epilepsy the direct effects of the drug used are hard to be deciphered from the effects of altered epileptic control or the underlying findings of the condition per se. Yet, antiepileptic drugs modulating GABA and glutamate neurotransmission and altering sodium and calcium channels functions seem to have direct effects on sleep other than their anticonvulsant actions. Newer AEDS though, such as levetiracetam have fewer negative effects and some such as gabapentin or lamotrigine even show some positive effects stabilizing sleep [11–13].

Sleep deprivation aggravates seizures and is known to be an activator of EEG epileptiform discharges, which has led to its use as a diagnostic aid [14]. Yet, since sleep deprivation usually occurs during physical overactivity or psychological distress, one cannot be certain whether the seizure provoking characteristics are due specifically to the lack of sleep per se or are caused by mechanisms underlying stress non-specifically [15].

Because of this reciprocal relationship between sleep and epilepsy, research related to the basic mechanisms of sleep–wake regulation and the anatomic regions involved, is essential for our understanding of the mechanisms that govern epileptogenesis and its clinical expression. Electroencephalographic findings, seizure types and frequency, vary depending on the sleep–wake cycle.

Epileptic seizures and EEG discharges alter the architecture of sleep [16]. They can influence both sleep's macrostructure, such as increasing stage shifts and the number and the duration of awakenings while reducing and fragmenting REM sleep, but also sleep's microstructure, such as the Cyclical Alternating Patterns (CAPs) and even phasic events such as K-Complexes and sleep spindles [17].

Epileptic seizures tend to occur during NREM sleep [18] suggesting their relation to hypersynchronization. Furthermore they tend to occur during CAP phases (mainly CAP-A) indicating their correlation to arousals or sudden changes in excitability during sleep [19]. Indeed sudden provoked awakening during sleep is noted as a precipitant of seizures [20].

3.2 Sleep Physiology

In order to shed light on the relationship between sleep and epilepsy, is clearly important to understand the dynamics underlying sleep and epilepsy's EEG correlates.

Sleep consists of distinct stages which result from a complex interplay between brain regions that are responsible for the transition from the waking to the sleeping state and vice versa [21]. Sleep onset is controlled by two factors (Fig. 3.1a): a circadian influence by the hypothalamic suprachiasmatic pacemaker nucleus and homeostatic influences exerted in hypothalamus by substances like adenosine accumulating in hypothalamus with increasing wake time. According to the proposed "flip-flop" scheme (Fig. 3.2a), this hypothalamic activation inactivates the centers of the upper brainstem which are known to tonically maintain wakefulness, i.e. the ascending reticular activating system containing aminergic (noradrenaline,

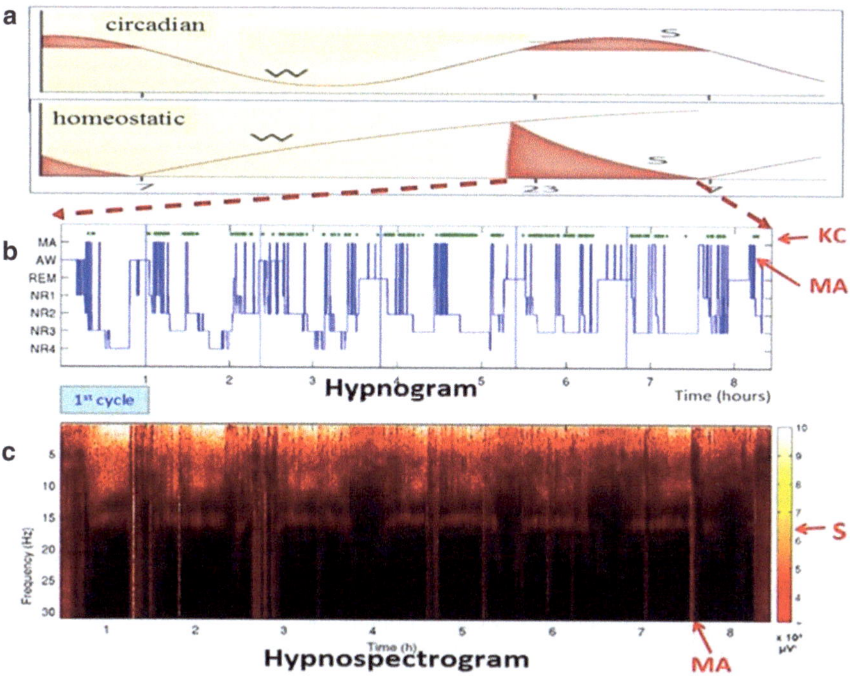

Fig. 3.1 Sleep macro- and microstructure. (**a**) Sleep time (S) is controlled by both circadian and homeostatic factors, the latter building up as waking (W) progresses and diminishing slowly after sleep onset. (**b**) On the basis of EEG, EOG and EMG features a whole night's sleep is described by a hypnogram (*blue*) denoting about 6 cycles of about 90 min duration, each containing a NREM and a REM part. NREM includes four stages (N1–N4). Elements of sleep microstructure are shown as K-complexes (KC, *green dots*), microarousals (MA) or rare awakenings (AW). (**c**) Time frequency analysis (hypnospectrogram color coding EEG power) for the same sleep reveals the frequent presence of spindles (S, 12–15 Hz) mainly during stage 2 of NREM sleep

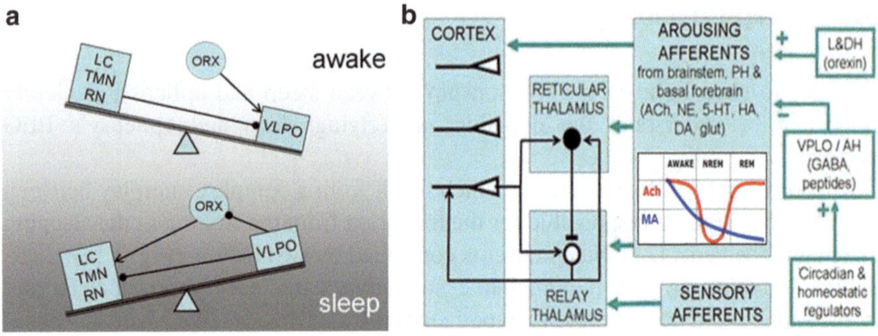

Fig. 3.2 Mechanisms promoting sleep. (**a**) According to a seesaw model sleep starts when hypothalamic centers like nucleus posterolateralis (VLPO) inactivate those centers of brainstem which maintain wakefulness like locus coeruleus (LC, releasing norepinephrine), tuberomamillary n. (TMN, releasing histamine), raphe n. (RN, releasing serotonin) and others. Additionally VLPO inactivates orexinergic (ORX) neurons which also promote arousal. *Arrows* and *filled circles* denote respectively excitatory and inhibitory connections. (**b**) Simplified diagram of brain circuits involved in sleep onset. AH and L&DH indicate respectively anterior, lateral and dorsal hypothalamus. Ach: acetylcholine, NE: norepinephrine, 5-HT: serotonin, HA: histamine, DA: dopamine, glut: glutamate. While all monoamines (MA, *blue curve*) decrease throughout sleep, Acetylcholine (*red curve*) is the only neuromodulator supporting REM sleep. The induced by hypothalamus diminution of ascending inputs to thalamus and cortex promotes oscillatory activity in the frequency of spindles or delta waves which blocks sensory processing from reaching the cortex

serotonin and dopamine) and cholinergic neurons, which project diffusely to the entire nervous system (Fig. 3.2b). The homeostatic factor constituting the sleep pressure will gradually decrease as sleep progresses and activation of brainstem arousing centers will start inactivating the hypothalamic ones leading to awakening (Fig. 3.2a). For the duration of sleep, the patterning of sleep stages is additionally influenced by a daylong oscillation of arousal with a period of about 90 min and by the interaction of specific REM-on and REM-off neurons in the brainstem to produce 5–6 cycles in a night's sleep (Fig. 3.1b), each of which consist of NREM and REM periods [22]. The above sleep promoting centers are located in the anterior hypothalamus preoptic area and contain mostly neurons releasing as neurotransmitters GABA and the neuropeptide galanin. In contrast, posterior hypothalamus is the home of two types of neurons which functionally belong to the activating ones like those in the brainstem: and respectively release histamine and orexin. It is quite impressive that the distinction between hypothalamic sleep promoting anterior and wake promoting posterior areas, was proposed as early as 1920 by Constantin von Economo, based on the lesions he observed in patients with lethargic encephalitis.

Additionally, 'external' influences (i.e. circadian rhythms and sensory stimuli on the Reticular Activating System—RAS) can 'switch' between sleep and wakefulness.

Wakefulness is maintained through the tonic activity in the RAS from the upper brain stem and the thalamocortical and cortical projections from the posterior hypothalamus and the basal forebrain. The transition to sleep is caused by both a withdrawal of external stimuli from the RAS and an activation of the hypothalamus' preoptic area [23]. This leads to a widespread increase in GABAergic activity that gives rise to the relative synchronous oscillations of NREM sleep such as sleep spindles, delta activity and the slow cortical oscillations.

At the microstructure level of NREM sleep organization, the cyclic alternating pattern (CAP) alternates with periods of quiescence (non-CAP periods) (Fig. 3.4c); NREM sleep is thus divided into periods showing CAP and periods not showing CAP. During the CAP periods, phases containing arousal phenomena and therefore indicate increased vigilance (phase A of CAP) alternate with phases without arousal phenomena, which correspond to reduced vigilance (phase B of CAP) [24].

Sleep spindles (Figs. 3.1c and 3.4a) generated by the reticular nucleus of the thalamus (RE) and its connections to the dorsal thalamus (Fig. 3.2b), consist of a 10–14 Hz activity lasting 2–3 s occurring especially in stages 2 and 3 of NREM sleep. The spindle oscillation is transferred to the thalamocortical relay cells, traveling to the cortex, generating rhythmic excitatory postsynaptic potentials and a synchronous activity over widespread cortical regions that can be observed at the macroscopic/EEG level [25]. Inhibitory RE to RE neurons collaterals desynchronize spindle activity limiting both the amplitude and the duration of sleep spindles [26].

Modeling of spindles from the EEG and MEG signal was remarkably unsuccessful in illuminating the nature of the generators that appeared to be dispersed in different areas with little consistent regularity within and across events. This was rather surprising given the rather widespread and consistent spindle imprint on the EEG (in terms of signal morphology of the sigma-band-pass filtered signal of central and dorsal EEG sensors. One plausible explanation was that the problem arose because the source localization algorithms did not have sufficient power to disentangle activity from a mixture of superficial and deep generators. In recent years a series of studies focused on the problem of identifying the generators of spindles from EEG, MEG, simultaneous EEG and MEG [27, 28], EEG triggered fMRI and invasive measurements in humans [29, 30] begin to converge to a rather surprising but consistent picture: spindles are sporadic events that at least in the cortex occur asynchronously over wide areas and with high variability from event to event. Our own tomographic analysis confirms the high variability of generator loci from spindle to spindle but provides a unified picture of the evolution of activity from the awake state through core periods of light sleep and the periods before and during spindles and K-complexes (Ioannides et al., in preparation). All of the established features of spindles are recovered by the tomographic analysis, but they appear strongly when the appropriate comparison is made between conditions. An example is demonstrated in Fig. 3.3: comparing the spectral density of individual events during and just before spindles shows the anterior foci of the low sigma activity and the more posterior foci in the high sigma periods. In addition and in a more ventral level increases during the spindles are seen for the entire sigma band in posterior cingulate and in the right caudate [31].

Spindles appear to play at least three roles. (a) shaping the early organization of sensory-motor cortical circuits [32] (b) maintaining sleep by increasing the threshold

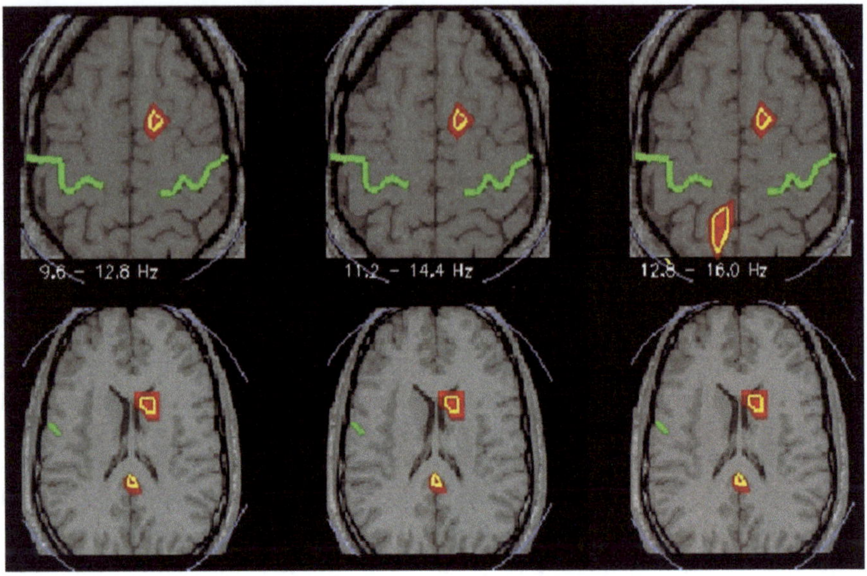

Fig. 3.3 Spectral statistics of MEG tomographic data showing statistical significant changes of activity common to all four subjects in three narrow frequency bands, alpha and low sigma band (9.6–12.8 Hz) statistical parametric, main sigma band (11.2–14.4 Hz) and high sigma band (12.8–16 Hz). *Yellow* outline marks common for all 4 subjects statistics at $p<0.05$. The *green line* indicates the central sulcus. The left hemisphere is on the *left*. The MRIs at the more dorsal/upper (15) and more ventral/lower (11) level are slightly off scale to fit the space provided

perceiving sensory stimuli [33] and (c) helping memory consolidation through a cortico-hippocampal interaction [34].

The delta oscillation (1–4 Hz) has both a thalamic [35] and a cortical component [36] and seems to result from the same circuit that generates sleep spindles. However, spindle and delta activity exclude each other within the thalamocortical neuron, with spindle activity arising at relatively depolarized resting membrane potentials and delta activity arising at more hyperpolarized membrane potentials [37].

Slow cortical oscillations at frequencies bellow 1 Hz, observed during slow wave sleep, are generated in the cerebral cortex with different cortical areas synchronizing in a widespread manner through corticocortical connections. K complexes (Fig. 3.1b), a fluctuating arousal occurring at periodic intervals during NREM sleep, can be triggered by both external sensory stimuli as well as internal very slow sleep oscillations [38]. They appear to have a very dynamic relationship to spindles and to theta bursts (Fig. 3.4a, b) [39, 40].

K complexes, sleep spindles and arousals but also epileptiform discharges and seizures (more about this later) seem to occur during the A phase of CAP and seem to be inhibited during the B phase of CAP (Fig. 3.4c) [41].

During REM, the selective reactivation of the RAS cholinergic system with neurons releasing serotonin or norepinephrine on the thalamocortical system remaining

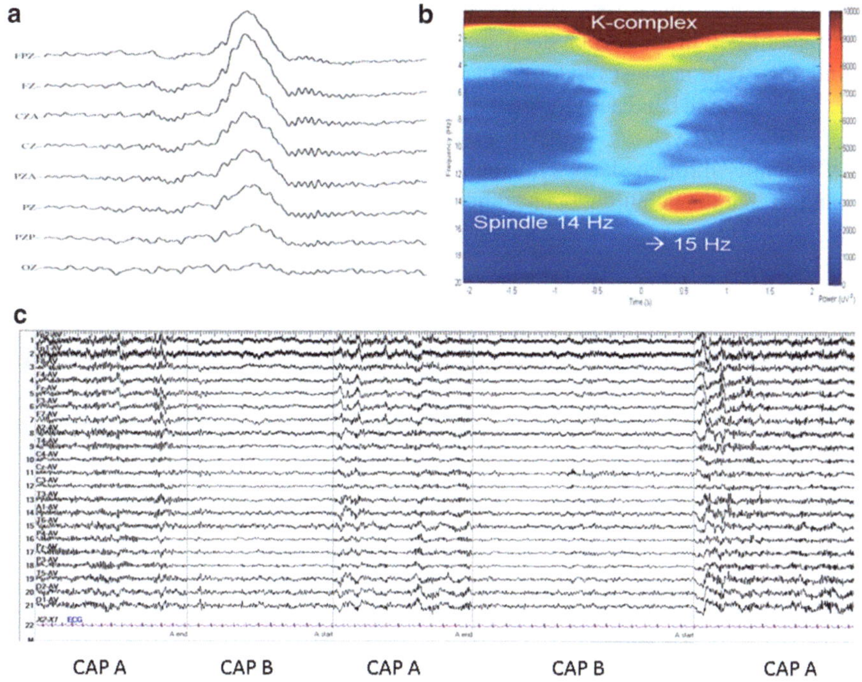

Fig. 3.4 Elements of sleep microstructure putatively involved in seizure expression. (**a**) K-complex followed by spindles in central EEG electrodes FPZ to OZ. (**b**) Averaged spectrogram of many K-complexes. Note the interruption of preceding spindle, development of a theta burst and the increased frequency of the spindles following the K-complex. (**c**) Cycle alternating patter in EEG during NREM with most power concentrating in periods "A"

inactive, leads to the observed EEG activity which is similar to what it is observed during wakefulness but with muscle atonia. Lastly, the transition to waking, is caused by the activation of the cholinergic as well as monoaminergic nuclei in the brainstem (Fig. 3.2b) and a resultant activation of the thalamocortical cells [42].

Thanks to the tomographic analysis of MEG data the changes during sleep can be studied non-invasively using MEG. The first tomographic analysis of whole night MEG data revealed the complexity of the processes involved [43] but also provided hints of a global organization and a relatively smooth transition from one stage to the next when the quiet rather than active periods of each sleep stage were compared [44]. The most consistent change running through all sleep staged was identified in near-midline areas on the left hemisphere, one in the frontal lobe and the other in the posterior mid-parietal area. In these two areas the gamma band activity was higher during REM sleep than other sleep stages and active wakefulness. A meta-analysis of recent neuroimaging studies showed that these two areas are at the centre of foci of increased activity identified in experiments with increased resting state activity compared to task periods (areas belonging to the "default system") and areas identified in experiments requiring understanding own and other peoples intention and introspection (areas of the "Theory of Mind" system). It seems significant that the areas identified in the earlier sleep study, the default mode areas and the Theory of mind area were not just randomly placed with respect to each other. The areas of increased gamma band activity were at the centre and not overlapping with the areas of the two other systems, suggesting they play a pivotal role in maintaining the person's identity through sleep. This makes us dare pushing beyond the suggestion of Domhoff that the default network might be involved in dreaming [45, 46]; we could suggest that the areas identified as super-active in the gamma band during sleep, and especially the dorsal medial prefrontal cortex are acting like a "Dream Box" releasing consciousness while the rest of the brain is largely subdue. Our recent research, focusing on light sleep suggests that an area close to the "Dream Box" areas is also important for spindles (Ioannides et al., in preparation).

3.3 The Interictal State and Sleep

Abnormal electrical discharges occurring in the time between seizures are referred to as 'interictal spikes'. The summation of this abnormal electrical activity along with the signs and symptoms, within a broad psychiatric and behavioral range, is referred to as the 'Interictal State' [47].

Interictal spikes consist of abnormal discharges of an inappropriately synchronized population of neurons from focal brain areas. Unlike seizures, interictal discharges do not spread across large brain areas and they do not cause clinical symptoms [48].

Generalized spike-wave discharges in idiopathic generalized epilepsy aggravate as NREM sleep progresses and after awakening, while they diminish during REM sleep. These discharges tend to occur during CAP-A phases of NREM sleep. A study on patients with JME showed that their distribution in CAP phases (A or B) may relate to (or even predict) seizure control. Increased epileptic pressure may cause disruption of the inhibitory mechanisms of phase B, increase the CAP rate by contributing to more A phases, and thereby foster more epileptiform discharges through the CAP A window. In other words there seems to be an epileptic positive feedback with clinical correlates: increased seizure activity is associated with enhanced intrusion of spike wave activity into phase B of CAP sleep, increased CAP rate, more epileptiform discharges and by implication higher probability of having more seizures. Increased electrographic awakenings fragment sleep and may independently contribute to the clinical deterioration by impairing sleep quality [49]. Similar findings were noted in childhood absence epilepsy [50]. Also, generalized spike–wave discharges alter morphologically during NREM sleep [51, 52].

Also, in partial epilepsies NREM activates interictal discharges while REM suppresses them without being clear though, which stages of NREM aggravate discharges; some studies suggest that interictal discharges increase during NREM stages 1–3 [53] and others suggest that they increase during NREM stages 3–4 instead [54].

Besides NREM's effect on the rate that interictal discharges occur, discharges seem to also be more widespread in terms of their partial extent during NREM than during wakefulness or REM sleep. Yet, it has been debated that interictal discharges observed during NREM are of a lower value to localizing the true epileptic focus in contrast to activity observed during REM and wakefulness [54], since there is a large percentage of novel epileptiform foci, unrelated to the true epileptic focus, observed during NREM even from the contralateral hemisphere.

Lastly, ripples (80–250 Hz) and fast ripples (>200–250 Hz), which seem to be generated within the neocortex or the hippocampus, and have been correlated to the interictal to ictal transition [55] also seem to be enhanced by NREM sleep [56]. Schevon et al. [57] suggested that fast ripples are produced by cortical domains near locally excitable clusters that produce microdischarges and are better correlated to interictal epileptiform events. High frequency oscillations, the increasingly recognized biomarkers of the epileptogenic zone [58, 59] are facilitated by slow waves in NREM sleep [60].

All mentioned above, indicate the relation of epileptiform activity to the increased neural synchronization that is observed during NREM sleep as it is to be discussed below.

3.4 Epileptic Seizures and Sleep

Similarly to interictal discharges, seizure activity is also correlated to sleep and it is modified both by the sleep wake cycle as well as the sleep stage. Early observations since the nineteenth century have tried to classify epilepsies into diurnal vs. nocturnal or diffuse [61] and Janz at 1962 [62] described the awakening epilepsies, which occur upon or soon after awaking and seem to be primary generalized in nature.

Idiopathic generalized epilepsies (IGE) include awakening epilepsies while only a minority of them may present with EEG abnormalities seen only during sleep; even then, seizures (absences, myoclonic and tonic-clonic) occur during wakefulness. Idiopathic focal epilepsies can occur during sleep in as much as 80 % of cases [63]. Frontal lobe seizures typically occur during NREM sleep, particularly during stage 2 (N2) (Fig. 3.5—nocturnal seizure with genital automatisms) while temporal lobe seizures are more likely to generalize during sleep [64]. In NFLE the vast majority of seizures will occur during sleep with NREM sleep enhancing focal epileptiform activity in association to the A phase of CAP A [65].

Epilepsy with continuous spike wave during slow wave sleep (CSWS) and Landau–Kleffner syndrome, are also strictly correlated to the sleep wake state, both showing a continuous spike wave activity during sleep [66].

Sleep stage is a key factor in seizure occurrence and characteristics. NREM sleep facilitates seizure onset and seizure spread whereas REM suppresses them [14]. Slow wave sleep is the state of maximum synchronicity in the brain while K complexes and sleep transients that are often correlated to epileptiform activity during lighter stages of sleep, are related to patterns of periodic arousal instability as it is

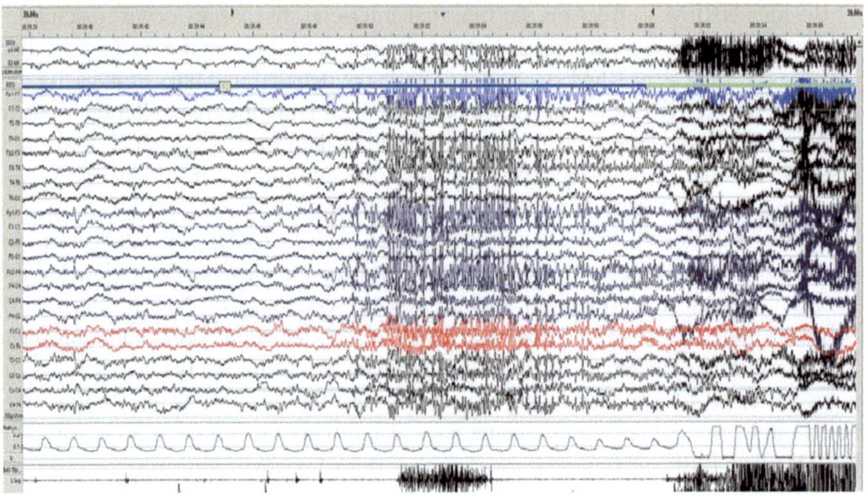

Fig. 3.5 Nocturnal frontal lobe seizure, seemingly arising from N3. The actual seizure discharge starts at the *blue arrow*, but sleep stage has already changed from N3 to lighter 3 s before (*green arrow*) as evidenced by the diffuse increase of faster rhythms. The patient engages in bilateral genital automatisms of which he has no recollection

described by the CAP [65], both indicating the relation to the hypersynchronization and hyperexcitability which characterize epileptiform activity.

Finally, sleep deprivation has been correlated to seizure inducing and precipitating epileptiform discharges possibly by inducing NREM sleep but also through affecting cortical excitability [14].

3.5 Mechanisms Underlying the Effect of Sleep on Epilepsy

3.5.1 Sleep Mechanisms Altering Brain Synchrony and Excitability

The theoretical background of the mechanisms involved in the interaction of sleep and epilepsy include the existence of shared neuronal circuits between sleep and epilepsy, the increased synchronization that is evident during certain sleep stages and seems to facilitate epileptiform activity, and various intrinsic characteristics of the epileptic focus which are influenced by sleep related activity.

Epileptiform activity is generally characterized by hypersynchronization and hyperexcitability. During sleep and wakefulness the level of neural synchronization varies. During NREM sleep there is increased neural synchronization compared to REM sleep or wakefulness. This suggests that different levels of neural synchronization can interfere with the abnormal synchronization occurring during ictal and interictal discharges, promoting it during NREM or suppressing it during REM or wakefulness [14].

Yet neural synchronization during NREM varies [67] and frequencies observed in the delta, theta and sigma band fluctuate during the course of sleep. There are conflicting data on what aspects of the neural activity during NREM promotes epileptiform discharges. The depth of sleep and the log delta power [68], the progression through deeper stages of NREM sleep or lighter stages of NREM sleep [69] and changes in the power in the sigma or theta band [53, 70–73], all have been correlated with the rate of discharges but fail to clearly predict which of the activity seen during NREM promotes discharges. A possible explanation of these varied data lies in the need of a necessary concurrence of neuronal activity of the epileptic loci with the activity of the rest of the brain temporally and spatially. Sigma band is best observed in the frontocentral regions; therefore it is expected to predict discharges coming from these areas [14]. Similarly the epileptic loci may carry some intrinsic rhythmicity itself which could fall within several frequencies normally occurring in the brain. Therefore, a similar extrinsic frequency of the brain would boost and be related to the intrinsic frequency of the epileptic loci. If the intrinsic frequency falls within sigma band for example, it is expected to be better correlated to power in the sigma band or sleep spindles [74].

A recent study showed that during NREM sleep epileptic spike discharges and high frequency oscillations appeared not continuously but with highest rate in association to larger slow waves and particularly during the highly synchronized transition from "down" to "up" states underlying these waves; not during the "up" states, which is the case with physiological activity [60]. The authors concluded that the activation of epileptic discharges during NREM sleep is not a state-dependent phenomenon but is predominantly associated with specific events, and apparently facilitated by increased synchronization rather than by increased excitability. Understanding the prime cause however remains as challenge, because dynamic bistability of neuronal membrane potentials widespread synchronization are mutually dependent and reinforced (see [3]).

On another note, intrinsic features of the epileptic focus may be responsible for the varied behavior regarding the enhancement or otherwise a relative immunity to the extrinsic synchronized rhythm which is present during NREM sleep [74]. There is data in literature where epileptic foci can be selectively enhanced by NREM in contrast to the rest of the brain or, in other cases, epileptic foci may lose the ability to be modulated by sleep related activity. This could explain the persistence of local hyperexcitability and hypersynchronization of a local area even in the absence of similar extrinsic activity. Yet, for the epileptiform activity to be spread to other regions, as seen to the intractable symptomatic and secondary generalized epilepsies, the synchronized activity that is present during NREM can facilitate the transmission of the epileptiform activity to the proximate normal cortex [74].

Additionally, in terms of generalized epileptiform discharges, it is speculated that sleep events and rhythms such as spindles, K complexes and delta oscillations share common or overlapping neuronal circuits involved in the generation of generalized discharges.

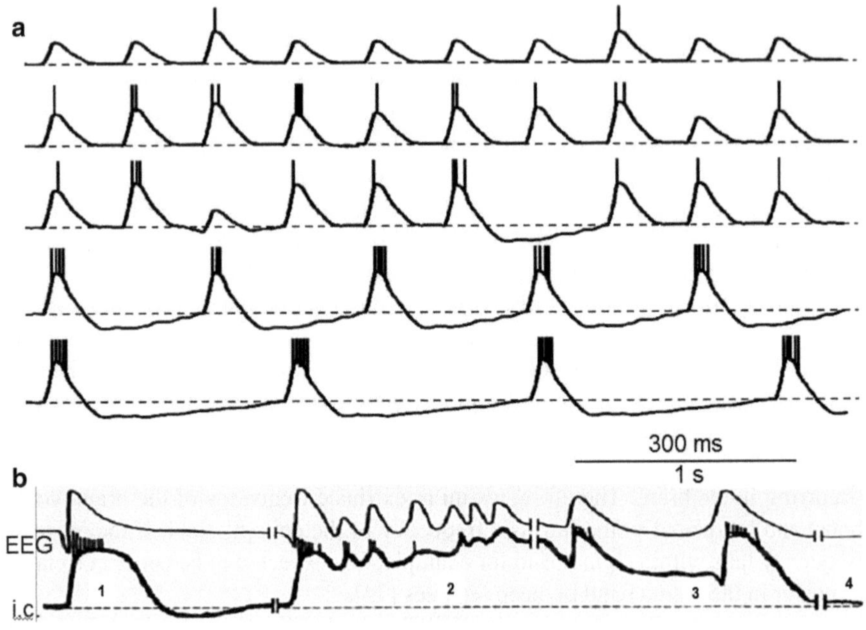

Fig. 3.6 (**a**) Diagram of suggested successive changes in membrane potential of PN during the transition from spindles to SWDs in the feline model absence seizures, as deduced from different experiments at the intracellular (i.c.) level, at successive times following small doses of barbiturates and penicillin. The observation that under these conditions spindles appear to be gradually replaced by SWDs of half or a third of the spindles frequency (in a step-like transition), lead to the following hypothesis: SWDs may be generated because the pyramidal neurons' usually subthreshold excitatory postsynaptic potentials (EPSPs) corresponding to each EEG spindle wave (*top trace*), upon an increase in cortical excitability, progressively (from *top* to *bottom traces*) become more and more effective in firing and so they activate cortical interneurons whose increasing inhibitory postsynaptic potentials annul the next EPSP and further on the next two EPSPs. (**b**) Diagram of intracellular (i.c.) events underlying EEG characteristics of generalized tonic-clonic seizures (produced by higher doses of penicillin in the same experiment), serves to show the fundamental difference in mechanisms underlying the two seizure models: Excitability of interictal spike (1) is not just increased passed threshold but clearly abnormal (paroxysmal depolarization shift). Inhibition, while instrumental in SWDs, is maintained only in (1) and is completely lost during the tonic (2) and clonic (3) phases. Finally there is postictal depression (4). Modified from [79]

Spike and wave discharges (SWDs), the electrographic hallmark of typical absence seizures, which are an integral component of several idiopathic generalized epilepsies [5], have been reported to occur preferentially during the light stages of NREM sleep, where the majority of sleep spindles are observed and in a reverse relationship to their rate throughout the night. Gloor in 1978 [75] proposed that the same TC circuit producing sleep spindles could generate SWDs in states of cortical hyperexcitability. The hypothesis found support and mechanistic explanation in a series of experiments in the model of feline generalized epilepsy with penicillin (Fig. 3.6a) and developed further on the basis of in vitro and in vivo experiments, especially those revealing the neuronal mechanisms of spindles generation [76] and those very fruitfully using rodent genetic models of absence seizures [77–83].

One of the most important recent discoveries in the field has been the identification of a cortical ‘initiation site’ of SWDs of rodent absence seizures [84]. Also high density EEG as well as MEG and fMRI studies in patients with different types of idiopathic generalized epilepsy have shown SWDs in discrete, mainly frontal and parietal cortical regions before they appear over the rest of the cortex [85–89]. This novel view is consistent with the above hypothesis (Fig. 3.6a) because they describe variable expressions of—the inherent in this hypothesis—cortical hyperexcitability/instability, not with the readiness of the thalamocortical system to so accurately synchronize the entire brain at specific frequency, which can be presumed to be a product of long latent period epileptogenic process on a specific genetic background. This epileptogenic process appears to cardinally involve changes in the thalamocortical mechanisms which generate sleep spindles.

These studies emphasize the importance of (a) mutual interaction between the sleep and epilepsy, (b) recognizing that different types of epilepsy may have fundamentally different mechanisms (compare Fig. 3.6a to Fig. 3.6b) and (c) the significance of observing the “bigger picture” in both time (i.e. CAP periods) and brain space, since both sleep and epilepsy by definition involve dynamic changes in large brain circuits.

Epileptiform discharges can be an expression of corrupted normal sleep rhymes [26]. Both sleep spindles and spike-wave discharges can be generated by modulating the degree of GABAergic inhibition in the thalamus [74].

The reticular nucleus (RE) intrinsic oscillatory properties generate the sleep spindles, which are then amplified by the reciprocal connections from the thalamocortical neurons to the RE. In addition downward projections from the cortical neurons to the thalamocortical and RE trigger and synchronize them across cortical neurons. GABAergic neurons inhibit the thalamortical neurons which project via glutamatergic synapses to cortical neurons and back to the RE. Also, inhibition among RE neurons limits synchronization within the RE thus limiting size and duration of spindles. RAS and sensory excitatory input can further modulate activity of the cortical, thalamocortical and reticular neurons [74].

In the theory of the spindle generating thalamocortical circuit, the reduced GABAergic inhibition within the reticular nucleus which would increase the degree of synchronization in the RE—thalamocortical neuron circuit or otherwise the increased excitation of RE and/or thalamocortical neurons, can lead to the replacement of sleep spindles by thalamic oscillations that are synchronized and epileptiform. Increased excitation to the RE can also explain the alteration of the frequency from the spindle frequency (~10 Hz) to the spike-wave frequency (~3 Hz) because of a possible alteration of GABAb to GABAa response to the RE-thalamocortical inhibitory synapse [74].

Indeed, for some cases, seizures are better correlated to sudden excitability during sleep as it is noted during CAPs and arousals or microarousals than increased synchronization during NREM sleep [19, 49, 50]. Sudden provoked awakening during sleep is a well described trigger of epileptiform discharges and seizures [20]. Also this includes seizures occurring upon awakening, seizures correlated with K complexes and seizures during the active A phase of the cyclic alternating patterns.

Sudden excitability can concern a local increase of excitability within the epileptic focus [19] or otherwise a diffuse excitation caused by glutamatergic and cholinergic projections upon arousal/ awakening.

3.5.2 Neuromodulation in Sleep and Epilepsy

Both pro- and anticonvulsive properties have been attributed to monoaminergic systems involved in sleep (Fig. 3.2b). Serotonergic and histaminergic systems involved in arousal are considered to have antiepileptic actions, while changes in the norepinephrine system have been implicated in absences [3, 90]. Melatonin, also a monoamine, has been found to be mostly anticonvulsive (see later discussion). The cholinergic system, which is involved in "REM-on" and arousal, has been shown to have anticonvulsive properties [3].

Adenosine, an endogenous hypnogen, is under investigation as a therapeutic target for seizure control. Adenosine is an inhibitory neuromodulator [91] which through A1 receptors that can act by blocking glutamate release but also independently of GABA and glutamatergic systems [92]. Its concentration as well as its receptor density has been shown to rise after seizures [93], with the latter remaining increased for a prolonged period of time. Endogenous adenosine blocks epileptiform discharges in human epileptogenic cortex maintained in vitro [94]. This endogenous anticonvulsive role is further increased by its neuroprotective properties [3].

On the other hand, the peptide orexin, which is involved in arousal from sleep (Fig. 3.1a, b), may favor epileptic activity. Orexin has been classically linked to the parasomnias (narcolepsy in particular) [95]. It has been reported to induce behavioral seizures, while an increase in its levels has been shown in both animal models of epilepsy [96] and in patients who experienced seizures during polysomnographic examination.

3.5.3 Circadian Rhythms and Epilepsy

Although the relationship between sleep and epilepsy is a well-established one, it is crucial to understand that the sleep-wake cycle (Fig. 3.1a) is, albeit the most preeminent and the easiest to identify, just one of many endogenously controlled physiologic processes, cycling at a 24-h period. Those circadian rhythms also include hormone production, temperature variation and cardiovascular pattern regulation, and are controlled by both photic and non-photic environmental time cues (zeitgebers) [95]. At the core of the circadian clock lie pacemaker neurons in the suprachiasmatic nucleus of the hypothalamus. These receive input from photosensitive ganglion cells in the retina (retinohypothalamic pathway), and are thus entrained to light–dark cycles [97]. Other afferent connections are the geniculohypothalamic and the median raphe serotonergic pathway, while efferent connections include output to the pineal (which produces melatonin), the subiculum and the hippocampus [98]. On a molecular level, the rhythm of SCN cells is regulated by a negative feedback loop of gene expression (such as CLOCK and BMAL1), is independent of action potentials and remains intact in zeitgeber deprivation. SCN neurons mainly use GABA and inhibit the neurons that they innervate [97, 98]. The molecular mechanisms of circadian rhythms have been closely linked with epilepsy. Membrane excitability is altered due to changes in potassium, regulated by clock gene products. The expression of hippocampal genes and ligand-binding activities (α1 adrenergic and benzodiazepine receptors) also oscillate depending on the SCN. In addition, transcription factors involved in epilepsy (such as those regulating the expression of the GABAα1 receptor subunit) have been shown to be important in circadian rhythms. Lastly, the SCN has been reported to be involved in a major signaling pathway of the acquired epilepsies (mTOR pathway, reviewed in [97]). Recently, a reduced seizure threshold was found in a knockout mouse model for the BMAL1 clock gene [99]. The link between circadian rhythms and epilepsy, is responsible for the effect of epilepsy on various circadian functions. People with epilepsy have limited diurnal pressure and heart-rate variability, a fact linked to sudden unexpected deaths (SUDEP). Recent findings suggest that SUDEP, results from depolarization spreading to and inactivating brainstem nuclei supporting vital cardiorespiratory functions [100].

Seizures have been shown to disrupt normal temporal rhythms of hormone production, with postictal elevations of cortisol, prolactin and growth hormone [95]. Melatonin, whose production is directly influenced by the light–dark cycle, has been extensively investigated regarding its relationship with seizures. Finally the monoamine melatonin reduces glutamatergic neurotransmission, while enhancing GABA-related activity. One of its metabolites (kynurenic acid), has also been shown to have antiepileptic properties, with others also being neuroprotective. While the majority of animal and patient studies have reported *in vivo* anticonvulsive effects, there are also data that contest the efficacy of melatonin as an add-on treatment for seizures [101].

3.6 Polysomnography in Long Term Monitoring of Epilepsy

Monitoring brain and body activities of patients with Epilepsy is a very important procedure that has proven to be very useful for clinical and research purposes. Polysomnography recordings is the most suitable mean to monitor these activities for a long period as it integrates many modalities such as Electroencephalogram (EEG), Electro-oculography (EOG), Electrocardiogram (ECG), Electromyogram (EMG). Each polysomnography modality is specialized to monitor effectively a single activity of the patient and many studies [102–105] have revealed modality specific patterns in sleep which could be used as biomarkers for epilepsy prognosis and diagnosis or could enable the accurate epileptic seizure detection.

3.6.1 Open Problems in Polysomnographic Monitoring of Epilepsy

Despite the great advancements in long term monitoring of epileptic patients there exist many unsolved problems until now.

First of all, there exist many different views on how much of the polysomnographic data should be recorded and stored in view of the space and power limitations of the existing devices. Some researchers and clinicians claim that only recording and keeping the data identified by the seizure and spike detection programs (usually supplemented by periodic sampling of the wake and sleep background activity) is sufficient. However, this may mean that atypical and subclinical seizures are not recorded. Moreover, any subtle changes (in e.g. EEG or ECG) that may occur before sustained ictal electrographic activity may be permanently lost. With the availability of large capacity and relatively inexpensive digital storage media, the continuous recording of the polysomnography signals and subsequent archiving of the data has become a more practical and economic option. However, new advancements in low power consumption devices and storage devices alongside with the incorporation of compression techniques in them are required to enable the efficient recording of polysomnographic "big data" [106].

Another problem that remains unsolved in spite of the efforts of the scientific community is the integration of different types of modalities. Integrating different types of modalities can be more effective than the utilization of any modality alone, for problems such as spatiotemporal pattern discovery. Effective integration aims to exploit the advantages of every modality in order to achieve accurate results, and is considered to be an essential task in modeling human brain activity. However, the integration of more than two modalities in a single monitoring device is not provided by most of the current monitoring devices and methods.

Existing monitoring devices for epileptic patients have emphasized so far to the problems of seizure prediction, and focus localization. However, much progress have been accomplished so far in the fields of algorithmic epilepsy differential diagnosis and therapy advice and these advancements should be included in monitoring devices to improve their functionality. A striking example in this direction is therapeutic devices for epileptic patients. Fisher [107] presented the so far limited available devices that not only attempt to detect seizures but also try to treat them using Responsive neurostimulation techniques or Optogenetics techniques (the latter in experimental animals). In this article it is mentioned that optimizing the place of devices in therapy for epilepsy will require further development and clinical experience.

Finally, another open issue is the incorporation of other non-standard modalities in monitoring devices for epileptic patients. For example, in the work of Beniczky et al. [108], a wireless device and a method were proposed for the detection of generalized results using only wrist accelerometer recordings. The experimental results indicated that this methodology achieved accuracy similar to a video–electroencephalography method, indicating that simpler alternative modalities may perform difficult tasks as good as more complex, time consuming and energy intensive techniques. Thus, an interesting future direction is the exploration of the potential of these simple alternative modalities and their incorporation in long term monitoring devices for epileptic patients.

3.6.2 *The ARMOR Approach*

As already mentioned in the present chapter sleep and epilepsy are highly related. In order to efficiently monitor patients with epilepsy it is extremely crucial to record and analyze polysomnographic signals during their sleep. It is even more important for every monitoring system not to disturb sleep or deteriorate its quality.

ARMOR platform, is a wireless, power efficient device which could incorporate many polysomnographic modalities. By design it does not disturb sleep significantly avoiding to use modalities such as video recordings which are believed to stress subjects. Multiple modalities recordings are integrated, streamed and stored to a specially designed database. The quality of the EEG signal of ARMOR platform is more than adequate to locate and analyze essential sleep microstructure events, such as sleep spindles, k-complexes and microarousals, alongside with other sleep macrostructure events such as sleep stages and CAPS. A variety of online algorithms have been incorporated to ARMOR platform to enable the location of sleep onset with the detection of alpha rhythm block and the detection of epileptic interictal spikes and seizures. Moreover, ARMOR polysomnographic sleep recordings can be further analyzed using ARMOR offline algorithms to answer more complex clinical questions. Another advancement of ARMOR platform is its advanced alert system which could help clinicians interfere during the subject's sleep when this is required, e.g. during an epileptic seizure. For all these reasons, the polysomnographic recording component of ARMOR project is an advancement compared to the state-of-the-art as it deals with most of the aforementioned open problems.

References

1. Kothare S, Kaleyias J (2010) Sleep and epilepsy in children and adolescents. Sleep Med 11(7):674–685
2. Tsiptsios DI, Howard RS, Koutroumanidis MA (2010) Electroencephalographic assessment of patients with epileptic seizures. Expert Rev Neurother 10(12):1869–1886
3. Kostopoulos GK (2009) Brain mechanisms linking epilepsy to sleep. In: Schwartzkroin PA (ed) Encyclopedia of basic epilepsy research, vol 3. Academic, Oxford, pp 1327–1336
4. Panayiotopoulos CP (2006) The epilepsies: seizures, syndromes and management. Bladon Medical Publishing, Oxfordshire, http://www.ncbi.nlm.nih.gov/books/NBK2606/
5. Panayiotopoulos CP (2010) A clinical guide to epileptic syndromes and their treatment. Springer, London
6. Gowers WR (1885) Epilepsy and other chronic convulsive diseases. William Wood and Company, New York

7. Bazil CW (2003) Epilepsy and sleep disturbance. Epilepsy Behav 4(suppl 2):S39–S45
8. Galer S, Urbain C, De Tiège X, Peigneux P, Van Bogaert P (2012) Impaired sleep-related consolidation of non-verbal declarative memory in benign childhood epilepsy: a preliminary study. Front Hum Neurosci. Conference Abstract: Belgian Brain Council. doi:10.3389/conf.fnhum.2012.210.00029
9. Derry CP, Duncan JS, Berkovic SF (2006) Paroxysmal motor disorders of sleep: the clinical spectrum and differentiation from epilepsy. Epilepsia 47(11):1775–1791
10. Manni R, Terzaghi M (2010) Comorbidity between epilepsy and sleep disorders. Epilepsy Res 90(3):171–177
11. Placidi F, Scalise A, Marciani MG, Romigi A, Diomedi M, Gigli GL (2000) Effect of antiepileptic drugs on sleep. Clin Neurophysiol 111(suppl 2):S115–S119
12. Legros B, Bazil CW (2003) Effects of antiepileptic drugs on sleep architecture: a pilot study. Sleep Med 4(1):51–55
13. Bazil CW (2003) Effects of antiepileptic drugs on sleep structure: are all drugs equal? CNS Drugs 17(10):719–728
14. Díaz-Negrillo A (2013) Influence of sleep and sleep deprivation on ictal and interictal epileptiform activity. Epilepsy Res Treat 2013, 492524. doi:10.1155/2013/492524
15. Malow BA (2004) Sleep deprivation and epilepsy. Epilepsy Curr 4(5):193–195. doi:10.1111/j.1535-7597.2004.04509.x
16. Vaughn B, D'Cruz O (2004) Sleep and epilepsy. Semin Neurol 24:301–313
17. Tezer FI, Remi J, Erbil N, Noachtar S, Saygi S (2014) A reduction of sleep spindles heralds seizures in focal epilepsy. Clin Neurophysiol 125:2207–2211
18. Herman S, Walczak T, Bazil C (2001) Distribution of partial seizures during the sleep-wake cycle: differences by seizure onset site. Neurology 56(11):1453–1459
19. Miano S, Paolino M, Peraita-Adrados R et al (2009) Prevalence of EEG paroxysmal activity in a population of children with obstructive sleep apnea syndrome. Sleep 32:522–529
20. Shvarts V, Chung S (2013) Epilepsy, antiseizure therapy, and sleep cycle parameters. Epilepsy Res Treat 2013, 670682. doi:10.1155/2013/670682
21. Olbrich E, Claussen JC, Achermann P (2011) The multiple time scales of sleep dynamics as a challenge for modelling the sleeping brain. Philos Trans A Math Phys Eng Sci 369(1952): 3884–3901
22. Saper CB (2013) The neurobiology of sleep. Continuum (Minneap Minn) 19(1 Sleep Disorders):19–31
23. Gvilia I (2010) Underlying brain mechanisms that regulate sleep-wakefulness cycles. Int Rev Neurobiol 93:1–21. doi:10.1016/S0074-7742(10)93001-8
24. Parrino L, Ferri R, Bruni O, Terzano MG (2012) Cyclic alternating pattern (CAP): The marker of sleep instability. Sleep Med Rev 16(1):27–45
25. Destexhe A, Sejnowski TJ (2010) Sleep and sleep states: thalamic regulation. In: Squire LH (ed) Encyclopedia of neuroscience. Academic, Oxford, pp 973–976
26. Beenhakker MP, Huguenard JR (2009) Neurons that fire together also conspire together: is normal sleep circuitry hijacked to generate epilepsy? Neuron 62(5):612–632
27. Urakami Y (2008) Relationships between sleep spindles and activities of cerebral cortex as determined by simultaneous EEG and MEG recording. J Clin Neurophysiol 25:13–24
28. Dehghani N, Cash SS, Chen CC, Hagler DJ, Huang M, Dale AM, Halgren E (2010) Divergent cortical generators of MEG and EEG during human sleep spindles suggested by distributed source modeling. PLoS One 5, e11454
29. Andrillon T, Nir Y, Staba RJ, Ferrarelli F, Cirelli C, Tononi G, Fried I (2011) Sleep spindles in humans: insights from intracranial EEG and unit recordings. J Neurosci 31:17821–17834
30. Frauscher B, von Ellenrieder N, Dubeau F, Gotman J (2015) Scalp spindles are associated with widespread intracranial activity with unexpectedly low synchrony. Neuroimage 105:1–12
31. Urakami Y, Ioannides AA, Kostopoulos GK (2012) Chapter 4 Sleep spindles-as a biomarker of brain function and plasticity. In: Dr. Ihsan Mohammad Abud A (ed) Advances in clinical neurophysiology, InTech. p 73–108. ISBN: 978-953-51-0806-1, doi:10.5772/48427.

http://www.intechopen.com/books/advances-in-clinical-neurophysiology/sleep-spindles-as-a-biomarker-of-brain-function-and-plasticity
32. Khazipov R, Sirota A, Leinekugel X, Holmes GL, Ben Ari Y, Buzsaki G (2004) Early motor activity drives spindle bursts in the developing somatosensory cortex. Nature 432:758–761
33. Colrain IM (2005) The K-complex: a 7-decade history. Sleep 28:255–273
34. Stickgold R, Walker MP (2013) Sleep-dependent memory triage: evolving generalization through selective processing. Nat Neurosci 16:139–145
35. McCormick DA, Pape HC (1990) Properties of a hyperpolarization-activated cation current and its role in rhythmic oscillation in thalamic relay neurones. J Physiol 431:291–318
36. Ball GJ, Gloor P, Schaul N (1977) The cortical electromicrophysiology of pathological delta waves in the electroencephalogram of cats. Electroencephalogr Clin Neurophysiol 43(3): 346–361
37. Nunez A, Amzica F, Steriade M (1992) Voltage-dependent fast (20-40Hz) oscillations in long-axoned neocortical neurons. Neuroscience 51:7–10
38. Steriade M, Amzica F (1998) Coalescence of sleep rhythms and their chronology in corticothalamic networks. Sleep Res Online 1:1–10
39. Kokkinos V, Kostopoulos GK (2011) Human non-rapid eye movement stage II sleep spindles are blocked upon spontaneous K-complex coincidence and resume as higher frequency spindles afterwards. J Sleep Res 20(1 Pt 1):57–72
40. Kokkinos V, Koupparis AM, Kostopoulos GK (2013) An intra-K-complex oscillation with independent and labile frequency and topography in NREM sleep. Front Hum Neurosci 7:163
41. Terzano MG, Parrino L, Anelli S, Halasz P (1989) Modulation of generalized spike-and-wave discharges during sleep by cyclic alternating pattern. Epilepsia 30(6):772–781
42. Brown RE, Basheer R, McKenna JT, Strecker RE, McCarley RW (2012) Control of sleep and wakefulness. Physiol Rev 92(3):1087–1187
43. Ioannides AA, Corsi-Cabrera M, Fenwick PB, Del Rio PY, Laskaris NA, Khurshudyan A, Theofilou D, Shibata T, Uchida S, Nakabayashi T, Kostopoulos GK (2004) MEG tomography of human cortex and brainstem activity in waking and REM sleep saccades. Cereb Cortex 14:56–72
44. Ioannides AA, Kostopoulos GK, Liu LC, Fenwick PBC (2009) MEG identifies dorsal medial brain activations during sleep. Neuroimage 44:455–468
45. Domhoff GW (2011) The neural substrate for dreaming: Is it a subsystem of the default network? Conscious Cogn 20:1163–1174
46. Fosse R, Domhoff GW (2007) Dreaming as non-executive orienting: A conceptual framework for consciousness during sleep. In: Barrett D, McNamara P (eds) The new science of dreaming: content, recall, and personality correlates, vol 2. Praeger, Westport, CT, pp 49–78
47. Mula M, Monaco F (2011) Ictal and peri-ictal psychopathology. Behav Neurol 24(1):21–25
48. Staley KJ, Dudek FE (2006) Interictal spikes and epileptogenesis. Epilepsy Curr 6(6):199–202. doi:10.1111/j.1535-7511.2006.00145.x
49. Bonakis A, Koutroumanidis M (2009) Epileptic discharges and phasic sleep phenomena in patients with juvenile myoclonic epilepsy. Epilepsia 50(11):2434–2445
50. Koutroumanidis M, Tsiptsios D, Kokkinos V, Kostopoulos GK (2012) Focal and generalized EEG paroxysms in childhood absence epilepsy: topographic associations and distinctive behaviors during the first cycle of non-REM sleep. Epilepsia 53(5):840–849
51. Sato S, Dreifuss FE, Penry JK (1973) The effect of sleep on spike wave discharges in absence seizures. Neurology 23(12):1335–1345
52. Shouse MN, Scordato JC, Farber PR (2004) Sleep and arousal mechanisms in experimental epilepsy: epileptic components of NREM and antiepileptic components of REM sleep. Ment Retard Dev Disabil Res Rev 10:117–121
53. Ferrillo F, Beelke M, De Carli F, Cossu M, Munari C, Rosadini G, Nobili L (2000) Sleep-EEG modulation of interictal epileptiform discharges in adult partial epilepsy: a spectral analysis study. Clin Neurophysiol 111(5):916–923

54. Sammaritano M, Gigli GL, Gotman J (1991) Interictal spiking during wakefulness and sleep and the localization of foci in temporal lobe epilepsy. Neurology 41(2):290–297
55. Timofeev I, Steriade M (2004) Neocortical seizures: Initiation, development and cessation. Neuroscience 123(2):299–336
56. Bagshaw AP, Jacobs J, LeVan P et al (2009) Effect of sleep stage on interictal high-frequency oscillations recorded from depth macroelectrodes in patients with focal epilepsy. Epilepsia 50:617–628
57. Schevon CA, Trevelyan AJ, Schroeder CE, Goodman RR, McKhann G, Emerson RG (2009) Spatial characterization of interictal high frequency oscillations in epileptic neocortex. Brain 132(11):3047–3059
58. Dümpelmann M, Jacobs J, Schulze-Bonhage A (2015) Temporal and spatial characteristics of high frequency oscillations as a new biomarker in epilepsy. Epilepsia 56(2):197–206
59. Salami P, Lévesque M, Benini R, Behr C, Gotman J, Avoli M (2014) Dynamics of interictal spikes and high-frequency oscillations during epileptogenesis in temporal lobe epilepsy. Neurobiol Dis 67:97–106
60. Frauscher B, von Ellenrieder N, Ferrari-Marinho T, Avoli M, Dubeau F, Gotman J (2015) Facilitation of epileptic activity during sleep is mediated by high amplitude slow waves. Brain 2015:1–13
61. Langdon-Down M, Brain W (1929) Time of day in relation to convulsions in epilepsy. Lancet 2:1029–1032
62. Janz D (1962) The grand mal epilepsies and the sleeping-waking cycle. Epilepsia 3:69–109
63. Panayiotopoulos CP, Michael M, Sanders S, Valeta T, Koutroumanidis M (2008) Benign childhood focal epilepsies: assessment of established and newly recognized syndromes. Brain 131:2264–2286
64. Crespel A, Coubes P, Baldy-Moulinier M (2000) Sleep influence on seizures and epilepsy effects on sleep in partial frontal and temporal lobe epilepsies. Clin Neurophysiol 111(suppl 2):S54–S59
65. Halasz P, Kelemen A, Szucs A (2012) Physiopathogenetic interrelationship between nocturnal frontal lobe epilepsy and NREM arousal parasomnias. Epilepsy Res Treat 2012, 312693, 8 pages. doi:10.1155/2012/312693
66. Singhal N, Sullivan J (2014) Continuous spike-wave during slow wave sleep and related conditions. ISRN Neurol 2014, 619079, 6 pages. doi:10.1155/2014/619079
67. Dang-Vu TT, Schabus M, Desseilles M, Albouy G, Boly M, Darsaud A, Gais S, Rauchs G, Sterpenich V, Vandewalle G, Carrier J, Moonen G, Balteau E, Degueldre C, Luxen A, Phillips C, Maquet P (2008) Spontaneous neural activity during human slow wave sleep. Proc Natl Acad Sci U S A 105(39):15160–15165
68. Malow BA, Lin X, Kushwaha R, Aldrich MS (1998) Interictal spiking increases with sleep depth in temporal lobe epilepsy. Epilepsia 39(12):1309–1316
69. Minecan D, Natarajan A, Marzec M, Malow B (2002) Relationship of epileptic seizures to sleep stage and sleep depth. Sleep 25(8):899–904
70. Nobili L, Baglietto MG, Beelke M, De Carli F, De Negri E, Tortorelli S, Ferrillo F (2000) Spindles-inducing mechanism modulates sleep activation of interictal epileptiform discharges in the Landau-Kleffner syndrome. Epilepsia 41(2):201–206
71. Nakabayashi T, Uchida S, Maehara T, Hirai N, Nakamura M, Arakaki H, Shimisu H, Okubo Y (2001) Absence of sleep spindles in human medial and basal temporal lobes. Psychiatry Clin Neurosci 55(1):57–65
72. Douw L, van Dellen E, de Groot M, Heimans JJ, Klein M, Stam CJ, Reijneveld JC (2010) Epilepsy is related to theta band brain connectivity and network topology in brain tumor patients. BMC Neurosci 11:103. doi:10.1186/1471-2202-11-103
73. Bersagliere A, Achermann P, Lo Russo G, Proserpio P, Nobili L (2013) Spindle frequency activity may provide lateralizing information in drug-resistant nocturnal mesial frontal lobe epilepsy: a pilot study on the contribution of sleep recordings. Seizure 22(9):719–725
74. Sinha S (2011) Basic mechanisms of sleep and epilepsy. J Clin Neurophysiol 28(2):103–110. doi:10.1097/WNP.0b013e3182120d41

75. Gloor P (1978) Generalized epilepsy with bilateral synchronous spike and wave discharge. New findings concerning its physiological mechanism. Electroencephalogr Clin Neurophysiol Suppl 34:S245–S249
76. Steriade M, Llinas RR (1988) The functional states of the thalamus and the associated neuronal interplay. Physiol Rev 68(3):649–742
77. Maheshwari A, Noebels JL (2014) Monogenic models of absence epilepsy: windows into the complex balance between inhibition and excitation in thalamocortical microcircuits. Prog Brain Res 213:223–252
78. Avoli M (2012) A brief history on the oscillating roles of thalamus and cortex in absence seizures. Epilepsia 53(5):779–789
79. Kostopoulos GK (2000) Spike-and-wave discharges of absence seizures as a transformation of sleep spindles: the continuing development of a hypothesis. Clin Neurophysiol 111(suppl 2):S27–S38
80. Coenen AM, Van Luijtelaar EL (2003) Genetic animal models for absence epilepsy: a review of the WAG/Rij strain of rats. Behav Genet 33(6):635–655
81. Avanzini G, Vergnes M, Spreafico R, Marescaux C (1993) Calcium dependent regulation of genetically determined spike and waves by the reticular thalamic nucleus of rats. Epilepsia 34(1):1–7
82. McCormick DA, Contreras D (2001) On the cellular and network bases of epileptic seizures. Annu Rev Physiol 63:815–846
83. Kostopoulos GK, Psarropoulou CT (1992) Possible mechanisms underlying hyperexcitability in the epileptic mutant mouse tottering. J Neural Transm Suppl 35:109–124
84. Meeren H, van Luijtelaar G, Lopes da Silva F, Coenen A (2005) Evolving concepts on the pathophysiology of absence seizures: the cortical focus theory. Arch Neurol 62(3):371–376
85. Miao A, Tang L, Xiang J, Guan Q, Ge H, Liu H, Wu T, Chen Q, Yang L, Lu X, Hu Z, Wang X (2014) Dynamic magnetic source imaging of absence seizure initialization and propagation: a magnetoencephalography study. Epilepsy Res 108(3):468–480
86. Tenney JR, Fujiwara H, Horn PS, Jacobson SE, Glauser TA, Rose DF (2013) Focal corticothalamic sources during generalized absence seizures: a MEG study. Epilepsy Res 106(1–2):113–122
87. Carney PW, Masterton RA, Flanagan D, Berkovic SF, Jackson GD (2012) The frontal lobe in absence epilepsy: EEG-fMRI findings. Neurology 78(15):1157–1165
88. Moeller F, LeVan P, Muhle H, Stephani U, Dubeau F, Siniatchkin M, Gotman J (2010) Absence seizures: individual patterns revealed by EEG-fMRI. Epilepsia 51(10):2000–2010
89. Tucker DM, Brown M, Luu P, Holmes MD (2007) Discharges in ventromedial frontal cortex during absence spells. Epilepsy Behav 11(2–3):546–557
90. Noebels JL (1984) A single gene error of noradrenergic axon growth synchronizescentral neurones. Nature 310(5976):409–411
91. Phillis JW, Kostopoulos GK (1975) Adenosine as putative transmitter in the cerebral cortex. Studies with potentiators and antagonists. Life Sci 17(1975):1085–1094
92. Masino SA, Kawamura M Jr, Ruskin DN (2014) Adenosine receptors and epilepsy: current evidence and future potential. Int Rev Neurobiol 119:233–255
93. Angelatou F, Pagonopoulou O, Maraziotis T, Olivier A, Villemeure JG, Avoli M, Kostopoulos G (1993) Upregulation of A1 adenosine receptors in human temporal lobe epilepsy: a quantitative autoradiographic study. Neurosci Lett 163:11–14
94. Kostopoulos G, Drapeau C, Avoli M, Olivier A, Villemeure JG (1989) Endogenous adenosine can reduce epileptiform activity in the human epileptogenic cortex maintained in vitro. Neurosci Lett 106(1–2):119–124
95. Hofstra WA, de Weerd AW (2009) The circadian rhythm and its interaction with human epilepsy: a review of literature. Sleep Med Rev 13(6):413–420
96. Goudarzi E, Elahdadi Salmani M, Lashkarbolouki T, Goudarzi I (2015) Hippocampal orexin receptors inactivation reduces PTZ induced seizures of male rats. Pharmacol Biochem Behav 130:77–83

97. Connors BW, Bear MF, Paradiso MA (2006) Neuroscience: exploring the brain, 3rd edn. Lippincott Williams & Wilkins, Baltimore
98. Cho CH (2012) Molecular mechanism of circadian rhythmicity of seizures in temporal lobe epilepsy. Front Cell Neurosci 6:55
99. Gerstner JR, Smith GG, Lenz O et al (2014) BMAL1 controls the diurnal rhythm and set point for electrical seizure threshold in mice. Front Syst Neurosci 8:121
100. Aiba I, Noebels JL (2015) Spreading depolarization in the brainstem mediates sudden cardio-respiratory arrest in mouse SUDEP models. Sci Transl Med 7(282):282ra46. doi:10.1126/scitranslmed.aaa4050
101. Banach M, Gurdziel E, Jędrych M, Borowicz KK (2011) Melatonin in experimental seizures and epilepsy. Pharmacol Rep 63(1):1–11
102. Massé F, Van Bussel M, Serteyn A, Arends J, Penders J (2013) Miniaturized wireless ECG monitor for real-time detection of epileptic seizures. ACM Trans Embed Comput Syst 12(4):102. doi:10.1145/2485984.2485990
103. Meier R, Dittrich H, Schulze-Bonhage A, Aertsen A (2008) Detecting epileptic seizures in long term human EEG: a new approach to automatic online and real time detection and classification of polymorphic seizure patterns. Clin Neurophysiol 25(3):119–131
104. Van de Vel A, Cuppens K, Bonroy B, Milosevic M, Jansen K, Van Huffel S, Vanrumste B, Lagae L, Ceulemans B (2013) Non-EEG seizure-detection systems and potential SUDEP prevention: state of the art. Seizure 22(5):345–355
105. Cavazos J, Girouard M, Whitmire L (2015) Novel ambulatory EMG-based GTC seizure detection device for home & hospital use (I6-4B). Neurology 84(14 Supplement)
106. Redline S, Dean D, Sanders MH (2013) Entering the era of "big data": getting our metrics right. Sleep 36(4):465–469
107. Fisher RS (2012) Therapeutic devices for epilepsy. Ann Neurol 71:157–168. doi:10.1002/ana.22621
108. Beniczky S, Polster T, Kjaer T, Hjalgrim H (2013) Detection of generalized tonic-clonic seizures by a wireless wrist accelerometer: a prospective, multicenter study. Epilepsia 54(4):e58–e61. doi:10.1111/epi.12120

Chapter 4
Source-Estimation from Non-invasive Recordings of Brain Electrical Activity in Sleep and Epilepsy

Andreas A. Ioannides, Lichan Liu, Vahe Poghosyan, Khalid Hamandi, and George K. Kostopoulos

Abstract Epilepsy and sleep are characterized by spontaneously occurring large graphoelements in electroencephalography (EEG) and magnetoencephalography (MEG) recordings, such as ictal and interictal epileptiform discharges in epilepsy and K-complexes (KC) in sleep. Localization of the neural sources of these graphoelements, which is of immense clinical and research importance, requires application of electromagnetic source analysis methods. A number of such methods are available; however their ability to localize *widespread synchronous cortical sources*, such as the sources of KCs and widespread epileptiform discharges, is contested. Here, we used KC as an exemplar of large graphoelements with such sources to test the performance of a diverse set of commonly employed source analysis methods. We analyzed segments of sleep MEG data with clear KCs using equivalent current dipole models, beamformer methods, linear distributed source methods, and a nonlinear distributed source method—magnetic field tomography (MFT). MFT provided the most robust and steady localization across KCs, which was also highly consistent with the intracranial findings: strong and widespread activations were reliably found in superior aspects of bilateral frontal cortex. Conversely, the localizations provided by the other methods were very variable across KCs and were all inconsistent with the intracranial findings: in many cases, the KCs were incorrectly localized in deep medial brain structures. Our current and earlier results showing

---

A.A. Ioannides (✉) • L. Liu • V. Poghosyan
Laboratory for Human Brain Dynamics, AAI Scientific Cultural Services Ltd., Nicosia, Cyprus
e-mail: a.ioannides@aaiscs.com

K. Hamandi
The Alan Richens Welsh Epilepsy Centre, University Hospital of Wales, Cardiff, UK

Cardiff University Brain Research Imaging Centre, School of Psychology, Cardiff University, Cardiff, UK

G.K. Kostopoulos
Neurophysiology Unit, Department of Physiology, Medical School, University of Patras, Patras, Greece

N.S. Voros, C.P. Antonopoulos (eds.), *Cyberphysical Systems for Epilepsy and Related Brain Disorders*, DOI 10.1007/978-3-319-20049-1_4

the excellent localization accuracy of MFT for focal as well as extended brain sources and the smart uses of MEG and EEG in epilepsy, demonstrate that the MFT analysis of MEG signals may be a powerful tool for the studies of epilepsy, epilepsy monitoring and in pre-surgical evaluation of patients.

4.1 Introduction

Epilepsy and sleep are characterized by hypersynchrony and spontaneously occurring high amplitude neural activity leading to large graphoelements, such as ictal and interictal epileptiform discharges in epilepsy and K-complexes (KC) in sleep. Section 4.1.1 reviews the history and recent advances in understanding KCs and their relationship to epilepsy. Accurate identification of neural processes underlying epileptiform discharges and K-complexes (KC) and their putative relationship is of great importance for both clinical reasons and basic research. Currently, the only noninvasive techniques that can provide time-resolved measurements of these characteristic waveforms are electroencephalography (EEG) and magnetoencephalography (MEG). These methods are therefore ideal for monitoring some of the neural mechanisms that come into play in epilepsy and sleep.

MEG and EEG measure the instantaneous electromagnetic signals (magnetic field just outside the head and electric potential between scalp electrodes) produced by coordinated electrical activity of neuronal assemblies in the brain. To identify the locations of these neuronal assemblies and to quantify their activity from the recorded MEG and EEG signals, i.e. to perform MEG/EEG source analysis, the electromagnetic inverse problem needs to be solved. A number of methods are available to this end. Section 4.1.2 provides brief introductions to EEG, MEG and electromagnetic inverse solvers, while Sect. 4.1.3 provides a brief overview of the source analysis in studying the neural correlates of the large epileptic (in the childhood focal epilepsies) and sleep graphoelements.

Noninvasive source localization of epileptic graphoelements, if available, can provide valuable information during the therapeutic planning and it is an important component of pre-surgical evaluation of patients with epilepsy, since it is ordinarily much cheaper and safer than invasive studies. Localization of KC generators is also necessary in many clinical and basic research applications. Thus the ability to accurately and reliably identify the sources of these large graphoelements is of immense clinical and research importance.

Recently, Wennberg and Cheyne [1] have assessed the utility of several widely used MEG/EEG source analysis methods (equivalent current dipole, ECD; low-resolution electromagnetic tomography, LORETA; and standardized LORETA, sLORETA) for localizing the sources of KCs by comparing the results directly to the intracranial recordings in the same patients. All the tested methods had *failed* to provide physiologically valid results, localizing the 'true' extended superficial cortical generators to deep medial structures. Motivated and intrigued by their results, we have tested several other commonly employed source analysis methods for their ability to localize neural sources of individual KCs. As the largest potential in the healthy EEG, KC is an ideal exemplar of large graphoelements with widespread synchronous cortical sources, such as sleep slow waves and widespread epileptiform discharges (e.g. generalized spike/wave complexes).

In the current study we have analyzed segments of sleep MEG data with clear KCs, identified from the concurrently recorded EEG signals (C3 and C4 channels according to International 10-20 system), using a number of commonly employed MEG source analysis methods: two ECD models—single and multiple dipole models, two beamformer methods—synthetic aperture magnetometry (SAM [2]) and linear constrained minimum variance (LCMV [3]), two linear distributed source methods—weighted minimum norm estimate (wMNE [4]) and exact LORETA (eLORETA [5]), and one non-linear distributed source method—magnetic field tomography (MFT [6, 7]). The performance of the methods was assessed qualitatively based on the detailed intracranial cortical localization results reported by Wennberg [8], i.e. "widespread synchronous superficial cortical activity arising maximally within the superior medial and lateral frontal cortices."

Moreover, in the current study, we have used MFT to analyze the neural sources of sensory-evoked KCs and examine the involvement of sensory-specific cortices in their generation. This part of the work was motivated by the recent evidence suggesting such involvement, in contrast to earlier widely accepted view that sensory-evoked KCs display largely stereotypic, sensory-modality-independent EEG responses.

4.1.1 K-Complex and Its Relationship with Epilepsy

4.1.1.1 Background

In 1937 Loomis and colleagues named as KC the EEG response they noticed to be evoked upon a knock on the door of a sleeper [9]. Since this first description the studies of KCs increasingly justify its name with regards to its complex underlying mechanisms and its roles in physiology and pathophysiology (for reviews see [10, 11]). There is a striking resemblance of spontaneously occurring KCs to those evoked by auditory stimulation and until recently it was thought that the same mechanisms underlie both. However, more recent evidence [12, 13] suggests that the neural generators of spontaneous and evoked KCs may at least partially differ.

4.1.1.2 Description of KC

The KCs are easily distinguishable, large multicomponent EEG graphoelements, which may occur spontaneously or can be triggered by sensory stimuli. The spontaneous KCs occur with a frequency between 1 and 2 per minute in stage 2 of non rapid eye movement (NREM) sleep and relatively more frequently in the first sleep cycles and in the ascending part of each sleep cycle. Of the several components, the most prominent is the negative high voltage peak (greater than 100 μV) lasting for more than 0.5 s, which is followed by a lower voltage but often longer lasting positivity which in about 70 % of cases overlaps with bursts of spindles. Prior to the large negative component, even in the spontaneously occurring KCs, one often observes traces similar to those preceding KCs evoked by external stimuli such as sounds, touches on the skin [14] and internal ones such as inspiratory interruptions [15]. They are generated in widespread cortical locations [16] though they tend to predominate over the frontal parts of the brain [17]. They are accompanied by autonomic discharges identical to those seen for arousals.

4.1.1.3 Neuronal Mechanisms Underlying the KC

It is proposed that different modalities of experimental or natural sensory evoked potentials may act as traveling cortico-cortical excitations, which upon reaching the fronto-central region, depending on the reactivity of cortex in different NREM sleep levels, elicit a biphasic slow wave response, the KC [13]. Animal and human intracranial field and single neurons recordings [16] have demonstrated that KCs are generated by the occurrence in widespread cortical areas of outward dendritic currents from the middle (III) to the upper (I) layers of the cerebral cortex, accompanied by a decrease in broadband (including gamma) EEG power. These neuronal network silence periods are defined as 'down-states' [16]. The rebound excitation upon the ending of the 'down-state' activates thalamic circuits contributing to the production and synchronization of sleep oscillations such as spindles and delta waves [18]. KCs may appear as solitary events in light seep but as NREM sleep progresses they can be repeated and included in CAP-A1 periods suggesting that they may be nested in, respectively, slow or even infra-slow cortical oscillations [11].

KCs appear in the same sleep stage and often in time association to spindles, which has possible implications to the mechanisms underlying both [19–22]. During the negative wave of KCs any already running spindle is invariably interrupted [21], most often replaced by a brief oscillation in the upper theta range whose maximal power shifts antero-posteriorly and spectral frequency tends to increase towards the alpha band. Finally KCs are most often followed by spindles which however are invariably faster than the sporadic fast centroparietal spindles observed in the same subject. Interestingly no correlation was found between any of the characteristics of KCs with those of the spindles which follow them [19, 21] arguing against a spindle triggering role of KC [18] and in favor of a third, possibly subcortical, trigger of both [20].

Sensory-Evoked KCs

Until recently, it was widely accepted that sensory-evoked KCs display largely stereotypic, sensory-modality-independent EEG topologies. However, recently Riedner and colleagues [12] using source analysis of high-density EEG have observed "aspects of the evoked K-complex response that were not stereotypic". More recently, Laurino and colleagues [13] have found that the electrodes with the shortest KC P200 latency were located over the stimulation-related primary sensory cortices. However, a clear demonstration of involvement of sources in the primary sensory cortices in the generation of KC components was lacking. Using state-of-the-art tomographic source analysis of MEG signals recorded from sleeping individuals receiving auditory and somatosensory stimulation, we found that the earliest brain activations within the sensory-evoked KCs are localized in the stimulation modality-specific primary sensory cortices [23].

4.1.1.4 Association of KC to Epilepsy

The mutual interference between sleep and epilepsy has been known since the antiquity (see references in [24]). It is long established that synchronous oscillations of cortical neurons underlying slow waves during NREM sleep promote seizure propagation, in contrast to desynchronization of the EEG, which seems to impede seizure propagation during REM sleep and wakefulness [25]. Lighter stages of NREM sleep seem to best promote seizures, while deeper stages of NREM sleep seem to best activate interictal epileptiform discharges. In patients with primary generalized epilepsy, the occurrence of spikes, polyspikes, and spike-and-wave discharges is enhanced during the stage 2 of NREM sleep. This is particularly noticed in conjunction with a major graphoelement of NREM, the KCs, and has been termed 'dyshormia' by Niedermeyer [26]. The latter is defined as faulty or deviant arousal, during which the epileptic KC is found localized mid-anterior-frontally rather than at the vertex.

In individuals with idiopathic generalized epilepsy, KC induced synchronization can trigger spike-and-wave discharges. This tends to happen most during the shift between wakefulness and NREM, and between NREM and REM sleep [27]. In autosomal dominant nocturnal frontal lobe epilepsy, KCs are almost invariably present at the start of seizures [28]. It may be concluded that the slow (<1 Hz) oscillation of NREM, and in particular spindles, KCs and delta waves, share some features that may contribute to the aggravation of epileptic phenomena. These effects may be related to the dynamic bistability of neuronal membrane potentials and neuronal readiness for bursting and widespread synchronization.

The thalamocortical mechanisms elaborating sleep spindles have been implicated in the development of EEG spike-and-wave discharges underlying absence seizures (evidence reviewed in [24]). A gating function in these discharges has been alternatively ascribed to KCs [11, 29]. In 1981 Halász demonstrated manifold similarities between the KC and the spike-wave pattern of the EEG in generalized epilepsy [30], which he regarded as an epileptic 'caricature' of the sleep induction momentum reflected in the KC phenomenon. A possibly fruitful examination of the common ground between the above two independently advanced putative explanations of spike-and-wave discharges as related to either spindles or KCs could be helped by recent findings of robust dynamic relationships between these two graphoelements of NREM [19–21].

Regarding focal epilepsies, a long standing hypothesis is that KCs may serve as vehicles for seizure discharge [31]. Nocturnal frontal lobe epilepsy (NFLE) and KCs share important characteristics like (a) both occur mostly during stage 2 of NREM sleep, (b) both have periodic rhythms appearing repeatedly during a sleep cycle, and (c) both have their highest EEG activity in frontal and vertex regions [32]. KC activity increases in NFLE, especially prior to a clinical seizure. According to Si and colleagues [32], this reflects an unstable sleep condition. All these suggest a correlation between KC and epileptic activities including seizures and EEG spike discharges.

4.1.1.5 Mechanisms Underlying the Effect of KC on Seizures

The challenge emerges to elucidate the mechanisms underlying the dual nature of a KC because both may be related to seizure expression as they deal with the two conditions most relevant to seizure expression: excitability and synchronization. KC initial parts behave like an arousal pattern that can be triggered by acoustic and other stimuli. This may be a state of phasic hyperexcitability. At the same time KCs change in parallel with the delta activity that means dampening along the sleep process and their rate is related to the depth of the actual sleep cycle, directly related to the degree of synchronization [26]. The observed dynamics between KC and theta as well as spindles oscillations may prove useful markers in the search of further associations between KCs and manifestations and mechanisms of epilepsy.

The slow negative wave of a KC is proposed to represent an isolated cortical down state, i.e., a period of synaptic quiescence of cortical neurons [16]. At the same time, however, sensory evoked responses are more easily produced during down as compared to up states, they propagate relatively further in cortex through recurrent excitation and they bias cortical neurons toward the up state [33]. Animal experiments [29] and human studies [10, 11, 16, 34], show that besides the main hyperpolarization (EEG surface negative) phase, KCs are preceded and followed by smaller depolarizing (surface positive) components, indicative of neuronal excitation. Also the observed intra-KC-theta-oscillation [20] may reflect subcortically evoked cortical responses [34] rather than cortical spontaneous activity, as is the case for KC negativity. Thus, during the occurrence of KCs, a window of opportunity for subcortical afferents to reach cortex and bias its neurons toward the up state may be offered. KCs may be closely repeated and be grouped in CAP activation (CAP-A) periods [35] which reflect sleep instability, a major determinant of seizure onset [36–38]. As an arousing reaction (see above) a KC can induce sleep instability [29, 39] and an unstable state may be important for the physiological roles of KCs but also promote abnormal events, including epileptic seizures, especially those usually preceded by EEG arousal.

4.1.2 EEG, MEG and Electromagnetic Inverse Solvers

4.1.2.1 EEG

The EEG measures electric field potential generated on scalp surface by the neural currents using electrodes attached to the scalp. Because a potential can only be measured as a difference, at least two electrodes are necessary to obtain one value, i.e. two electrodes form one channel. There are several ways of forming the EEG channels for visualization and analysis, which are referred to as *montages*. The *common reference montage* measures the potentials at all electrodes with respect to a single common reference electrode. There is no standard position for this reference; the placement of the reference depends on the intentions of the study; a popular reference is "linked ears," which is a physical or mathematical average of electrodes attached to earlobes or mastoids. In the *average reference montage* each channel is referenced against the average of all channels. The *bipolar montage* uses a distinct, typically neighboring, reference electrode for each channel. An example of such a montage is the sequential montage, where each channel is formed by the difference between two adjacent electrodes. The entire montage consists of a series of these channels (e.g. 'Fp1-F3', 'F3-C3' etc.). In the *Laplacian montage* each channel represents the difference between an electrode and a weighted average of the surrounding electrodes. Each of these montages has its advantages and disadvantages. Almost always the data are recorded using a common reference montage, because later it can be re-referenced to any other montage. EEG has wide range of applications in neurology and cognitive neuroscience. It is a key clinical tool in studying of epilepsy, sleep and sleep disorders.

4.1.2.2 MEG

The MEG measures magnetic field produced by the neural currents inside the head using very sensitive SQUID-based sensors (SQUID—Superconducting Quantum Interference Device). EEG scalp voltages are on the order of tens of microvolts and thus readily measured using relatively low-cost scalp electrodes and amplifiers. In contrast, magnetic fields generated by neural currents are very weak; they are of the order of femtoteslas (about eight orders of magnitude smaller than the earth's magnetic field). Therefore, expensive and complex hardware is necessary for recording usable magnetic fields from the brain. Different types of MEG sensors exist: *magnetometers* have a single coil for picking the magnetic flux; *gradiometers* have two or more coils, which are designed to reduce noise by diminishing signals coming from farther away. Various gradiometer configurations are used in different MEG systems, with most common being the first-order series *planar* (two coils in the same plane) and symmetric series *axial* (two coils along the same axis) gradiometers. Each of the sensor types has its advantages and disadvantages. Moreover, it is possible to construct synthetic (software-based) higher-order gradiometers from the recorded MEG signals.

4.1.2.3 Main Differences Between MEG and EEG

MEG is significantly more expensive than EEG, due to costly equipment, which includes shielded rooms, cryostats, and SQUIDs. However, it has several key advantages: magnetic fields are less distorted than electric fields by the skull and other tissues intervening between the brain and scalp, which results in a better spatial resolution of the MEG even with simple models for the head conductivity; MEG is reference-free and measures absolute, rather than relative change at each sensor, this confers advantages in source localization and connectivity analyses, while scalp EEG relies on a reference, which makes interpretation of the data difficult; MEG measurements are easier to perform, with no electrodes attached to the skin, this is of special importance when dealing with patients.

One of the main differences between MEG and EEG that is often cited is that MEG is mainly sensitive to tangentially oriented sources and is insensitive to radial sources, whereas EEG can detect sources of all orientations. A radially oriented current source produces no magnetic field outside a *spherically symmetric* volume conductor. However, since the human head is not exactly spherically symmetric, the radial orientation is not well defined, and an approximately radial source in the brain is not necessarily silent in MEG. Similarly the convoluted folding of the human brain and spatial extent of firing neural ensembles are unlikely to produce pure radial sources. Therefore, it is possible that all source orientations generate a magnetic field that is detectable by MEG. Moreover, only very few (<5 %) cortical sources are expected to have the orientation of the lowest sensitivity (close to radial) in MEG [40].

4.1.2.4 Electromagnetic Inverse Problem

The goal of solving the inverse problem is to estimate the location, direction and strength of the primary current density that generate the measured electromagnetic signals. This problem of source localization is an ill-posed inverse problem: there are an infinite number of solutions (configurations of intracranial current sources) that explain the measured data (electric potentials or magnetic fields recorded at the scalp) equally well. The problem arises partly because of silent sources. Silent sources are generators that even with arbitrary large source strength produce no electromagnetic signals. For example, a radial source in a spherical conductor is a silent source (for MEG) because the symmetry properties of the basic mathematical equations that govern the electromagnetic field associated with this configuration demands that no external magnetic field can be present outside the spherical bounding surface. In descriptive terms, the total flow of current that contributes to the magnetic signal consists of the original radial source and the vortex like currents at the inner surface of the resistive spherical boundary; these two contributions exactly cancel each other. A second reason for non-uniqueness is the fact that any given set of measurements can be reconstructed from measurements placed in a closed surface inside the head. Since there are an infinite set of such surfaces there are an infinite number of possible source configurations that can explain the data. Only by introducing a priori assumptions about the nature of the sources and the volume conductor can the inverse problem be solved. These a priori assumptions are crucial, since they determine whether the solution is limited to only explaining the data or if the solution actually gives neurophysiological information about where the signals were generated in the brain. Thus, the accuracy and *validity* of the current source estimates depend on the biological correctness of the assumptions and priors adopted in the solver. Although the neuroelectromagnetic inverse problem is indeed formidable in its mathematical form, in practice reasonable a priori constraints can lead to very accurate reconstructions.

There are three basic approaches that encompass most neuroelectromagnetic inverse solvers: (1) equivalent current dipole (ECD) models, (2) spatial filters or beamformers, and (3) distributed source models.

Dipolar Models

The basic a priori assumption underlying ECD model is that the measured electromagnetic signals are generated by *one or a few, focal* current sources, which can be modeled as current dipoles. To solve this overdetermined (there are less unknowns than measured data) problem, ECD algorithms minimize a data-fit cost function defined in a multidimensional space of parameters. Usually, the algorithms estimate five nonlinear parameters per dipole: the location (three coordinates in a 3D space), and the orientation (two angles necessary to define dipole orientations in 3D space). The amplitudes of dipoles are linear parameters estimated directly from the data.

One general risk of these methods is that they can converge to local minima, resulting in the algorithm accepting an 'incorrect' location. However, even if the algorithm converges to the global minimum, it would be incorrect to assume that the mathematically absolute minimum is the 'correct' solution. Among all solutions compatible with the data, the global minimum solution is at best only slightly more likely than the others [41]. It would thus be presumptuous to automatically equate a 'correct' solution with the absolute minimum of the chosen cost function. In general, the current dipole approach is inadequate in most cases, because its assumptions (few, focal sources) are rarely true: in most cases multiple, widely distributed neural sources are active in the brain. Moreover, these methods require assumptions regarding the number and approximate location of the active brain sources, which seldom can be correctly estimated.

Spatial Filters or Beamformers

Rather than attempting to identify discrete sets of sources by adjusting their non-linear location parameters, spatial filters or beamformers independently and systematically scan for dipoles within a predefined grid inside the brain. Here the goal is to estimate the activity at a source point or region while avoiding the crosstalk from other regions so that these affect as little as possible the estimate at the region of interest. At each point of the brain grid, a narrow-band spatial filter is formed and the contribution to data from an elementary source model (e.g. a single or a triplet of current dipoles) is evaluated, while contributions from other brain regions are ideally muted, or at least attenuated. The main caveat of these algorithms is that their results are not an estimation of the current density everywhere in the brain. They represent a score map of a source model—generally a current dipole—that is evaluated at the points of a predefined spatial lattice, which sometimes leads to misinterpretations. The localization issue now becomes a signal detection problem within the score map. The most widely used spatial filter or beamformer methods include LCMV, SAM and DICS.

Distributed Source Models

Instead of performing low-dimensional nonlinear optimization or spatial scanning, distributed inverse solvers estimate brain electrical activity in each point of pre-defined source space, which can be a 3D grid and/or mesh (e.g. points in grey matter or entire brain). All points within the source space are considered as a possible location of a current source, thus there is no a priori assumption about the number of dipoles in the brain. The goal is to find a unique configuration of activity at these pre-defined source points that explains the measured data. Unfortunately, this problem is highly underdetermined (there are substantially more unknowns than measured data). To solve it, different a priori assumptions/constraints need to be introduced to identify the 'optimal' or 'most likely' solution. The distributed inverse solvers differ in their choice and implementation of these constraints. Some are purely mathematical, while others incorporate biophysical or physiological knowledge. It is important to note that the validity of the a priori constraint defines the validity of the inverse solution. The most widely used distributed inverse solvers include the *linear* minimum norm estimate (MNE) and its derivatives such as weighted MNE, LORETA/sLORETA/eLORETA and dSPM.

4.1.2.5 Magnetic Field Tomography (MFT) and Electric Field Tomography (EFT)

MFT and EFT are distributed source analysis methods for MEG and EEG respectively. They rely on a *non-linear* algorithm to estimate the neural activity in the brain with minimal a priori assumptions [6, 7]. Currently, MFT is the most robust and accurate method for the MEG source localization, as has been repeatedly demonstrated in comparison with fMRI localization [42], in computer simulations and real data precision tests [43], in the tests with realistic head-shaped phantom [44] etc. MFT has been used in a large number of research studies providing significant novel findings [45, 46], including overturning a long accepted view on neural correlates of attention [47, 48], differences in activity and connectivity patterns of brain activations in schizophrenic and control subjects [49] and new insights about basic state properties of sleep activity patterns at different sleep stages [50, 51]. MFT estimates can be effectively combined with cytoarchitectonic probability maps, mapping brain activity within individual cytoarchitectonic areas [52] and tracing the evolution of stimulus-evoked activity through different cytoarchitectonic areas [43, 47]. Moreover, MFT can be used to localize deep subcortical sources in thalamus, amygdala, cerebellum and brainstem [44, 49, 50, 53].

Below we summarize some of the key features of MFT/EFT:

- In MFT/EFT, *only the direction* of the primary current density is expressed by a linear expansion in lead fields. The strength of the source currents is determined more explicitly from the measured signals by solving a highly nonlinear system of equations. This is in contrast to linear distributed inverse solvers where *both direction and strength* of current source density are expressed in a single linear expansion in the lead fields. It is easy to see that MFT constraint is consistent with the laws of physics while linear constraints are not. A priori constraints (like any linear one) that fix the general properties of both the direction and strength of the unknown source configurations assume (implicitly or explicitly) more than what the laws of Physics (i.e. Maxwell's equation) can provide. Only the non-linear constraint of MFT makes assumptions that are consistent with what information is a priori available. The MFT constraint was arrived at first heuristically [5] and only later demonstrated theoretical that it has optimal properties for tomographic analysis [6]. In broad terms, MFT/EFT relies less on a priori assumptions (that are nor supported by either physics or biology) than the linear methods. To compensate for relying less on the raw lead fields MFT makes more use of <u>all</u> the available information in the measured data and this entails solving a non-linear system for each set (time-slice) of data. In contrast, linear methods do not fully exploit the information in the measured signals because of their over-reliance on the raw lead fields, which are reflected in the blurred nature of their solutions or in the artificial additional constraints that are then imposed to make the solution unique.
- A key a priori assumption imposed by MFT/EFT is the low likelihood of very large source currents; this is reasonable from the point of view of neurophysiology. This supports the neurophysiological validity of its solutions, that is, one can be reasonably confident that the source currents estimated by the algorithm reflect true neural activity in the brain. Methods using purely mathematical constraints to solve the ill-posed inverse problem run the risk of obtaining solutions that are limited to only explaining the data without providing valid neurophysiological information.
- The space of MFT/EFT solutions (source space) includes the entire brain, including the brainstem, cerebellum, subcortical structures and the cortex. No prior assumptions are made regarding the likely location or orientation of the source current densities. This is in contrast to many current methods, which a priori restrict the sources to the cortex or grey matter, and restrict the orientations of the source currents to be along the norm of the cortical surface. These assumptions

are only partially justified and severely restrict the validity of their obtained solutions. There is strong evidence that MEG and EEG record substantial signals from subcortical sources and white matter [44, 45, 49, 50, 53]. The fact that the MFT solutions are localized in the gray matter and coincide with true gold standard, if such is available, can then be justifiably taken as a true measure of the localization power of the method.

4.1.3 *Source Analysis of Large Epileptic and Sleep Graphoelements*

4.1.3.1 Source Analysis of Epileptic Graphoelements

The most widely used MEG/EEG source analysis method in epilepsy is the equivalent current dipole (ECD) model. ECD methods model the brain activity with one or few point-like generators, which in many cases of *focal* epilepsy provide adequate localization for restricted epileptogenic cortical zones. However, ECD methods have several well-documented deficiencies, such as they require subjective judgments (e.g. about the number and initial positions of dipoles, the acceptable goodness-of-fit, GOF etc.) and in some cases they may be spatially and temporally inaccurate, especially when the signal-to-noise ratio (SNR) is low [54, 55]. Furthermore, even in the best cases with high SNR and a single dominant focal generator, when ECD provides an excellent fit, the rest of the brain cannot be ignored and the condition needs to be explained in terms of abnormal neural networks rather than changes in a focal brain region [56, 57]. Therefore, more advanced source analysis methods are necessary for understanding the neural processes underlying epileptic discharges. Recent evidence suggests that *distributed source models* are more valuable than ECD for this purpose [55, 58–62].

MEG in Idiopathic Focal Epilepsies (IFE)

The idiopathic focal epilepsies (IFE) form a group of childhood onset epilepsy syndromes characterized by focal onset seizures and characteristic defining EEG/MEG waveforms. They include one of the commonest childhood epilepsy syndromes, Benign Epilepsy with Centro-Temporal Spikes (BECTS), also known as rolandic epilepsy, characterised by childhood focal hemifacial and laryngeal onset sensorimotor seizures and characteristic high amplitude centrotemporal (or rolandic) spikes on EEG/MEG. The other commonly recognized IFE being Panayiotopoulos syndrome, with autonomic seizures, unresponsiveness and high amplitude, typically occipital EEG/MEG discharges.

MEG studies of IFE have focused on spike localization, their relationship to clinical features, and lately on time-frequency analyses of oscillatory rhythms.

An early MEG study in five patients found the spike ECDs localized to the same area as lower lip stimulation [63]. Subsequently, a study in seven cases reported spikes with an anterior positivity in the superior rolandic (hand motor) region in four patients and in the inferior rolandic (oromotor) region in three patients [64], related to seizure semiology: orofacial seizure manifestations in patients with inferior rolandic spikes and hand manifestations in those with superior rolandic spikes. Increased fast wave activity was reported in five patients with neuropsychological deficits [64]. In one case-study unilateral spikes were localized to bilateral operculum [65]. Combined EEG/MEG source localization identified interictal epileptic discharges (IED) in the lower somatosensory cortex, co-located with tongue movement fMRI activation [66].

Later studies using whole-head MEG systems localized IEDs 10–20 mm anterior and lateral to the hand somatosensory cortex (localized by median nerve stimulation in the same session) [67]. Dipole analysis of bilateral discharges showed two ECD sources in homotopic motor areas. A single ECD explained most of the unilateral spikes in a pre-central location, and over 98 % of spikes were seen simultaneously in EEG and MEG; suggesting a stable tangential dipole source. For bilateral IED, the temporal difference between bilateral foci was 15–21 ms. The same group correlated the location of IED sources with sensory responses [68], finding IED sources closer to SII than SI. Further analysis in the time frequency domain using the Morlet Wavelets showed power increase in the 0.5–40 Hz range in the side of the spike (most prominent in the alpha band) and increase in the range 0.5–25 Hz on the homologous area of the other hemisphere [69].

A study using the spatiotemporal multiple signal classification (MUSIC) analysis in five cases found that a single dipolar source was sufficient to explain the spiking activity in two cases, and in three cases complex sources were resolved that started in the more superior areas (finger/hand) and propagated and extend down along the precentral sulcus to the mouth/tongue area [70]. With a modest but nevertheless larger series a study of 15 patients with BECTS found three main types of spikes according to ECD analysis: (i) superiorly oriented spike MEG dipoles in the opercular area, (ii) anteriorly oriented spike dipoles in the rolandic area, and (iii) laterally oriented spike dipoles in the interhemispheric area [71]. Perhaps in the most detailed descriptive study of BECTS IED to date from 17 patients, Pataraia et al. [72] found right spikes in 6, left in 9, and bilateral in 2 patients. Examination of isopotential and isofield maps over 250 ms before and after the main negative peak of the spike using principle component analysis and spatiotemporal dipole modeling suggested that spikes were generated by a single tangential dipolar source located in the precentral gyrus with the positive pole directed frontally and the negative pole directed centro-temporally with a stable dipole over the entire (500 ms) time window analysed, with no differences in spike location, orientation or stability over time.

A correlation between cognitive deficits and spike location is described in 20 children with IFE with decreased language tests in those with left perisylvian spikes, and lower information processing in those with occipital spikes [73]. No relationship between spike rate and psychological deficits was found. One publication reports MEG findings in Panayiotopoulos syndrome (PS) [74]: 13 patients where studied with ECD analysis. Dipoles were localized along the parieto-occipital or calcarine sulcus in 11 of 13 patients, and in the rolandic area in 2 with atypical PS and rolandic IEDs.

Few MEG studies have been reported in Landau-Kleffner Syndrome (LKS) or CSWS, typically in conjunction with pre-surgical evaluation for multiple sub-pial transection. A study of four children with LKS found that the earliest spike activity originated in the intrasylvian cortex, spreading to contralateral sylvian cortex over 20 ms in one patient. Secondary spikes occurred within 10–60 ms in ipsilateral peri-sylvian, temporo-occipital, and parieto-occipital areas [75]. Others have also reported more widespread spikes in LKS: from 19 patients, 13 of whom had peri-sylvian MEG spikes, bilateral in 10 and unilateral in 3, 4 also had non-sylvian spikes in frontal or parietal areas [76]. In a larger cohort of 28 patients with LKS, 80 % had bilateral epileptic discharges generated in the auditory and language related perisylvian cortex, approximately 20 % had a unilateral perisylvian pacemaker that triggered secondary bilateral synchrony of spikes [77]. Analyzing MEG along with FDG Positron Emission Tomography (PET) findings in six children with CSWS, a study found that spike wave onset corresponded to areas of PET hypermetabolism during wakefulness, in superior temporal gyrus in LKS and centro-parietal regions in atypical rolandic epilepsy—areas of spike propagation predominantly showed PET hypermetabolism [78].

The first and only MFT analyses of interictal epileptic activity were over two decades ago using the 35-channel, flat dewar Siemens KRENIKON system. The MFT results obtained from the average MEG signals of 39 ‘equivalent’ sharp waves identified by template matching technique found two distinct dominant focal sources, one superficial and the other relatively deep; the relative amplitude of the two generators changed over time. The ECD solution appeared to move over time between the two sources identified by MFT with excellent agreement between the two methods only when the MEG signals were dominated by the contribution from one or other of these two sources [79]. A follow-up MFT analysis revealed a much richer dynamics that were masked by the averaging, but with the same dominant generators as the ones identified by the MFT analysis of the average MEG signals [80].

If a single or fixed number of ECDs are postulated then the appropriate fixed single foci will be identified. Under these circumstances the modeling is judged to be successful if the solutions are plausible and the data are fitted well. This seems to be the case for rolandic spikes. However, MUSIC analysis demonstrated that at least in some cases a succession of generators is involved, even for rolandic spikes [70]. It might then appear that the results change depending on what method one uses. In reality, each model has limitations but also flexibility to allow for useful generalizations. For example by allowing multiple and independent ECD solutions (for example for data from different times, periods or subjects) the ECD model can provide a distribution of solutions rather than a single location, as was done for example in a comparison of 10 PS and 10 children with other type of epilepsy [81].

4.1.3.2 Source Analysis of Sleep Graphoelements

Several studies have used source analysis to estimate neural generators of EEG and MEG signals during KCs. Using various methods, KC-related brain activity has been identified in deep central temporal [82–84], parietal [85, 86] and frontal [86] areas. Surprisingly, all these localizations are inconsistent with the 'true' cortical localization obtained from intracranially recorded KCs, which indicate that KCs are generated by widespread synchronous cortical activity arising maximally within the superior medial and lateral frontal cortices [1, 8]. To assess the validity of MEG/EEG source localization for large extended cortical sources, such as generators of KCs, Wennberg and Cheyne [1] have applied variety of popular MEG and EEG source analysis methods (ECD, sLORETA and LORETA) and compared the results directly to the intracranial cortical localization within the same patients. They found that all the tested methods (ECD, sLORETA and LORETA, each with various head models) provided physiologically invalid source estimates in all patients, localizing the 'true' extended superficial cortical generators of KCs to deep medial structures.

Here we use MEG data from the sleep study performed at RIKEN Brain Science Institute (Japan). Earlier reports using the same data focused on eye movements during REM sleep [50] and on quiet periods of different sleep stages [51]. Here we focus on the source identification during K-complexes testing a wider range of methods than the Wennberg and Cheyne study [1], including MFT.

4.2 Methods

Experimental procedures used in this study have been described in detail earlier [51]. Here we will summarize only the key points.

4.2.1 Subjects

Four healthy right-handed male subjects participated in the sleep experiment, after a night of acclimatization. For the details on acclimatization night see [51]. RIKEN's (the institution where the experiment was conducted) ethics committee approved the study, and all the subjects gave their informed written consent after all procedures were explained to them before the experiment.

4.2.2 Data Acquisition and Pre-processing

MEG was recorded throughout the night using a 151 gradiometer whole-head system (CTF/VSM Omega System, Canada) at a sampling rate of 625 Hz and bandpass of 0–208 Hz. In synchrony with MEG the following auxiliary signals were recorded:

scalp EEG from C3 and C4 locations referenced to A2 and A1 respectively, vertical and horizontal electrooculogram (EOG) and electromyogram (EMG) from the chin.

The continuous recording was interrupted every 3 min to record the subject's head location relative to MEG sensors, using head localization coils attached to the subject's head. Co-registration of the MEG sensors with the individual high-resolution anatomical MRIs was accomplished using a standard procedure described in [51].

Offline, the MEG was converted to a third order synthetic gradient and bandpass filtered from 0.625 to 200 Hz, with notch filters at 50 Hz and its harmonics. Independent components analysis was used to remove eye blink, cardiac and movement artifacts from the data.

Commonly used procedures were employed to score the sleep stages and visually mark stage 2 sleep KCs using EEG C3-A2 and C4-A1 channels [87]. For details on sleep stage scoring and identification of KCs see [51]. In this study, the epochs containing KCs were extracted using a time window from −2 to 2 s with respect to the KC onset.

4.2.3 Sensory Stimulation During Sleep

All subjects participated also in an additional MEG experiment where they received auditory and somatosensory stimulations during sleep. Stimuli of three intensities (subthreshold, at perception threshold and suprathreshold) were delivered to the subjects with an inter-stimulus-interval of 2 s, in 3 min. blocks. Sounds were pure tones delivered binaurally. Somatosensory stimuli were electrical pulses delivered one at a time to the left and right index fingers and toes. Here we report the results from the suprathreshold stimulations only.

4.2.4 Source Analysis

Source analysis of MEG signals for individual epochs was performed using seven different methods: two ECD models—single and multiple dipole models, two beamformer methods—SAM [2] and LCMV [3], two linear distributed source methods—wMNE [4] and eLORETA [5], and one non-linear distributed source method—MFT [6, 7]. All the methods except MFT were applied using the Fieldtrip toolbox [88]. The source space was defined as a regular 3D grid with a 5 mm resolution and the lead fields (forward computations) were performed using a realistically shaped single shell volume conduction model [89].

The dipole fits (single and multiple ECDs) were performed using all 151 MEG sensors and using only the 50 sensors best covering the top of the brain. The dipole fits using all sensors could not satisfactorily explain the data at any latency throughout the 4 s epochs; the best goodness-of-fit (GOF = 1—Residual Variance) across all studied KCs and time points was below 60 %, typically in the best case of two dipoles it was around 50 %. The dipole fit was performed with 1–5 ECDs; the best results in terms of GOF were obtained with 2 ECDs. Therefore in this chapter we report dipole modeling results obtained using 50 selected sensors and 2 ECDs; in all such cases the dipole models fit the data with a GOF of 90 % or higher.

The beamformers (SAM and LCMV) and linear distributed source methods (wMNE and eLORETA) used in the current study are known to have a bias towards superficial brain sources, because of the attenuation of the lead fields with the increasing source depth. To compensate for this bias, we have applied these methods using depth normalized lead fields. In each case the depth weighting parameter p was varied from 0.4 to 0.9. The different depth weightings did not significantly affect the localization results; here we report the results obtained with $p = 0.5$, which is the default depth normalization parameter used in Fieldtrip.

Similar to depth weighting, each method was applied with various noise regularization parameters, λ. In the cases of SAM and LCMV, the relative level of the noise that is used to calculate λ was varied from 5 to 25 %, for eLORETA λ values between 0.04 and 0.09 were used, and for wMNE we varied the signal-to-noise ratio (SNR) values between 3 and 7. The regularization parameter λ can be calculated from the relative noise level and SNR based on the estimates of source and noise covariance [90]. For each epoch, the noise covariance was calculated using a 2 s window preceding the onset of KC; the data covariance was calculated using the full 4 s epoch. The different regularization parameters used here did not significantly affect the localization results; here we report the results obtained with the values of 15 % for SAM and LCMV, 0.05 for eLORETA and 5 for wMNE. These are the values commonly found in literature and/or used as default in Fieldtrip.

MFT source analysis was performed using in-house developed software and an already standardized procedure routinely used in many prior studies, e.g. [42–44, 47, 48, 91]. Five partially overlapping source spaces, covering the left and right hemispheres, and front, back and top of the brain were defined as regular 3D grids with ~7 mm resolution. For each source space the forward computations were performed using a distinct spherical volume conduction model and the 90 MEG sensors best covering the source space under examination.

4.3 Results and Discussion

4.3.1 Localization of Neural Sources of Spontaneous KCs Using Different Methods

Figure 5.1 shows a typical spontaneous KC used in the current study. Here we report the source localization results from the peak latency of the main negative wave.

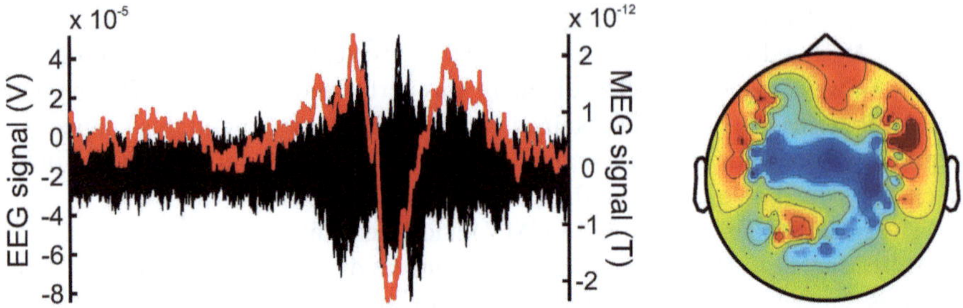

Fig. 5.1 A typical K-complex (KC) examined in the current study. On the left side a data epoch (4 s long) containing a clear KC is shown. The 'butterfly' plot of 151 MEG signals (*black*) is shown together with the EEG signal from C3 electrode (*red*). The ordinate axes on the left and right sides of the plot show the EEG and MEG signal magnitudes respectively. On the right side of the figure, a topographic map of the MEG signal distribution over the head at the time of the EEG (C3) main negative peak is shown

All methods *except MFT* provided very variable localization across different KCs, all of which were inconsistent with the intracranial findings [1, 8]: "widespread synchronous superficial cortical activity arising maximally within the superior medial and lateral frontal cortices." In many cases, the KCs were localized in deep medial brain structures as reported also by Wennberg and Cheyne [1]. Conversely, the strongest activations identified by MFT were consistently present during main negative wave of nearly all studied KCs. Moreover, the estimates provided by MFT were most consistent with the intracranial findings [1, 8]; strong and widespread activations were reliably found in superior aspects of bilateral frontal cortex. Figure 5.2 shows 'best' result obtained for each method that was most consistent with the intracranial findings [1, 8].

The MFT results are consistent the results of Wennberg and Cheyne [1] in that the most common source analysis methods used nowadays incorrectly estimate the neural sources of KCs. However, we strongly disagree with their more generalized conclusion that "existing noninvasive source localization techniques may not provide valid solutions for large extended cortical sources such as the human K-complex." The results of our current study and numerous earlier studies [42–53, 79, 80, 91] show that MFT, a noninvasive source localization technique, provides physiologically valid and accurate estimates for extended as well as focal brain sources.

The ECD model by its nature (one or few point-like generators) is not suitable for the localization of large extended sources, while LORETA, sLORETA, eLORET, wMNE and beamformers (SAM and LCMV) are linear inverse problem solvers, which too heavily rely on raw lead fields. Such an overreliance on lead fields, in our opinion, is the main reason for the tendency of linear models to provide deep medial solutions for widespread, bilaterally synchronous neural sources, clearly evident in the case of KC localizations. Our current findings show that MFT is a suitable tool for noninvasive source localization of extended cortical sources, such as the sources of sleep slow waves and widespread epileptiform discharges (e.g. generalized spike/wave complexes).

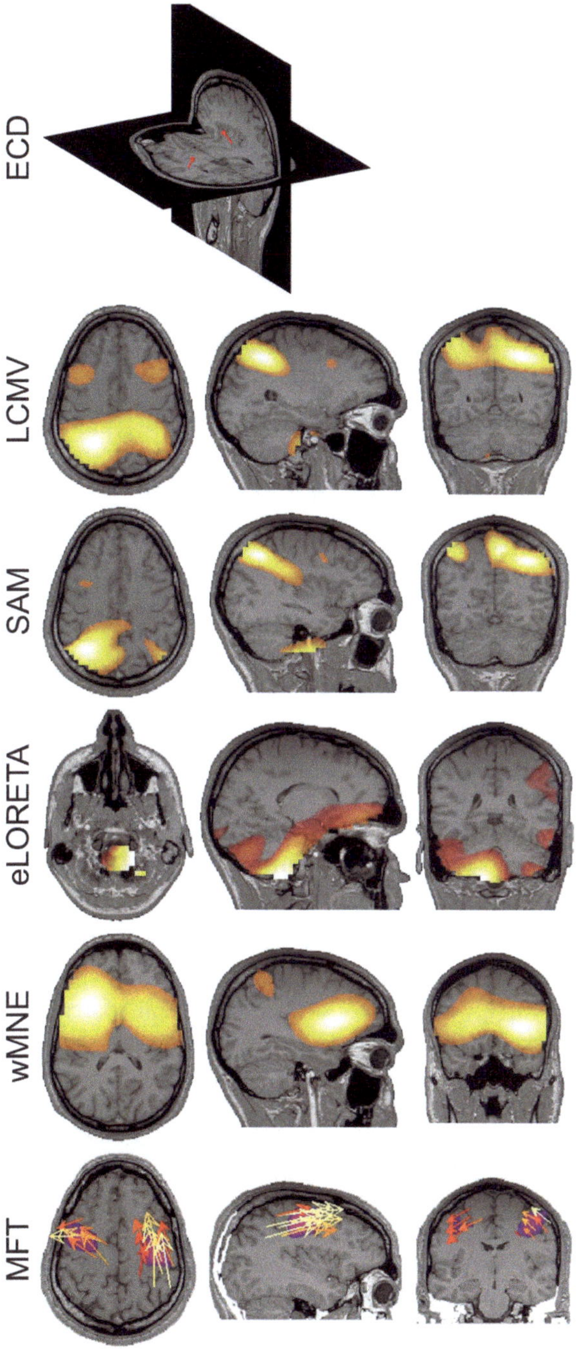

Fig. 5.2 Source localizations from the MEG data at the time of the EEG main negative peak. The 'best' localizations, i.e. closest to the intracranial findings [8], among the studied KCs, provided by different methods are shown. From top to bottom the results obtained from ECD model (2 ECDs, 50 MEG sensors), LCMV, SAM, eLORETA, wMNE and MFT are shown. For all methods, except ECD, the strongest activations are shown on coronal (*left*), sagittal (*middle*) and axial (*right*) MRI slices. For LCMV, SAM, eLORETA and wMNE the modulus of the estimated moment is plotted. For MFT the current density vector (*yellow* and *orange arrows*) and its modulus (*pink blobs*) are plotted. In all cases only the values above 50 % of the instantaneous maximum are shown

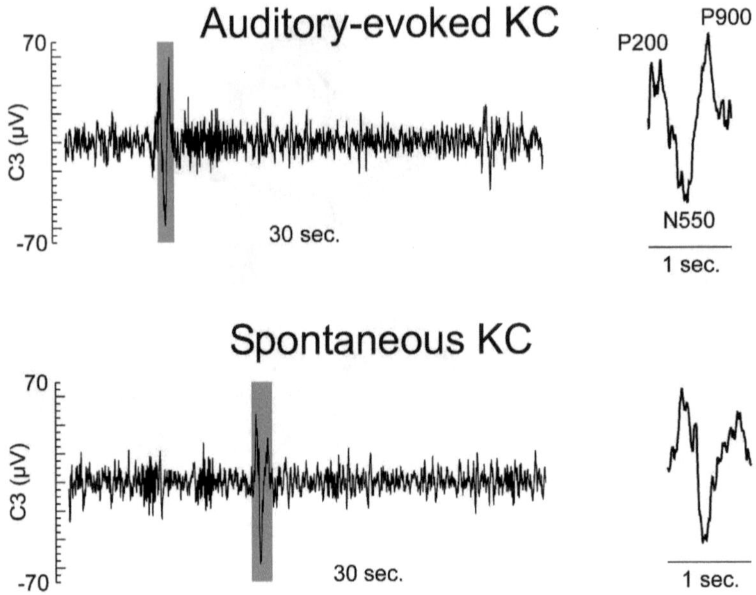

Fig. 5.3 Typical sensory-evoked (*upper row*) and spontaneous (*lower row*) KCs found in the current study. EEG signals from C3 electrode are shown in epochs of 30 s containing clear auditory-evoked and spontaneous KCs. The *grey shaded* period marks the KCs identified by a specialist. The *insets* on the right of each plot show the zoomed view of the marked KC (~1 s, *shaded* period). The conventional names of the tri-phasic evoked KC components are indicated next to each peak

4.3.2 Sensory Cortical Sources of the Auditory and Somatosensory Evoked KCs

Figure 5.3 shows a typical (auditory) evoked KC used in the current study together with a spontaneous KC recorded in the same subject.

MFT was used to localize the neural sources of auditory and somatosensory evoked KCs from recorded MEG signals. At the earliest latencies within KCs, around the onset of P200 and 150–260 ms post-stimulus, brain activations were typically localized in the stimulation-related primary sensory cortex (Fig. 5.4).

That is binaurally presented sounds elicited strong activations in bilateral primary auditory cortex (Fig. 5.4, upper row), while activation in contralateral primary somatosensory cortex were found in response to electrical pulses delivered to index fingers and toes (Fig. 5.4, lower row). Following these responses in the primary sensory cortices, the activity rapidly spread to bilateral frontal cortex, peaking in frontal cortical brain areas found to underlie the spontaneous KCs.

These novel results clearly demonstrate that the earliest component of sensory-evoked KC is not stereotypic, but depends on the stimulated sensory modality. Moreover, using a robust and accurate source analysis method, MFT, we localized the earliest brain activation within the sensory-evoked KC in the stimulation modality-specific primary sensory cortex.

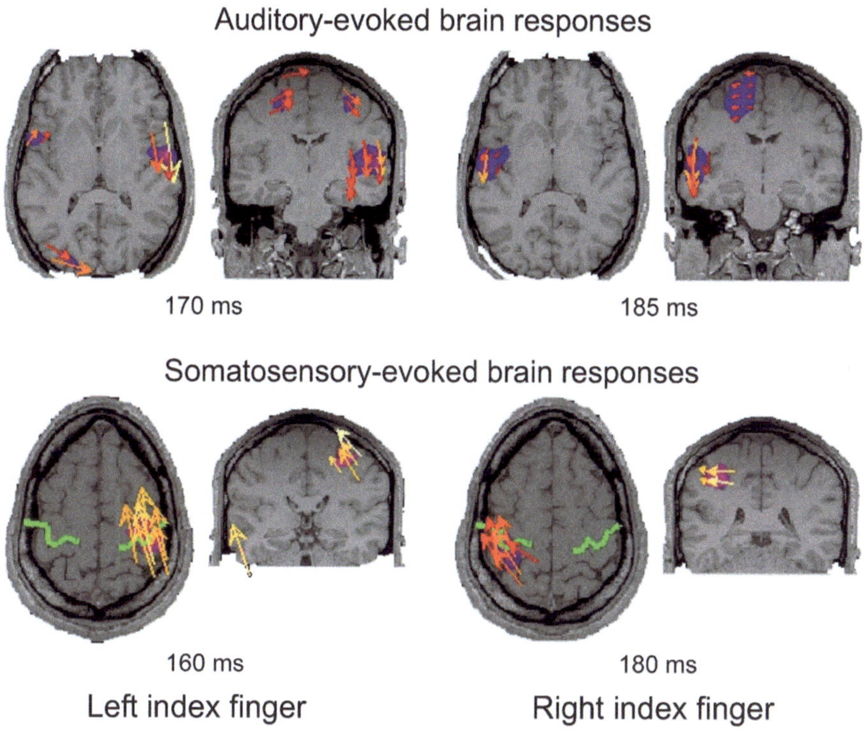

Fig. 5.4 Typical localization of auditory (*upper row*) and somatosensory (*lower row*) evoked brain response around the onset of KCs. The strongest responses are shown on axial (*left*) and coronal (*right*) MRI slices at peak post-stimulus latencies indicated below the images. The current density vector (*yellow* and *orange arrows*) and its modulus (*pink blobs*) are plotted. In all cases only the values above 50 % of the instantaneous maximum are shown

4.4 Conclusions

A careful study of sleep and especially of KCs may be very helpful in general epileptology research as well as clinic, while it is mandatory for the diagnosis and follow up of some particular nocturnal epilepsies. KCs are a cardinal feature to be taken into account in sleep stage scoring (hypnogram) and thereby realizing sleep quality metrics important for establishing possible effects of epilepsy and its medication on the patient's sleep. It would further help distinguishing certain types of seizures like NFLE or the effect of sleep deprivation on others. Sleep studies of epilepsy constitute an important indication for ambulatory EEG home monitoring and will be greatly helped by developments like this of ARMOR.

The current results, together with the earlier studies [42–45, 47–50, 52, 53, 91] showing the excellent localization accuracy of MFT for focal as well as extended sources, and the smart uses of MEG and EEG in epilepsy discussed in Chap. 13, demonstrate that the MFT analysis of MEG signals can be a powerful tool for studying epilepsy and in pre-surgical evaluation. To realize MFT's great potential for epilepsy, lengthy clinical studies are needed. The results suggest that it may be possible to eventually replace the invasive intracranial EEG procedures that are now necessary in epilepsy with MFT or some other (non-linear) method for localization of sources from MEG and possibly EEG data can.

References

1. Wennberg R, Cheyne D (2013) On noninvasive source imaging of the human K-complex. Clin Neurophysiol 124:941–955
2. Vrba J, Robinson SE (2001) Signal processing in magnetoencephalography. Methods 25:249–271
3. Van Veen BD, Van Drongelen W, Yuchtman M, Suzuki A (1997) Localization of brain electrical activity via linearly constrained minimum variance spatial filtering. IEEE Trans Biomed Eng 44:867–880
4. Liu AK, Dale AM, Belliveau JW (2002) Monte Carlo simulation studies of EEG and MEG localization accuracy. Hum Brain Mapp 16:47–62
5. Pascual-Marqui RD (2007) Discrete, 3D distributed, linear imaging methods of electric neuronal activity. Part 1: exact, zero error localization. arXiv:0710. 3341
6. Ioannides AA, Bolton JPR, Clarke CJS (1990) Continuous probabilistic solutions to the biomagnetic inverse problem. Inverse Probl 6:523–542
7. Taylor JG, Ioannides AA, Mueller-Gaertner HW (1999) Mathematical analysis of lead field expansions. IEEE Trans Med Imaging 18:151–163
8. Wennberg R (2010) Intracranial cortical localization of the human K-complex. Clin Neurophysiol 121:1176–1186
9. Loomis AL, Harvey EN, Hobart GA (1937) Cerebral states during sleep, as studied by human brain potentials. J Exp Psychol 21:127
10. Colrain IM (2005) The K-complex: a 7-decade history. Sleep 28:255–273
11. Halasz P (2005) K-complex, a reactive EEG graphoelement of NREM sleep: an old chap in a new garment. Sleep Med Rev 9:391–412
12. Riedner BA, Hulse BK, Murphy MJ, Ferrarelli F, Tononi G (2011) Temporal dynamics of cortical sources underlying spontaneous and peripherally evoked slow waves. Prog Brain Res 193:201–218
13. Laurino M, Menicucci D, Piarulli A, Mastorci F, Bedini R, Allegrini P, Gemignani A (2014) Disentangling different functional roles of evoked K-complex components: mapping the sleeping brain while quenching sensory processing. Neuroimage 86:433–445
14. Roth M, Shaw J, Green J (1956) The form, voltage distribution and physiological significance of the K-complex. Electroencephalogr Clin Neurophysiol 8:385–402
15. Webster KE, Colrain IM (1998) Multichannel EEG analysis of respiratory evoked-potential components during wakefulness and NREM sleep. J Appl Physiol 85:1727–1735
16. Cash SS, Halgren E, Dehghani N, Rossetti AO, Thesen T, Wang C, Devinsky O, Kuzniecky R, Doyle W, Madsen JR, Bromfield E, Eross L, Halasz P, Karmos G, Csercsa R, Wittner L, Ulbert I (2009) The human K-complex represents an isolated cortical down-state. Science 324:1084–1087
17. McCormick L, Nielsen T, Nicolas A, Ptito M, Montplaisir J (1997) Topographical distribution of spindles and K-complexes in normal subjects. Sleep 20:939–941
18. Amzica F, Steriade M (1997) Cellular substrates and laminar profile of sleep K-complex. Neuroscience 82:671–686
19. Koupparis AM, Kokkinos V, Kostopoulos GK (2013) Spindle power is not affected after spontaneous K-complexes during human NREM sleep. PLoS One 8, e54343

20. Kokkinos V, Koupparis AM, Kostopoulos GK (2013) An intra-K-complex oscillation with independent and labile frequency and topography in NREM sleep. Front Hum Neurosci 7:163
21. Kokkinos V, Kostopoulos GK (2011) Human non-rapid eye movement stage II sleep spindles are blocked upon spontaneous K-complex coincidence and resume as higher frequency spindles afterwards. J Sleep Res 20:57–72
22. Mak-McCully RA, Deiss SR, Rosen BQ, Jung KY, Sejnowski TJ, Bastuji HÃ, Rey M, Cash SS, Bazhenov M, Halgren E (2014) Synchronization of isolated downstates (K-complexes) may be caused by cortically-induced disruption of thalamic spindling. PLoS Comput Biol 10, e1003855
23. Poghosyan V, Ioannides AA (2014) Sensory cortical sources of the auditory and somatosensory evoked K-complexes. In: Organization for Human Brain Mapping 2014 Annual Meeting, p 4207
24. Kostopoulos GK (2009) Brain mechanisms linking epilepsy to sleep. In: Schwartzkroin PA (ed) Encyclopedia of basic epilepsy research. Academic, Oxford, pp 1327–1336
25. Shouse MN, Scordato JC, Farber PR (2004) Sleep and arousal mechanisms in experimental epilepsy: epileptic components of NREM and antiepileptic components of REM sleep. Ment Retard Dev Disabil Res Rev 10:117–121
26. Niedermeyer E (2008) Epileptiform K complexes. Am J Electroneurodiagnostic Technol 48:48–51
27. Steriade M, Amzica F (1997) Coalescence of sleep rhythms and their chronology in corticothalamic networks. Sleep Res online 1:1–10
28. El Helou J, Navarro V, Depienne C, Fedirko E, LeGuern E, Baulac M, An-Gourfinkel I, Adam C (2008) K-complex-induced seizures in autosomal dominant nocturnal frontal lobe epilepsy. Clin Neurophysiol 119:2201–2204
29. Amzica F, Steriade M (2002) The functional significance of K-complexes. Sleep Med Rev 6:139–149
30. Halasz P (1980) Generalized epilepsy with spike-wave paroxysms as an epileptic disorder of the function of sleep promotion. Acta Physiol Acad Sci Hung 57:51–86
31. Wieser HG (1990) Temporal lobe epilepsy, sleep and arousal: stereo-EEG findings. Epilepsy Res Suppl 2:97–119
32. Si Y, Liu L, Li Q, Mu J, Tian LY, Chen JN, Zhou D (2010) Features of the K-complex waves in refractory nocturnal frontal lobe epilepsy. Epilepsy Res 92:219–225
33. Petersen CC, Hahn TT, Mehta M, Grinvald A, Sakmann B (2003) Interaction of sensory responses with spontaneous depolarization in layer 2/3 barrel cortex. Proc Natl Acad Sci 100:13638–13643
34. Kohsaka S, Sakai T, Kohsaka M, Fukuda N, Ariga T (2012) Activation of the brainstem precedes and outlasts the K-complex in humans. Neuroscience 202:243–251
35. Terzano MG, Monge-Strauss MF, Mikol F, Spaggiari MC, Parrino L (1997) Cyclic alternating pattern as a provocative factor in nocturnal paroxysmal dystonia. Epilepsia 38:1015–1025
36. Bonakis A, Koutroumanidis M (2009) Epileptic discharges and phasic sleep phenomena in patients with juvenile myoclonic epilepsy. Epilepsia 50:2434–2445
37. Kokkinos V, Koupparis AM, Tsiptsios D, Kostopoulos GK, Koutroumanidis M (2013) Spatiotemporal profiles of focal and generalised spikes in childhood absence epilepsy. Epileptic Disord 15:14–26
38. Koutroumanidis M, Tsiptsios D, Kokkinos V, Kostopoulos GK (2012) Focal and generalized EEG paroxysms in childhood absence epilepsy: topographic associations and distinctive behaviors during the first cycle of non-REM sleep. Epilepsia 53:840–849
39. Niiyama Y, Satoh N, Kutsuzawa O, Hishikawa Y (1996) Electrophysiological evidence suggesting that sensory stimuli of unknown origin induce spontaneous K-complexes. Electroencephalogr Clin Neurophysiol 98:394–400
40. Hillebrand A, Barnes GR (2002) A quantitative assessment of the sensitivity of whole-head MEG to activity in the adult human cortex. Neuroimage 16:638–650
41. Michel CM, Murray MM, Lantz GÃ, Gonzalez S, Spinelli L, de Peralta RG (2004) EEG source imaging. Clin Neurophysiol 115:2195–2222

42. Moradi F, Liu LC, Cheng K, Waggoner RA, Tanaka K, Ioannides AA (2003) Consistent and precise localization of brain activity in human primary visual cortex by MEG and fMRI. Neuroimage 18:595–609
43. Poghosyan V, Ioannides AA (2007) Precise mapping of early visual responses in space and time. Neuroimage 35:759–770
44. Papadelis C, Poghosyan V, Fenwick PBC, Ioannides AA (2009) MEG's ability to localise accurately weak transient neural sources. Clin Neurophysiol 120:1958–1970
45. Ioannides AA, Fenwick PBC, Liu LC (2005) Widely distributed magnetoencephalography spikes related to the planning and execution of human saccades. J Neurosci 25:7950–7967
46. Ioannides AA (2006) Magnetoencephalography as a research tool in neuroscience: state of the art. Neuroscientist 12:524–544
47. Ioannides AA, Poghosyan V (2012) Spatiotemporal dynamics of early spatial and category-specific attentional modulations. Neuroimage 60:1638–1651
48. Poghosyan V, Ioannides AA (2008) Attention modulates earliest responses in the primary auditory and visual cortices. Neuron 58:802–813
49. Ioannides AA, Poghosyan V, Dammers J, Streit M (2004) Real-time neural activity and connectivity in healthy individuals and schizophrenia patients. Neuroimage 23:473–482
50. Ioannides AA, Corsi-Cabrera M, Fenwick PBC, Del Rio PY, Laskaris NA, Khurshudyan A, Theofilou D, Shibata T, Uchida S, Nakabayashi T, Kostopoulos GK (2004) MEG tomography of human cortex and brainstem activity in waking and REM sleep saccades. Cereb Cortex 14:56–72
51. Ioannides AA, Kostopoulos GK, Liu LC, Fenwick PBC (2009) MEG identifies dorsal medial brain activations during sleep. Neuroimage 44:455–468
52. Papadelis C, Eickhoff SB, Zilles K, Ioannides AA (2011) BA3b and BA1 activate in a serial fashion after median nerve stimulation: direct evidence from combining source analysis of evoked fields and cytoarchitectonic probabilistic maps. Neuroimage 54:60–73
53. Ioannides AA, Fenwick PBC (2005) Imaging cerebellum activity in real time with magnetoencephalographic data. Prog Brain Res 148:139–150
54. Ioannides AA (1995) Estimates of 3D brain activity millisecond by millisecond from biomagnetic signals: method (MFT), results and their significance. In: Eiselt E, Zwiener U, Witte H (eds) Quantitative and topological EEG and MEG analysis. Universitatsverlag Druchhaus-Maayer GmbH, Jena, Germany, pp 59–68
55. Tanaka N, Stufflebeam SM (2014) Clinical application of spatiotemporal distributed source analysis in presurgical evaluation of epilepsy. Front Hum Neurosci 8:62
56. Spencer SS (2002) Neural networks in human epilepsy: evidence of and implications for treatment. Epilepsia 43:219–227
57. Richardson MP (2012) Large scale brain models of epilepsy: dynamics meets connectomics. J Neurol Neurosurg Psychiatry 83:1238–1248
58. Shiraishi H, Ahlfors SP, Stufflebeam SM, Takano K, Okajima M, Knake S, Hatanaka K, Kohsaka S, Saitoh S, Dale AM, Halgren E (2005) Application of magnetoencephalography in epilepsy patients with widespread spike or slow-wave activity. Epilepsia 46:1264–1272
59. Tanaka N, Cole AJ, von Pechmann D, Wakeman DG, Hamalainen MS, Liu H, Madsen JR, Bourgeois BF, Stufflebeam SM (2009) Dynamic statistical parametric mapping for analyzing ictal magnetoencephalographic spikes in patients with intractable frontal lobe epilepsy. Epilepsy Res 85:279–286
60. Tanaka N, Hamalainen MS, Ahlfors SP, Liu H, Madsen JR, Bourgeois BF, Lee JW, Dworetzky BA, Belliveau JW, Stufflebeam SM (2010) Propagation of epileptic spikes reconstructed from spatiotemporal magnetoencephalographic and electroencephalographic source analysis. Neuroimage 50:217–222
61. Tanaka N, Peters JM, Prohl AK, Takaya S, Madsen JR, Bourgeois BF, Dworetzky BA, Hamalainen MS, Stufflebeam SM (2014) Clinical value of magnetoencephalographic spike propagation represented by spatiotemporal source analysis: correlation with surgical outcome. Epilepsy Res 108:280–288
62. Kanamori Y, Shigeto H, Hironaga N, Hagiwara K, Uehara T, Chatani H, Sakata A, Hashiguchi K, Morioka T, Tobimatsu S, Kira J (2013) Minimum norm estimates in MEG can delineate the

onset of interictal epileptic discharges: a comparison with ECoG findings. Neuroimage Clin 2:663–669
63. Minami T, Gondo K, Yamamoto T, Yanai S, Tasaki K, Ueda K (1996) Magnetoencephalographic analysis of roland discharges in benign childhood epilepsy. Ann Neurol 39:326–334
64. Kamada K, Möller M, Saguer M, Kassubek J, Kaltenhäuser M, Kober H, Uberall M, Lauffer H, Wenzel D, Vieth J (1998) Localization analysis of neuronal activities in benign rolandic epilepsy using magnetoencephalography. J Neurol Sci 154:164–172
65. Morikawa T (2001) Rolandic discharges in benign childhood epilepsy with centrotemporal spikes, and in other forms of partial epilepsies. Epileptic Disord 2:23–28
66. Van der Meij W, Huiskamp GJM, Rutten GJM, Wieneke GH, Van Huffelen AC, Van Nieuwenhuizen O (2001) The existence of two sources in rolandic epilepsy: confirmation with high resolution EEG, MEG and fMRI. Brain Topogr 13:275–282
67. Lin YY, Chang KP, Hsieh JC, Yeh TC, Yu HY, Kwan SY, Yen DJ, Yiu CH, Hari R (2003) Magnetoencephalographic analysis of bilaterally synchronous discharges in benign rolandic epilepsy of childhood. Seizure 12:448–455
68. Lin YY, Shih YH, Chang KP, Lee WT, Yu HY, Hsieh JC, Yeh TC, Wu ZA, Ho LT (2003) MEG localization of rolandic spikes with respect to SI and SII cortices in benign rolandic epilepsy. Neuroimage 20:2051–2061
69. Lin YY, Hsiao FJ, Chang KP, Wu ZA, Ho LT (2006) Bilateral oscillations for lateralized spikes in benign rolandic epilepsy. Epilepsy Res 69:45–52
70. Huiskamp G, van der Meij W, van Huffelen A, van Nieuwenhuizen O (2004) High resolution spatio-temporal EEG-MEG analysis of rolandic spikes. J Clin Neurophysiol 21:84–95
71. Ishitobi M, Nakasato N, Yamamoto K, Iinuma K (2005) Opercular to interhemispheric source distribution of benign rolandic spikes of childhood. Neuroimage 25:417–423
72. Pataraia E, Feucht M, Lindinger G, Aull-Watschinger S, Baumgartner C (2008) Combined electroencephalography and magnetoencephalography of interictal spikes in benign rolandic epilepsy of childhood. Clin Neurophysiol 119:635–641
73. Wolff M, Weiskopf N, Serra E, Preissl H, Birbaumer N, Kraegeloh-Mann I (2005) Benign partial epilepsy in childhood: selective cognitive deficits are related to the location of focal spikes determined by combined EEG/MEG. Epilepsia 46:1661–1667
74. Kanazawa O, Tohyama J, Akasaka N, Kamimura T (2005) A magnetoencephalographic study of patients with Panayiotopoulos syndrome. Epilepsia 46:1106–1113
75. Paetau R, Granstroem M-L, Blomstedt G, Jousmaeki V, Korkman M, Liukkonen E (1999) Magnetoencephalography in presurgical evaluation of children with the Landau-Kleffner syndrome. Epilepsia 40:326–335
76. Sobel DF, Aung M, Otsubo H, Smith MC (2000) Magnetoencephalography in children with Landau-Kleffner syndrome and acquired epileptic aphasia. Am J Neuroradiol 21:301–307
77. Paetau R (2009) Magnetoencephalography in Landau-Kleffner syndrome. Epilepsia 50:51–54
78. De Tiege X, Trotta N, Bourguignon M, Marty B, Wens V, Nonclercq A, Goldman S, Van Bogaert P (2013) Neurophysiological activity underlying altered brain metabolism in epileptic encephalopathies with CSWS. Epilepsy Res 105:316–325
79. Ioannides AA, Hellstrand E, Abraham-Fuchs K (1993) Point and distributed current density analysis of interictal epileptic activity recorded by magnetoencephalography. Physiol Meas 14:121–130
80. Bamidis PD, Hellstrand E, Lidholm H, Abraham-Fuchs K, Ioannides AA (1995) MFT in complex partial epilepsy: spatio-temporal estimates of interictal activity. Neuroreport 7:17–23
81. Yoshinaga H, Koutroumanidis M, Kobayashi K, Shirasawa A, Kikumoto K, Inoue T, Oka M, Ohtsuka Y (2006) EEG dipole characteristics in Panayiotopoulos syndrome. Epilepsia 47:781–787
82. Yoshida H, Iramina K, Ueno S (1996) Source models of sleep spindles using MEG and EEG measurements. Brain Topogr 8:303–307
83. Iramina K, Ueno S (1996) Source estimation of spontaneous MEG activity and auditory evoked responses in normal subjects during sleep. Brain Topogr 8:297–301
84. Ueno S, Iramina K (1990) Modeling and source localization of MEG activities. Brain Topogr 3:151–165

85. Lu ST, Kajola M, Joutsiniemi SL, Knuutila J, Hari R (1992) Generator sites of spontaneous MEG activity during sleep. Electroencephalogr Clin Neurophysiol 82:182–196
86. Numminen J, Makela JP, Hari R (1996) Distributions and sources of magnetoencephalographic K-complexes. Electroencephalogr Clin Neurophysiol 99:544–555
87. Rechtschaffen A, Kales A (1968) A manual of standardized terminology, techniques and scoring system for sleep stages of human subjects. US Dept. of Health, Education, and Welfare, Bethesda
88. Oostenveld R, Fries P, Maris E, Schoffelen JM (2010) FieldTrip: open source software for advanced analysis of MEG, EEG, and invasive electrophysiological data. Comput Intell Neurosci 2011:1–8. doi:10.1155/2011/156869
89. Nolte G (2003) The magnetic lead field theorem in the quasi-static approximation and its use for magnetoencephalography forward calculation in realistic volume conductors. Phys Med Biol 48:3637
90. Lin FH, Witzel T, Hamalainen MS, Dale AM, Belliveau JW, Stufflebeam SM (2004) Spectral spatiotemporal imaging of cortical oscillations and interactions in the human brain. Neuroimage 23:582–595
91. Ioannides AA, Poghosyan V, Liu L, Saridis GA, Tamietto M, Op de Beeck M, De Tiège X, Weiskrantz L, de Gelder B (2012) Spatiotemporal profiles of visual processing with and without primary visual cortex. Neuroimage 63:1464–1477

Chapter 5
Current Practices in Epilepsy Monitoring; Future Prospects and the ARMOR Challenge

Andreas A. Ioannides, Vahe Poghosyan, and George K. Kostopoulos

Abstract Methods and hardware for EEG recording are advancing rapidly with novel solutions approaching the performance of EEG recordings with conventional electrodes. The prime mover of the new solutions is the electronic game industry. Although there are some common desirable characteristics needed by EEG measurements in both the game industry and for home monitoring of epilepsy there are differences too, primarily in the demands for high signal quality, and this is where the new solutions are naturally still insufficient for clinical applications. Advances in the signal analysis are more mature with the use of tomographic estimates of activity from MEG but also from EEG data opening new ways for advancing the capability and usefulness of home monitoring for epilepsy management. The combination of the advances in EEG hardware and data analysis together with genetic and anatomical information for individual subjects coupled to powerful data mining techniques for "big data" is likely to revolutionize the monitoring and management of epilepsy.

5.1 General Introduction

The aim of the ARMOR project was to design a more holistic, personalized, medically efficient and economical system for home monitoring of epileptic patients. In this chapter we review current practices in epilepsy monitoring and use the ARMOR experience and the challenges that came along during its lifetime as litmus tests for the various options that currently appear as promising prospects for monitoring epilepsy in the future.

A.A. Ioannides (✉) • V. Poghosyan
Laboratory for Human Brain Dynamics, AAI Scientific Cultural Services Ltd., Nicosia, Cyprus
e-mail: a.ioannides@aaiscs.com

G.K. Kostopoulos
Neurophysiology Unit, Department of Physiology, Medical School, University of Patras, Patras, Greece

N.S. Voros, C.P. Antonopoulos (eds.), *Cyberphysical Systems for Epilepsy and Related Brain Disorders*, DOI 10.1007/978-3-319-20049-1_5

Epilepsy is a chronic disorder that is characterized by hyper synchronous and high amplitude surges in the electrical activity of the brain that can occur as isolated events, spikes and sharp waves in one or more places in the brain, or as seizures, an avalanche of electrical surges that can take hold of large brain areas in one or both hemispheres that are often, but not always, accompanied by extreme uncontrollable facial and other body part distortions and/or motor actions. These are obviously disturbing for both the patient and other people that witness it and can often be dangerous because of resulting injuries due to fall. When epilepsy is suspected the first task of the clinician is to determine whether the symptoms that give rise to the suspicion are truly due to epilepsy or they are due to other physiological or psychogenic origin. If epilepsy is indeed the underlying condition then it must be managed. Managing epilepsy involves stopping seizures, or at least controlling them finding the optimal way of doing so and if this involves medication reducing unwanted side effects and upsetting the patient's normal life. Part of epilepsy management is documenting what happens daily what happens in terms of behavior, feelings and of course electrical events in the brain. This monitoring part is obviously central in epilepsy but it is difficult to achieve because epileptic events do not come on demand, so they may be missed during one-off or intermittent examinations. The ARMOR aim is to provide near continuous monitoring in the subject's home environment so that the above problems can be efficiently overcome. Epilepsy is primarily an electrical phenomenon that can best be monitored by either electroencephalography (EEG) or magnetoencephalography (MEG). Obviously MEG is not suitable for continuous monitoring at home as intended in ARMOR but its indirect contribution can be important, as described in Chap. 13. In this chapter we discuss the challenges posed by the ARMOR requirements for unobtrusive home monitoring with EEG by reviewing current practices and capabilities and showing how these must be modified to make ARMOR monitoring an efficient and commercially viable proposition.

The remainder of this chapter is divided naturally into three self-contained parts. In following, Sect. 5.2, the hardware requirements are discussed. This is a key area that was deliberately left outside the original ARMOR project because it is a rapidly developing field where not only clinical concerns are paramount but also economic issues play a role because the main driver of improved devices comes from large commercial initiatives outside medicine (tied to the rapidly expanding game industry). As (the first phase of) ARMOR draws to a close the time is right to review from the point of view of future ARMOR services what EEG hardware is now available and what is likely to be available in the not too distant future. In Sect. 5.3 we examine the current practices in the analysis of the EEG and discuss the advances accomplished through recent advances in source estimation and other EEG analysis methods that were adapted and applied specifically for ARMOR purposes. The chapter concludes with the final Sect. 5.4 that discusses the new opportunities that are now opening up thanks to the advances already described in the previous sections and other related scientific and technological breakthroughs.

5.2 Demand and Prospects of Novel Sensors for Long Term EEG Monitoring of Epilepsy at Home

5.2.1 Introduction

Electroencephalography (EEG) introduced in 1929 by Hans Berger was quickly established as a technique for non invasive continuous recording of the electrical activity and response of the brain. Since then, there has been little change in the physical principles that sustain the signal acquisition probes, the EEG electrodes. EEG records the time-varying magnitude of electric fields emanating from the brain. The great value of EEG is that it most directly reflects real brain function in real time. The process of synaptic communication in the brain generates ionic currents that can be thought of as elementary dipoles mostly along the apical dendrites of cortical pyramidal neurons (see [1]). This active electric current generation is passively volume conducted through the intervening medium of widely different resistive and capacitive compartments: brain tissue, meninges, skull and skin. The expected distortion and diminution of the ionic flow produces a change in the electrical potential topography on the scalp. It is these changes in electrical potential on the scalp that the EEG records non-invasively. The underlying physics is described by Maxwell equations that dictate that the changes in ionic currents at the source location propagate and appear as changes in the electrical potential on the scalp with the speed of light; therefore for all intents and purposes the scalp electrical potential is an instantaneous reflections of the primary neuronal ionic currents. The mathematical format of the Maxwell equations describes a linear relationship between the vector ionic current and the scalar electrical potential. A direct consequence of linearity is that ionic currents with similar strength but opposite direction from (nearly) the same location in the brain will cancel each other out while similar ionic currents in the same direction will produce maximum combined effect. The combined consequence of the nature of the generation of the ionic currents, the instantaneous summation and the linearity of the Maxwell's equations are that the EEG signal is dominated by synchronous cortical activity from pyramidal neurons at the gyri of the cortex; in contrast the contribution from opposing walls in the sulci largely cancels out.

Although EEG signals have been firmly related to body functions and cognitive processes, routine use of EEG recording is limited by the relatively low power signals that are often obscured by transient signal noises and large-amplitude artifacts. Typical range of EEG signals is roughly 1–100 μV in the low frequency range that most routine EEG operates (less than 100 Hz). Signal quality therefore continues to be a priority challenge, which advanced technology is attempting to overcome by developing new acquisition devices. Signal quality and signal stability are of particular interest to ARMOR as a project aiming to long-term EEG monitoring.

Fig. 5.1 EEG electrodes. From left to right, (1) conventional EEG electrode filled with conductive gel, (2) 64 electrode cap used at UOP-NU, (3) comb-like dry-contact electrode, (4) active electrode

The key element of a good EEG recording is therefore the actual measurement of the electrical potential on the scalp.

EEG electrodes are material objects made of electrical conductors and shaped to make good electrical contact with the skin. Connecting electrodes at two different parts of the scalp allows the measurement of the electrical potential difference between these two scalp location. In conventional scalp EEG, the recording is obtained by placing 21–256 Ag/AgCl electrodes on the scalp with a conductive gel, paste or saline filled sponge, usually after preparing the scalp area by light abrasion to reduce impedance due to dead skin cells to about 5 kΩ. The preparation procedure is lengthy (up to one hour), requires well trained staff (typically medical technical assistants), and wear time is limited by the stability of the electrolytes (gels) [2]. Furthermore the gel may spread outside the electrode contact, thus shortening the recording of neighboring electrodes. Electrodes are either attached independently to a wire or are embedded in caps or nets, which is particularly common when high-density arrays of electrodes are needed (Fig. 5.1).

As an alternative to these conventional gel-coupled or "wet electrodes", and because they cannot stay wet for longer than a few hours, gel-free "dry-contact electrodes" were developed. Initially made of metals they were rigid, but lately the use of other conducting materials gave them flexibility [3]. While dry-contact electrodes are already in the market, research is in progress for the development of non-contact electrodes, which are based on capacitance rather than current. Traditional electrodes are passive contacts and their (non-shielded) wire connection is usually the source of artifact due to power line interference. Partly to eliminate the latter the electrodes described below are active in the sense that they incorporate a miniature circuit for signal amplification (and not only). Especially when the contacts are equipped with advanced preprocessing and wireless transmitting capabilities they are called sensors in distinction to passive electrodes. Additional challenges emerging from clinical but recently mostly non clinical applications of EEG relate to the speed and easiness of positioning the sensors, the convenience to the recorded subject, the remote and long term recording etc. The novel recording systems face these challenges as well by integrating the advanced electrodes with wireless transmission of data and powerful algorithms for EEG signal analysis helped by personalized databases [4].

5.2.2 *The Challenges Faced by Conventional Electrodes EEG Recording*

5.2.2.1 Clinical Demands for EEG Are Increasing and Changing

- There is an increase of diagnosed brain disorders including many expressed electrophysiologically and some demanding long-term EEG monitoring i.e. epilepsy, sleep apnea and sleep quality etc. Although the diagnosis is still carried out mostly in hospitals by expert practitioners, there is a trend of non-invasive, unobtrusive and long-lasting patient's monitoring outside the clinical environment, as a solution to both diagnostic demands and cost-efficiency. This trend has surfaced over the last few years and led to the development of a number of non-invasive monitoring devices, which can also be personal and home-based (like the present project ARMOR)
- New analysis methods for tomographic estimates of brain activity based on EEG aim to eventually replace invasive depth recordings for the identification of epileptic foci and the study of drug resistant interictal activity and seizures
- Progress is made in new methods to evoke brain responses, to study EEG rhythms, cross-frequency relationships and brain connectivity based on raw EEG or eventually its tomographic solutions
- EEG is now assisting or integrated with other methods of brain imaging i.e. EEG-triggered-functional MRI, EEG-functional Infrared Spectroscopy, EEG combined with MEG etc.
- the need is increasingly recognized to incorporate EEG into "body area networks" of multimodal sensors of whole body activity including expressions from the autonomic nervous system and the motor system

5.2.2.2 New Non Medical EEG Applications Are Emerging

Brain Computer Interface, video games industry, neurofeedback, psychology and neuropsychology evaluation, neuromarketing research, pharmaceutical development, research for military applications etc. Also novel electrodes are evaluated for detection of drowsiness/fatigue, which is a well recognized safety concern for drivers and industrial workers who must stay alert and attentive for long periods of time [5].

5.2.2.3 In Many Ways EEG Acquisition Fails Short of the Demand

Application of conventional silver/silver-chloride (Ag/AgCl) electrodes for EEG requires accurate, time-consuming individual preparation of each electrode position and application of additive electrolyte materials. The later are gels containing NaCl and for optimal impedance (about 5 kΩ) and data quality the electrode has to be firmly pressed on the skin until establishing sufficient and enduring contact.

These processes and materials increase patient inconvenience and stress, risk of skin irritation and limit acquisition time—when the gel dries and hardens. The use of electrode gel entails serious problems that are especially pronounced in real-world settings when experts are not available. Furthermore, the wiring poses limitations for patient mobility and measurement environment. After EEG recording, electrode removal is difficult, often even painful, and time consuming. Hence, conventional electrode caps are inappropriate for emerging EEG applications especially those demanding long term monitoring.

5.2.3 Development of Dry-Contact Electrodes

In order to face the drying and hardening problems posed by wet electrodes several dry (that is gel free) EEG electrodes were proposed based mostly on the micro electromechanical systems (MEMS) technique [6–8]. Sullivan et al. [9] presented an architecture for high-density, dry-electrode EEG based around the concept of integrating the entire signal processing (amplification, filtering, digitizing) chain on top of a dry MEMS electrode. This enables electrodes to be easily daisy-chained and expanded with only one common-wire, significantly reducing the clutter associated with a conventional EEG system. It is easily wearable and provides access to the forehead locations without gels or other preparation. Monitoring of user attention or alertness is another area that has been explored as a candidate for dry-contact EEG. Several headsets have been developed with this in mind. In 2009 Tsai et al. presented a detailed study of using dry-contact EEG sensors to monitor for driver drowsiness.

However, the measurement of MEMS-based dry electrodes is usually a semi-invasive approach, and their manufacture cost are relatively expensive. Although dry electrodes were constructed also from different conductive materials like rubber, fabric, polymer foam etc. the skin-electrode interface impedance remained much higher than that of the wet electrodes due to the capacitor characteristics of this interface, while hairy skin presented a serious obstacle especially in motion [10]. To solve the latter issue several spike- or comb-shaped dry electrodes were developed for EEG recording [11].

The dry contact electrodes were also connected to a preprocessing circuit (g.SAHARA) i.e. a capacitor was in series connected to the input node of the operational amplifier to eliminate DC offset, although this contribute a phase distortion. Another proposal used a metal-pin electrode with spring load [12].

Park et al. [13] developed a simple and cheap one-channel wireless dry EEG device for drowsiness detection. A pronged dry-AgCl electrode without conductive gel but equipped with a miniature amplifier circuit along with a reference and ground are kept on the occipital region of the scalp by a headband harness and sends signals to a printed circuit board base unit which then wirelessly transfers the digital signal through bluetooth technology to a computer where it is analyzed. Thus it eliminates the discomfort of gel electrodes while obtaining strong signals from hair covered areas of the scalp and discriminates power in alpha, beta and theta bands (see also [14]).

Dry electrodes with metal pins overcome most of the problems of wet electrodes, but their rigidity causes discomfort and pain. To overcome this problem flexible polymer-based dry electrodes with high user comfort have been presented (see [15, 16] used electrodes fabricated from ethylene propylene diene monomer rubber matrix containing various additives for optimum conductivity, flexibility and fabrication yield. The hardness and elastic modulus both increase with higher carbon content; with ~45 % content of carbon being selected as best compromise between electrical and mechanical properties. The impedance of those electrodes measured on phantoms and human skin showed contact impedances ~10 times larger than gel electrodes. Although they were capable of recording strong signals such as ECG, for low-amplitude EEG, the electrodes needed to be coupled with a small active circuit. With this addition they showed promising results: alpha-waves could be observed clearly with correlation and coherence similar to gel electrode signals and only a slightly lower signal to noise ratio, while subjects reported no discomfort from the polymer electrodes. They thus concluded that the polymer-based dry electrodes are promising alternatives for the rigid dry electrodes in development and the conventional gel electrodes.

Although the above dry electrodes could increase the skin-electrode contact area in hairy scalp, their skin-electrode interface impedances were still relatively higher than that of the wet electrode. Attempting to solve this by applying higher pressure via some specific mechanical design caused discomfort. In order to improve the above issue, a different active comb-shaped dry electrode (active CSDE) was recently proposed [17] to measure EEG in hairy scalp. This CSDE separates the hair layer, and an active circuit was designed to reduce the signal attenuation and phase shift, and avoid the reduction of common mode rejection ratio (CMRR). In deviation from other active electrodes, a high input-impedance bias current path was used in the input node of the active circuit to protect against the parasitic voltage and ensure the operational amplifier working in the active region. By using the design of the active circuit, the proposed active CSDE could effectively measure EEG without applying a higher pressure and was validated in recording alpha rhythm and steady state visual evoked potentials. Finally compared to MEMS-based dry electrodes that need the complex and expensive manufacture procedure, that of the proposed active CSDE is simple, and it is suitable for mass-production.

From their inception to their current research use in BCI [18], several alternative constructs of dry EEG electrodes have been presented made of different materials. The field has now matured enough for assessing their advantages and drawbacks. Slater et al. [19] calculated that using a dry EEG electrode system can reduce by 80 % the time from start placing the electrodes to obtaining an interpretable EEG. Able preprocessing electronics mounted on the electrode provided an adequate EEG signal quality even though they showed impedances as high as 10 MΩ. A drawback of dry-contact electrodes is that they have to touch the skin, so they have high motion artefacts and sensitivity to skin condition. Recently Fiedler et al. [20] developed and compared novel EEG electrodes with titanium and polyurethane as base materials, which were coated with nanometer sized titanium nitride films. Furthermore gold multipin electrodes were arranged on printed circuit boards. Impedances were found significantly higher for all dry electrode types compared to wet electrodes, but still within the measurement range of today's standard biosignal amplifiers. It was found that the novel dry titanium and polyurethane based electrodes show biosignal quality equivalent to conventional electrodes.

Developers of novel or improved front-end circuits for biopotential recordings using dry electrodes face the challenge of validating their design. In a recent critical and in depth evaluation of problems in assessment of novel biopotential front-end with dry electrodes, Gargiulo et al. [21] concluded that although dry electrodes allow more user-friendly and pervasive patient-monitoring, proof is required that new devices can record with a quality at least comparable to existing medical devices. Aside from electrical safety requirement recommended by standards and concise circuit requirement, there is not yet a complete validation procedure able to demonstrate improved or even equivalent performance of the new devices. At present, dry-contact EEG systems are limited to only reliably acquiring forehead signals, due to the vastly varying thickness of human hair. For clinical applications to be truly viable, clinical procedures and validation must be established for this limited set of EEG signals first.

5.2.4 Development of Non-contact Electrodes

Dry electrodes require direct electrical contact with the skin which is best in hairless body areas, they are very sensitive to the condition of the skin and have high susceptibility to motion artifacts. An alternative road to solving these problems is taken by the developers of non-contact electrodes. Based on the capacitance coupling theory they developed electrodes which do not require ohmic connection to the body and thus are insensitive to skin condition. Since they are capable of measuring biopotentials through hair or clothing, they appear highly suitable for embedment inside garment for a completely unobtrusive patient-friendly monitoring.

Very essential progress in developing contactless biopotential recording systems has been made by the group of Sullivan, Cauwenbergs, Chi and others at the University of California at San Diego, USA—see Sullivan et al. [22]; Chi YM, Ph.D. Thesis [23]; Chi and Cauwenbergs [24] and patent US 8,694,084 B2 of 4 April 2014 "Non-contact biopotential sensor". In the above sources one can find the full schematics and design considerations for building high-quality, discrete-component non-contact sensors. Their detailed characterization of the discrete component sensors, led them to conclude that (a) they could not reliably achieve clinical standards and (b) cost-effective manufacturing of the sensors was precluded if they would maintain critical specifications, like high-input impedance and the required considerable manual tuning. They considerably improved on (b) by fully controlling the parasitics associated with the input node, but nevertheless, unresolved problems with noise and motion artifacts remained a difficulty. The signal from the non-contact sensor, while much more accurate than before, were still much more noisy. In contrast to previous views they found that the noise in their sensor was intrinsic to the interface and irreducible. This revelation helped them develop a fully custom analog sensor front-end, achieving input impedances and frequency responses far exceeding what was previously possible, all completely without the need for manual adjustment.

Recently Kitoko [25] used non-contact capacitive EEG processing integrated circuit (EPIC) electrodes in a headset system and showed effective EEG recordings and classification of the cognitive EEG signals as relaxation with eye-closed, mathematical calculation and writing letter tasks. An ability to acquire quality EEG signals at different conditions with minimal discomfort was demonstrated, along however with a lack of optimal effectiveness and efficiency of the proposed noncontact electrode in comparison with wet electrode. They found a roughly one order of magnitude difference between the wet and dry electrodes absolute impedances and a more negative impedance phase for non-contact EPIC electrode, meaning that the non-contact EPIC electrodes are more capacitive than bristle dry electrode and gold-wet electrodes for low frequencies. The critical aspect appeared to be the optimization evaluation of contact pressure at the interface between headset-based non-contact EPIC electrode and scalp to minimising movement of the electrode with respect to the skin, and this will directly affect the optimal quality of the recording EEG signals.

Evaluation: Non-contact electrodes allow user-friendly and pervasive patient monitoring with no need for conducting NaCl gel—as the wet electrodes—or even touching the skin—as the dry electrodes. They are relatively comfortable to the wearer, need no preparation and work through hair. However the fact that non-contact electrodes 45 years after their first reporting [26] are still not in the market, suggests that the approach faces serious obstacles. The main is the large and highly varying skin-electrode coupling impedance, which is mainly capacitive. From this stem all their technical difficulties and drawbacks of their use: high inherent susceptibility to motion/friction artifacts; poor settling times and excessive low frequency noise often exceeding the interface thermal noise. The large skin electrode coupling impedance demands amplifiers with ultrahigh input impedance, low noise and thus zero input currents and minimum input and guard capacitances. These requirements make the design of the front-end op-amp biasing a real challenge [27]. In short, non contact electrodes—due to above—cannot yet offer the quality EEG recordings needed for advancement to the market, let alone for clinical validation. Current research raises hopes for future improvements within the mechanical construction and signal processing domain. Combining the circuit innovations with new industrial design and algorithms, which will improve the signal to noise ratio and identify and reject artifacts, may finally enable non-contact sensing as a practical tool for mobile health (including long term monitoring of epilepsy), fitness and brain-computer interface applications.

5.2.5 Beyond the Electrode/Skin Interface

Regardless of the electromechanical type of the electrodes, advances in miniaturization of electronics will undoubtedly continue to incorporate into sensors more and more preprocessing functions beyond amplification, like filters, A/D converters etc., eventually enabling wireless transmission of well conditioned digital signals. Since the major obstacle in long term recordings is signal quality, it is essential that the sensors impedance (a measure of contact to the skin) would be measured at installation by delivering a constant current pulse through the electrode to ground and displaying the response on a local screen. Because the contact may deteriorate over time or even be completely interrupted, this test should be repeated at close intervals throughout the recording through the incorporation in the sensors of appropriate device for surveillance of whether the skin-electrode interface impedance and/or the input values exceed some set range of values. In the latter case they would send a note to an attendant to correct the contact and also they would make a mark on the record for the eventual reader of the EEG; otherwise it would be difficult to identify artifacts with certainty in an offline evaluation of the EEG record. Furthermore the designer of any multimodal monitoring system should take into account the ability of each sensor to be incorporated in a body network (issues of proper common grounding, avoidance of cross talk between sensor channels etc.). Finally in a multisite EEG system, sensors may include remotely controlled on/off switches, in order to conserve energy by utilizing the optimum number of EEG sensors needed according to online analysis.

5.2.6 Conclusions and Prospects

Recently, significant advances in technology have enabled new EEG applications, not only in neurology but also in unexpected fields of applications, from Brain Computer Interface (BCI) and neurofeedback to a booming games industry. In all these applications, new aspects such as usability and gel-free operation are first order priorities. This is a crucial demand in any long-term and/or non-hospital-based EEG, because the application of current electrodes establishing contact with the scalp through conducting gel are time and labor intensive and of short-lived quality. Thanks to new advances in materials and integrated electronic systems technologies, a new generation of dry-contact electrodes made of new conducting materials has been developed to fulfill the need, while research is under way to develop wearable non-contact electrodes based on capacitance coupling. In parallel efforts are made to transform electrodes into sensors, i.e. integrated acquisition preprocessing and wirelessly emitting systems embedded in a mobile headset. Placement of dry contact electrodes is quick and no skills demanding, while avoiding the deterioration of signal and skin irritation of wet electrodes. Qualitative benchmarks show that the signal quality they offer is not very far from that of conventional wet electrodes. Research is burgeoning in an effort to more effectively overcome their drift and noise problems as well as their difficulty to record in hairy skin and during motion. The latter two problems should theoretically be more easily faced with the non contact electrodes, which could also be woven into caps and therefore may eventually expand the range of applications. However the advantages non-contact electrodes have been shown only at the level of proof of concept and several issues remain to be tested and solved. They often introduce problems such as comfort and most importantly the still show low EEG signal quality. While several dry-contact EEG sensors are already in the market, the industry has only recently entered the field trying to select the types of electrodes showing the best signal, the least calibration variability, the smaller motion artefacts or which are the least cumbersome or use the less costly electronics. Non contact electrodes definitely work, i.e. they are fully capable of resolving EEG signals, but they are yet nowhere near the cost effectiveness or the medical reliability of the traditional gel-covered wet-contact Ag/AgCl electrodes.

The development of novel advanced technology and more effective and convenient to the patient EEG sensors would be a very suitable task beyond, and in the direction started by, the ARMOR project. The review above fully vindicates to leave outside the scope of ARMOR the choice and deployment of electrodes that was correctly judged to be outside the original call which invited only ready off-the-shelf components to be used. The latter have served well the development of the system, but for future development, it is becoming obvious that long-term, high quality and convenient to the patient EEG monitoring demands the incorporation of advanced technology EEG sensors. Although non contact EEG electrodes are still in a research stage, the market for dry EEG electrodes is making a very dynamic start although for non clinical applications that may soon be relevant to future continuations of ARMOR, pending appropriate validation.

5.3 EEG Data Analysis for Epilepsy

Typically, specialists visually inspect the recorded raw EEG data for epileptiform activity or less often use specialized software tools to this end. Regardless, they rely on the raw EEG data, which has a number of limitations. Most critically, the electric potentials measured at the scalp are considerably distorted (smeared) by volume conduction of the skull and other tissues intervening between the brain and scalp, therefore the raw EEG signals may give a misleading impression as to the location, time course or other properties of the neural sources of the epileptiform electrical activity, especially when the source activity is distributed, weak or is localized in deeper brain structures. Another important limitation of the raw EEG is the spatial sampling, which in routine scalp EEG is incomplete, as significant amounts of cortex, particularly in basal and medial areas of the hemispheres, are not covered by standard electrode placement.

To overcome the limitations of the raw EEG data, we suggest more common usage of advanced source analysis methods. Currently, EEG source analysis is used only during the pre-surgical evaluation and even then only a simple equivalent current dipole (ECD) models are usually employed. An overview of different electromagnetic source analysis methods and their applications in epilepsy and sleep studies is provided in the Chap. 4. Emphasize in the Chap. 4 was on magnetoencephalography (MEG) source analysis, however the same ideas apply also to EEG source analysis. One key difference is that accurate EEG source analysis requires a more detailed and accurate geometric and conductivity model of the head.

In the ARMOR project we have performed a series of EEG source analyses. In the next few sections we will briefly describe some of the key results obtained from these analyses. For other applications of electromagnetic source analysis that enable smart uses of MEG and EEG in epilepsy see Chap. 13.

5.3.1 Tomographic Analysis of Epileptic Spikes Identified by ARMOR Automatic Spike Detection Methods

EEG data (9 h sleep, ~100 spikes) from one epileptic patient recorded at St. Loukas Hospital, Thessaloniki, was used in this work. Spikes were marked manually by a specialist and were identified by the program (*detectSpikes*) developed at UoP-CEID as part of the ARMOR platform. The specialist marked 101 spikes; the program detected a total of 149 spikes, 98 of which corresponded to the manually marked ones. Thus assuming the manually marked spikes as the 'gold standard', the assessment of the results by UoP-CEID was: 98 true positives, 3 missed spikes ('false negative') and 51 false positives. To evaluate this assessment and the performance of the ARMOR's *detectSpikes* program, the EEG data were subjected to detailed tomographic source analysis at AAISCS.

The EEG data contained signals from 26 electrodes placed according to the international 10-20 standard (Fp1, F3, C3, P3, O1, F7, T3, T5, F9, T9, P9, Fz, Pz, Fp2, F4, C4, P4, O2, T4, T6, F10, T10, P10, F8, Cz, Oz) and sampled at 500 Hz. Before the tomographic analysis, these signals were re-referenced to the 'common average reference' and were high-pass filtered at 1 Hz, with notches at 50 Hz and its harmonics. EEG data segments of 1 s around the manually marked and automatically detected spikes (−0.5 to 0.5 s) were extracted for tomographic source analysis.

The MRI of the patient or the digitized positions of the electrodes were not available, therefore we used the MNI152 template MRI and estimated the positions of the scalp electrodes based on the names. Four-shell realistic head model (brain, CSF, skull and scalp) was created based on ICBM tissue probability maps (http://www.bic.mni.mcgill.ca/ServicesAtlases/ICBM152NLin2009) and symmetric boundary element method (BEM) was used to compute the EEG lead fields [28–30]. Electric field tomography (EFT) was applied independently to every timeslice (2 ms) of each spike-related EEG data segment (from −500 to 500 ms) to estimate the three-dimensional distribution of neural current sources throughout the brain. EFT is the EEG adaptation of the magnetic field tomography (MFT) algorithm [31, 32]; EFT was recently developed for a different project (DEFT, ΕΠΙΧΕΙΡΗΣΕΙΣ/ΠΡΟΪΟΝ/0311/42).

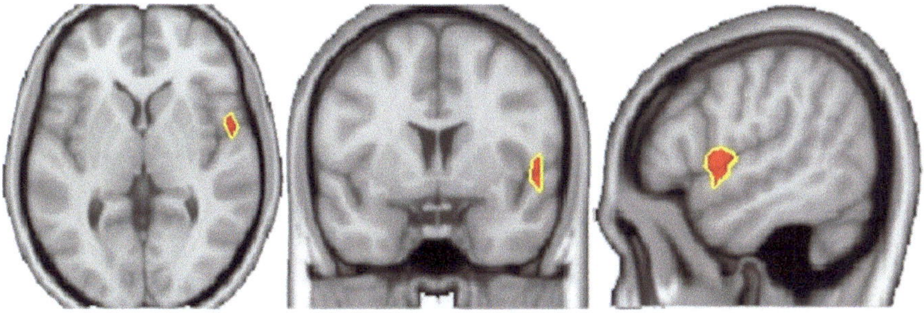

Fig. 5.2 Source of the spike. The activation identified based on manually marked spikes is shown on the MNI152 template MRI. From *left* to *right*, the axial, coronal and sagittal views are shown. The *yellow contour* on each view encompasses the region with SNR >8

It was modified for the ARMOR purposes, and especially for these data, accounting for the missing MRI and electrode positions.

Time-varying signal-to-noise ratio (SNR) maps were computed from the sets of EFT estimates corresponding to manually marked, 'true positive', 'false positive' and 'false negative' spikes [33]. The source of the spike was identified from the SNR map corresponding to the manually marked spikes (Fig. 5.2). Figure 5.2 shows the source extending over Temporal and Frontal lobes, with its centroid localized in the Right Superior Temporal Gyrus (BA 22, MNI coordinates: 53, 3, −1).

The activation time-course of this source in all selected data segments was computed using a standard procedure, routinely employed at AAISCS [34–37]. Figure 5.3 shows the source time courses in all segments around the manually marked spikes (101 spikes). As expected the spike is clearly identifiable, even on the level of single segments.

The *detectSpikes* program identified 51 spikes, which were not manually marked by the specialist. By assuming the manually marked spikes as the 'gold standard', these spikes were labeled as 'false positive'. However, our analysis shows that 41 (80 %) of these automatically detected spikes very likely are 'true positives' (Fig. 5.4). For ease of reference here we termed them as *false 'false positives'*.

It is clearly evident that the source time-courses in these segments contain spikes, which are very similar to the manually marked / true positive ones shown in Fig. 5.3. It is nevertheless worth noticing that the false 'false positive' spikes are more variable in their magnitude than the manually marked ones. That is the human specialist missed the more variable events, possibly because they were weaker in strength and/or other strong electrical activity interfered making it more difficult to stand out in the raw EEG.

The real false positive spike segments (10 of the 51 'false positive' segments) are distinctly different (Fig. 5.5).

The markers for the six of the 41 false 'false positive' spikes were set 20–25 ms before the spike (Fig. 5.6). The time shift is clearly visible in Fig. 5.6: the solid curve peaks at ~23 ms.

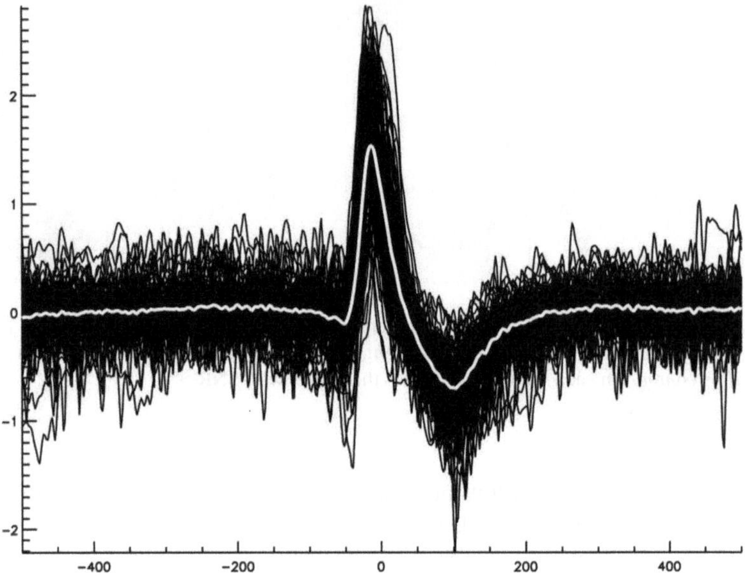

Fig. 5.3 Source activation time-courses in segments around the manually marked spikes. The time-courses extracted from the 101 segments around manually marked spikes are overplotted (*black curves*) together with the corresponding average time course (*thick white curve*). Abscissa is the time in milliseconds with respect to the spike marker. Ordinate is the neural current source magnitude in arbitrary units

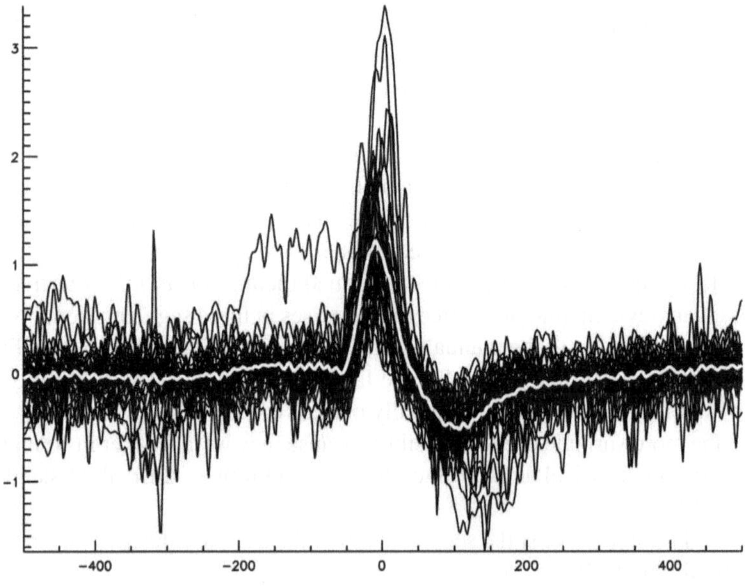

Fig. 5.4 Source activation time-courses in segments around the automatically detected false 'false positive' spikes. The time-courses extracted from the 41 segments around the automatically detected false 'false positive' spikes are overplotted (*black curves*) together with the corresponding average time course (*thick white curve*). Abscissa is the time in milliseconds with respect to the spike marker. Ordinate is the neural current source magnitude in arbitrary units

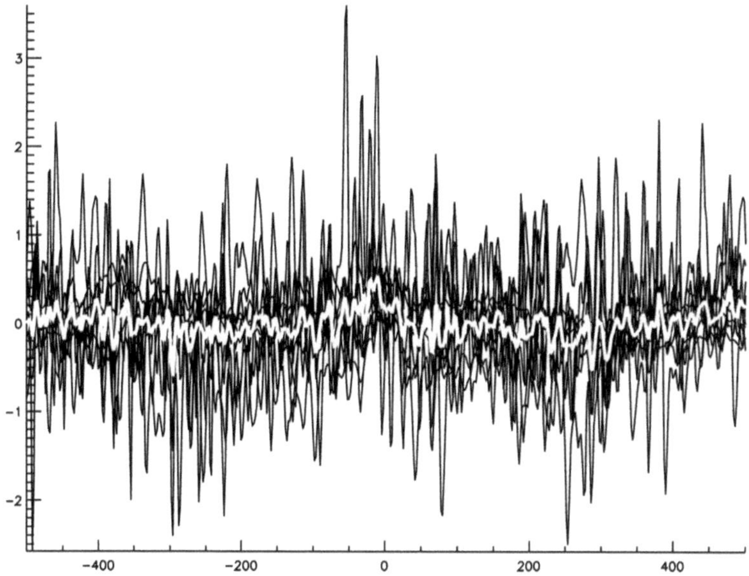

Fig. 5.5 Source activation time-courses in segments around the automatically detected real false positive spikes. The time-courses extracted from the 10 segments around the automatically detected real false positive spikes are overplotted (*black curves*) together with the corresponding average time course (*thick white curve*). Abscissa is the time in milliseconds with respect to the spike marker. Ordinate is the neural current source magnitude in arbitrary units

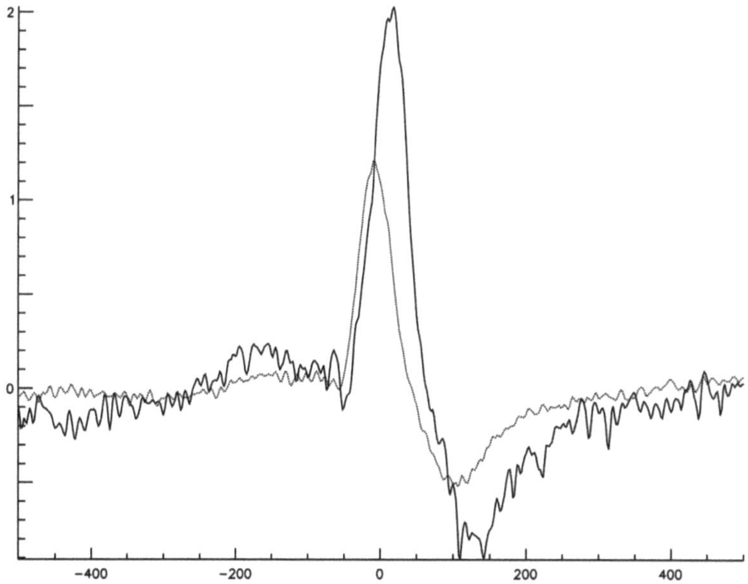

Fig. 5.6 Averaged source activation time-courses in segments around the automatically detected false 'false positive' spikes. The averaged time-courses from the 35 properly aligned (*dotted curve*) and 6 shifted (*solid curve*) false 'false positive' spike segments are overplotted. Abscissa is the time in milliseconds with respect to the spike marker. Ordinate is the neural current source magnitude in arbitrary units

In summary, the results obtained from the tomographic source analysis suggest that the automatic detection program performed much better than the human specialist. Specifically in 41 of the 51 'false positives' the source time-course contained clear spikes, which were very similar to the true positives. The tomographic analysis therefore suggests that these 41 events are very likely 'true positives', *i.e.* they are 'true spikes' that were missed in manual marking, possibly because the EEG signature was contaminated by other co-occurring events. In 6 of these 41 detected spikes, the program set the markers 20–25 ms before the peak of the spike.

5.3.2 Tomographic Analysis of Epileptic Spikes Recorded from an ARMOR Test Case Patient

Here we performed tomography-based analysis of EEG data obtained from one epileptic patient (Patient ID: 01off), who was included in the cohort of test cases for ARMOR evaluation. The patient suffers from intractable cryptogenic (normal brain MRI) frontal lobe seizures and repeated long EEG recordings (including a 24 h ambulatory EEG and all night full EEG montage polysomnography) failed to record seizures, although they demonstrated left frontal spikes. The patient needs longer EEG recordings to ascertain a frontal onset of his seizures and be enrolled in the pre-surgical evaluation process using intracranial recordings. The latter are deemed necessary, but their positioning needs to be guided somehow, as brain MRI is normal.

Twenty eight (28) epileptic spikes, which occurred in two ~10 s periods during the recording, were identified and marked by expert clinicians at KCL. The primary purpose of the tomographic analysis was to address the question whether the underlying brain activity was indeed focal, and if so, to localize it. In addition a more informative raw EEG signal representation was sought that reflected best the spike-related brain activity and could be used in future long-term measurements from this patient.

The EEG data contained signals from 21 scalp electrodes placed according to the international 10-20 standard and sampled at 200 Hz. Before the tomographic analysis, these signals were re-referenced to the 'common average reference' and were band-pass filtered between 1 and 45 Hz. EEG data segments of 1 s around the marked spikes (–0.5 to 0.5 s) were extracted for tomographic source analysis.

Neither the MRI of the patient nor the digitized positions of the electrodes were available. Therefore we used the MNI152 template MRI and estimated the positions of the scalp electrodes based on the 10-20 nomenclature. Four-shell realistic head model (brain, CSF, skull and scalp) was created based on tissue probability maps and symmetric boundary element method (BEM) was used to compute the EEG lead fields [28–30]. Electric field tomography (EFT) was applied independently to every timeslice (5 ms) of each spike-related EEG data segment (from –500 to 500 ms) to estimate the three-dimensional distribution of neural current sources throughout the brain.

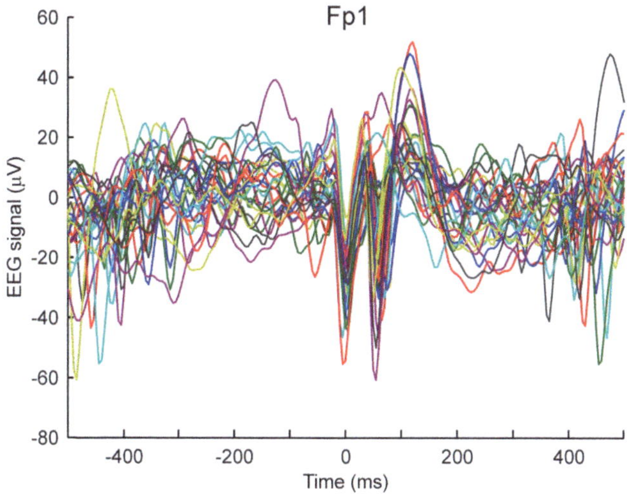

Fig. 5.7 EEG spikes. Signals over Fp1 electrode in all 28 extracted epochs (overplotted)

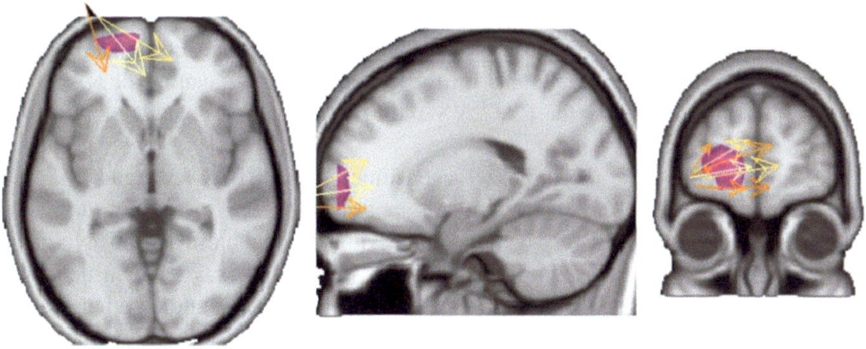

Fig. 5.8 Neural source of the spike. The averaged brain activation at the peak of the spike is shown on the MNI152 template MRI. From *left* to *right*, the axial, coronal and sagittal views are shown. The current density vector (*yellow* and *orange arrows*) and its modulus (*pink blobs*) are plotted. Only the values above 70 % of the instantaneous maximum are shown

5.3.2.1 Spike-Related Brain Activity

The spikes were clearly evident over the Fp1 electrode in all 28 extracted epochs and in most, but not all of the epochs the time course showed a double peak (Fig. 5.7).

The strongest brain activity at the first peak of the spike was localized in the Left Medial Frontal Gyrus (Left MFG BA 10, Fig. 5.8).

To investigate the temporal dynamics of the spike-related brain activity we computed a number of time courses: regional activation curves (RAC), which are

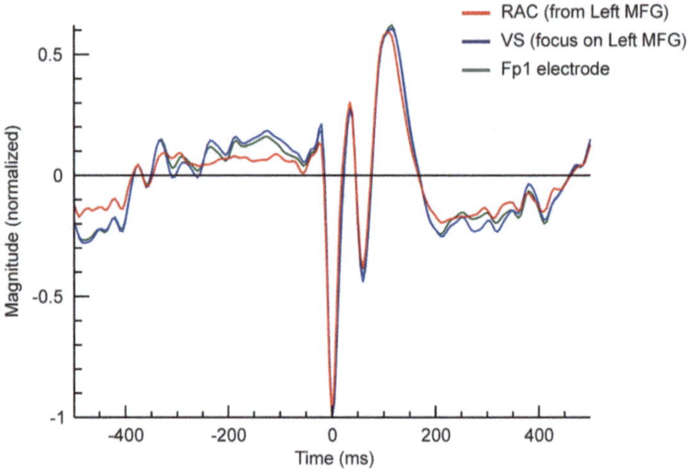

Fig. 5.9 Averaged time courses of spike-related brain activity. Each time course is normalized to its absolute maximum, because the physical units of the plotted measures are different

generated from EFT estimates and most faithfully describe the activation time course of a brain region; virtual signal (VS), which is estimated by means of simple lead field based spatial filter focused on the specific brain region; and signal recorded over the Fp1 electrode. For details on the generation and meaning of RACs see [34–37]. Figure 5.9 shows the RAC, VS and Fp1 time courses averaged across all identified spikes.

In this specific case, the three averaged time courses (RAC, VS and Fp1) are nearly identical. Thus the electric signals measured by Fp1 electrode during the identified spikes represent very well the underlying regional brain activity. Note that here the analysis was focused on the data epochs when spikes were identified from the EEG signals; it is possible that spikes in the Left MFG, and/or other places, occurred also in other time periods however they were not evident in the measured EEG signals. This could be because other concurrently active brain source masked it and/or the other generators are too deep.

It is worth noting that during normal brain activity the signals measured by a single electrode do not reflect the activity of a single brain source so faithfully. Brain activity is distributed and fairly similar across different areas. Superficial activity close to an electrode would give rise to a stronger signal than activity from deeper brain structures. In the current case, the neural source of the identified epileptic spikes was focal, strong and superficial, and therefore it stood out, especially in the EEG signals measured by the Fp1 electrode.

5.3.2.2 Selection of Optimal Electrode Positions

The data presented here were recorded from a patient using 21 EEG electrodes placed according to the international 10-20 standard, and the epileptic spikes were most evident over the Fp1 electrode. With the information provided, it is possible to

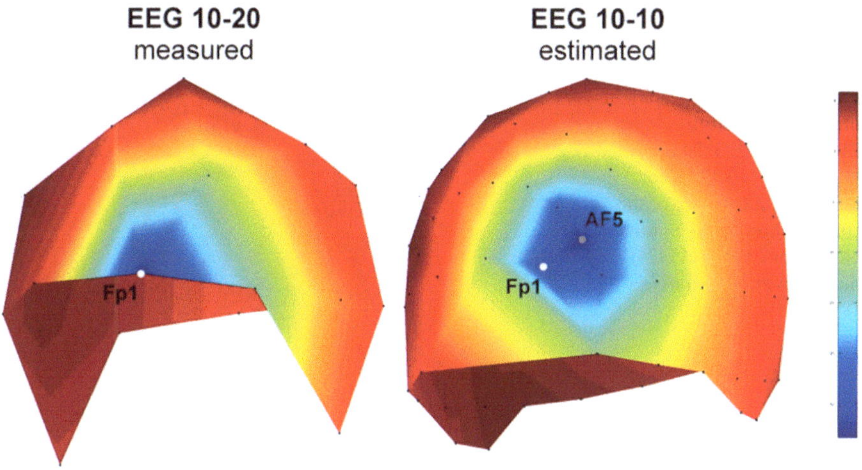

Fig. 5.10 Averaged EEG signal topographies at the peak of the spike. On the *left side*, the EEG topography of the measured 21-channel (10-20 system) EEG data is shown. On the *right side*, the corresponding topography of the estimated (from 21-channel EEG data) 94-channel (10-10 system) EEG data is shown. On both topographic, the Fp1 electrode is marked with a *white circle*. The *grey circle* on the 10-10 topography marks the AF5 electrode with the strongest absolute EEG signal in the estimated data

later monitor the patient using much fewer electrodes placed at optimal scalp positions. As part of the ARMOR project AAISCS has developed a new methodology for determining the optimal number and position of electrodes based on prior available data (see Chap. 13). Briefly, the methodology involves two main steps: predicting EEG signals at large number of unmeasured scalp positions from a set of MEG or low-density EEG measurements of relevant epileptic event and 2) selecting the few electrode positions, which in combination provide the most information about the event.

Here we used the recorded 21-channel (10-20 system) EEG data to predict the signals for the 94-channel (10-10 system) EEG (Fig. 5.10).

The predicted EEG signals at the peak of the spike were used to determine the optimal number and position of EEG electrodes for monitoring this specific brain event (see Chap. 13 of the ARMOR book). The algorithm settled for two such electrodes: AF5 and FC4 (Fig. 5.11). Thus according to our methodology it is best to monitor the specific brain event in this patient using only these two electrodes attached to the scalp.

To evaluate the advantage provided by the selection of the two optimal electrodes we compared 3 EEG signals: measured over Fp1, estimated at AF5 and a simple virtual signal computed as AF5—FC4 (Fig. 5.12).

Similar to the time courses presented in the Fig. 5.9, here the three averaged time courses (Fp1, AF5 and AF5-FC4) are nearly identical. Thus in this specific case no substantial advantage is gained from the optimal electrode selection. However, in many other cases this methodology provides significant benefits (see Chap. 13).

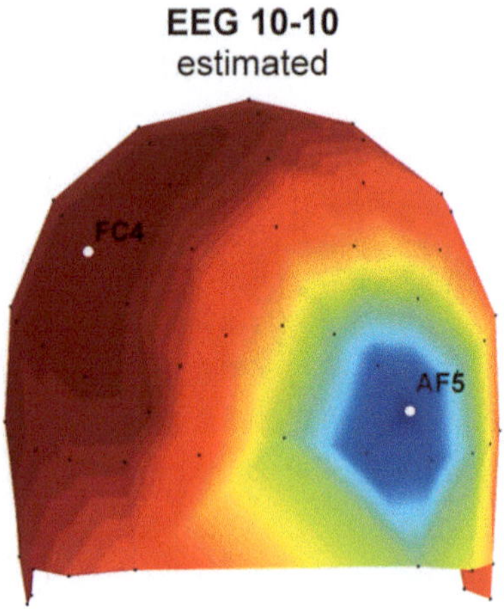

Fig. 5.11 Optimal electrode positions for monitoring the specific brain event. The averaged EEG topography of the estimated (from 21-channel EEG data) 94-channel (10-10 system) EEG data at the peak of the spike is shown. The optimal electrodes (AF5 and FC4) are marked with *white circle*

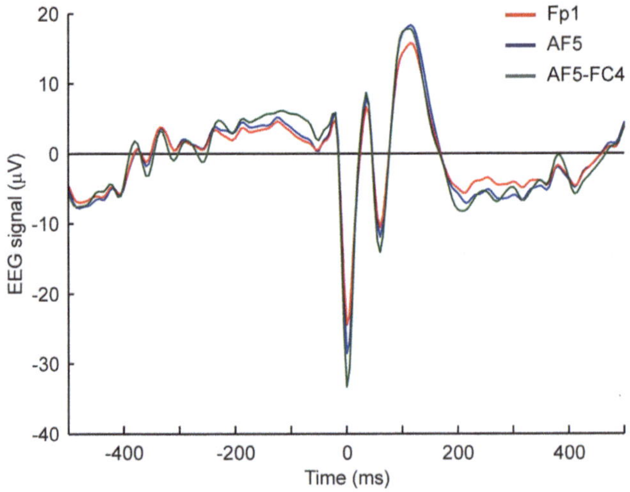

Fig. 5.12 Averaged time courses of spike-related EEG signals

5.3.2.3 Summary

We identified the brain sources underlying the marked epileptic spikes. The strongest brain activity at the peak of the spike was localized in the Left Medial Frontal Gyrus (Left MFG BA 10), consistent with the predictions from signal-based analysis and clinical report. However, because the MRI of the patient or the digitized positions of the electrodes were not available and templates were used instead, the high accuracy of the localization cannot be assured.

All the signal-based estimates (raw EEG signal representations) of spike-related brain activity used in the current analysis (measured signal over Fp1, predicted signal over AF5, predicted AF5-FC4 and VS) were very accurate for this patient. Usually this is not the case; especially when considering the signals measured by a single electrode. Such signals do not reflect the activity of a brain source so faithfully.

Normally the brain activity is distributed evenly across different areas that intermittently become more active for short time intervals (a few milliseconds). The current case was exceptional in this respect showing a persistent focal, strong and superficial generator activated at the peak of the identified spike.

From the point of view of ARMOR we need to ask: what if any new insights were furnished by the tomographic analysis of the identified spikes. First, the key question is adequately addressed: all indication are that the identified spikes are generated in one brain area. Second, it is evident that the identified epileptic activity can be monitored with few electrodes, making it an ideal case for home monitoring with ARMOR, if the clinical decision is to follow up this patient over a longer time period. However, the overall available information is insufficient for confident reliance on the localization provided by EFT. The uncertainty arises partly because the MRI and electrode locations were not available, but also because the electrodes did not fully cover the area of maximum EEG activity (it is evident in Fig. 5.10 that the most active electrode (Fp) is at the edge of the area covered by the 10-20 montage. The 10-10 montage would obviously allow for a more accurate reconstruction. If the clinical decision was to go for surgery a new EEG would be recommended or if possible an MEG recording.

5.4 Outlook

The results presented here make point towards exciting new prospects for monitoring epilepsy. Although the evolution of EEG recording is fast, the advances are closely aligned to the demands of the electronic game industry rather than clinical consideration. While the decision to stay clear from the development of new EEG hardware was both necessary and correct on general grounds, the situation is nevertheless sufficiently close to realization today that a new initiative that will involve development of new EEG sensors, specifically for ARMOR-like applications seems highly desirable. While progress in wearable devices for the game industry will continue it is unlikely that the signal quality will improve sufficiently to satisfy the ARMOR requirements. In the end it may be more effective to consider an initially expensive solution that will fit closer the ARMOR requirements and work towards making it more affordable through the development of significant home monitoring applications that will make mass production attractive.

The tomography based achievements accomplished within ARMOR open up new opportunities for home monitoring and even clinical applications for all aspects of epilepsy management: diagnosis, identification of epileptic foci, recording of rare events and evaluating changes induced by intervention (pharmaceutical or other).

A future deployment of ARMOR will be facilitated by the availability of detailed information about the subject to allow a truly personalized system optimized for each individual patient. Such additional information should include at least one recording with a high density EEG or preferable MEG and good quality anatomical information, including MRI (for the conductivity model definition) and ideally tractography data showing at the individual level the connectivity between areas. This will allow testing before deployment candidate foci to be monitored by integrate them into a wider network activity pattern that may be clinically more informative.

Finally, a future deployment of ARMOR will provide a huge volume of data that together without databases of epileptic activity and genetic information will allow "big data" methods to be used for a better understanding of the mechanism of epilepsy. The database of epilepsy data and the knowledge that can be extracted from it can then be used to improve the individual diagnosis and monitoring and possibly guide decisions about treatment in individual cases.

References

1. Ioannides AA, Kostopoulos GK (2008) Real-time tomography from magnetoencephalography (MEG)? Experimental and theoretical challenges, possible physiological basis and evidence from the analysis of real data. Int J Bioelectromagnetism 10:37–51
2. Tallgren P, Vanhatalo S, Kaila K, Voipio J (2005) Evaluation of commercially available electrodes and gels for recording of slow EEG potentials. Clin Neurophysiol 116:799–806
3. Leleux P, Badier J-M, Rivnay J, Benar C, Herve T, Chauvel P, Malliaras GG (2014) Conducting polymer electrodes for electroencephalography. Adv Healthc Mater 3:490–493
4. Casson AJ, Smith S, Duncan JS, Rodriguez-Villegas E (2008) Wearable EEG: what is it, why is it needed and what does it entail? Eng Med Biol Soc. 30th Annual International Conference of the IEEE, p 5867–5870
5. Arnin J, Anopas D, Horapong M, Triponyuwasi P, Yamsa-Ard T, Iampetch S, Wongsawat Y (2013) Wireless-based portable EEG-EOG monitoring for real time drowsiness detection. Eng Med Biol Soc. 35th Annual International Conference of the IEEE, p 4977–4980
6. Lopez-Gordo MA, Sanchez-Morillo D, Valle FP (2014) Dry EEG electrodes. Sensors 14:12847–12870
7. Hoffmann KP, Ruff R (2007) Flexible dry surface-electrodes for ECG long-term monitoring. Eng Med Biol Soc. 29th Annual International Conference of the IEEE, p 5739–5742
8. Chang CW, Ko LW, Lin FC, Su TP, Jung TP, Lin CT, Chiou JC (2010) Drowsiness monitoring with EEG-based MEMS biosensing technologies. GeroPsych 23:107
9. Sullivan TJ, Deiss SR, Jung TP, Cauwenberghs G (2008) A brain-machine interface using dry-contact, low-noise EEG sensors. IEEE International Symposium on Circuits and Systems, p 1986–1989
10. Lin YP, Wang Y, Wei CS, Jung TP (2014) Assessing the quality of steady-state visual-evoked potentials for moving humans using a mobile electroencephalogram headset. Front Hum Neurosci 8:182
11. Sellers EW, Turner P, Sarnacki WA, McManus T, Vaughan TM, Matthews R (2009) A novel dry electrode for brain-computer interface. In: Human-computer interaction. Novel Interaction Methods and Techniques. Springer, Berlin, pp 623–631
12. Toyama S, Takano K, Kansaku K (2012) A non-adhesive solid-gel electrode for a non-invasive brain-machine interface. Front Neurol 3:114
13. Park J, Xu L, Sridhar V, Chi M, Cauwenberghs G (2011) Wireless dry EEG for drowsiness detection. In: 33rd annual international conference of the IEEE EMBS, Boston, MA, USA, August 30–September 3, 2011

14. Guger C, Krausz G, Allison BZ, Edlinger G (2012) Comparison of dry and gel based electrodes for P300 brain-computer interfaces. Front Neurosci 6:60
15. Grozea C, Voinescu CD, Fazli S (2011) Bristle-sensors low-cost flexible passive dry EEG electrodes for neurofeedback and BCI applications. J Neural Eng 8:025008
16. Chen YH, Op de Beeck M, Vanderheyden L, Vanstreels K, Vandormael H, Van Hoof C (2014) Soft conductive polymer dry electrodes for high-quality and comfortable ECG/EEG measurements. Adv Sci Technol 96:102–107
17. Huang Y, Wu C, Wong A, Lin B (2015) Novel active comb-shaped dry electrode for EEG measurement in hairy site. IEEE Trans Biomed Eng 62:256–263
18. Zander TO, Lehne M, Ihme K, Jatzev S, Correia J, Kothe C, Picht B, Nijboer F (2011) A dry EEG-system for scientific research and brain-computer interfaces. Front Neurosci 5:53
19. Slater JD, Kalamangalam GP, Hope O (2012) Quality assessment of electroencephalography obtained from a “dry electrode” system. J Neurosci Methods 208:134–137
20. Fiedler P, Haueisen J, Jannek D, Griebel S, Zentner L, Vaz F, Fonseca C (2014) Comparison of three types of dry electrodes for electroencephalography. ACTA IMEKOl 3(3):33–37
21. Gargiulo GD, Bifulco P, Cesarelli M, Fratini A, Romano M (2014) Problems in assessment of novel biopotential front-end with dry electrode: a brief review. Machines 2:87–98
22. Sullivan D, Mizuseki K, Sorgi A, Buzsaki G (2014) Comparison of sleep spindles and theta oscillations in the hippocampus. J Neurosci 34(2):662–674
23. Chi YM (2011) Non-contact biopotential sensing. PhD thesis, UC San Diego electronic theses and dissertations. Permalink: https://escholarship.org/uc/item/4z4898fb. Local identifier: b7185097
24. Chi YM, Cauwenberghs G (2010) Wireless non-contact EEG/ECG electrodes for body sensor networks. 2010 IEEE international conference on body sensor networks, vol 52, pp 297–301
25. Kitoko V (2014) Improvement of EEG signal recording with non-contact electrodes in an headset system. Int J Biomed Eng Sci 1(3)
26. Lopez A, Richardson PC (1969) Capacitive electrocardiographic and bioelectric electrodes. IEEE Trans Biomed Eng 16(1):99
27. Valchinov ES, Pallikarakis NE (2004) An active electrode for biopotential recording from small localized bio-sources. Biomed Eng Online 3:25
28. Gramfort A, Papadopoulo T, Olivi E, Clerc M (2010) OpenMEEG: opensource software for quasistatic bioelectromagnetics. Biomed Eng Online 9:45
29. Gramfort A, Papadopoulo T, Olivi E, Clerc M (2011) Forward field computation with OpenMEEG. Comput Intell Neurosci 2011:923703
30. Kybic J, Clerc M, Abboud T, Faugeras O, Keriven R, Papadopoulo T (2005) A common formalism for the integral formulations of the forward EEG problem. IEEE Trans Med Imaging 24:12–28
31. Ioannides AA, Bolton JPR, Clarke CJS (1990) Continuous probabilistic solutions to the biomagnetic inverse problem. Inverse Probl 6:523–542
32. Taylor JG, Ioannides AA, Mueller-Gaertner HW (1999) Mathematical analysis of lead field expansions. IEEE Trans Med Imaging 18:151–163
33. Ioannides AA, Poghosyan V, Liu L, Saridis GA, Tamietto M, Op de Beeck M, De Tiège X, Weiskrantz L, de Gelder B (2012) Spatiotemporal profiles of visual processing with and without primary visual cortex. Neuroimage 63:1464–1477
34. Ioannides AA, Fenwick PBC, Liu LC (2005) Widely distributed magnetoencephalography spikes related to the planning and execution of human saccades. J Neurosci 25:7950–7967
35. Poghosyan V, Ioannides AA (2007) Precise mapping of early visual responses in space and time. Neuroimage 35:759–770
36. Poghosyan V, Ioannides AA (2008) Attention modulates earliest responses in the primary auditory and visual cortices. Neuron 58:802–813
37. Ioannides AA, Poghosyan V (2012) Spatiotemporal dynamics of early spatial and category-specific attentional modulations. Neuroimage 60:1638–1651

Chapter 6
Data Management Processes

Christos P. Antonopoulos and Nikolaos S. Voros

Abstract Complete end-to-end security support in the context of a complex CPS system in context of sensitive medical applications goes beyond pure technical countermeasures, hardware and techniques. Especially since highly sensitive medical data will be created, accesses, stored, processes and transferred it is of paramount importance to define accurate and comprehensive data management processes. Critical objectives achieved from accurate and comprehensive data management processes definition include:

Data privacy involving the right of any individuals to expect that personal information collected about them will be processed securely and will not be disseminated in any form without their written consent.

Data protection consists of a framework of security measures designed to guarantee that data are handled in such a manner as to ensure that they are safe from unforeseen, unintended, unwanted or malevolent use. Data protection is the technical mechanism to ensure data privacy.

Data management processes, as seen in this document, comprise be rules, actions to be taken in specific event as well as guidelines assuring adequate security level provision both in normal operation conditions (i.e. no threat is apparent) as well as in cases where specific security threat is identified.

A critical aspect necessitating the definition of such rules is the fact that the human factor is always in the loop in the context of respective systems' operation which may comprise the weak link from the security point view. Therefore, these rules mainly concern actions taken by humans concerning data management (e.g. password policy and account creation), algorithm configuration (e.g. acceptable encryption policy and security credential policy), equipment management policy (e.g. equipment disposal policy and removable media policy) etc. Another critical aspect is to highlight relevant legal and legislative requirements and extract significant principles, concerns and recommendations with respect to up to data literature.

C.P. Antonopoulos (✉) • N.S. Voros
Embedded System Design and Application Laboratory, Computer and Informatics Engineering Department, Technological Educational Institute of Western Greece, Antirio, Greece
e-mail: cantonopoulos@teiwest.gr; voros@teiwest.gr

N.S. Voros, C.P. Antonopoulos (eds.), *Cyberphysical Systems for Epilepsy and Related Brain Disorders*, DOI 10.1007/978-3-319-20049-1_6

6.1 Identifying the Need for Data Management Processes

In the context of ARMOR project the compilation of an Ethics Blueprint document was of cornerstone importance as a survey of the principles and legal requirements pertaining to work in ARMOR where biomedical data is incorporated into an ICT research and development project. In this section the most important ethics and data protection issues relevant to Cyberphysical systems for Epilepsy and related Brain Disorders and respective recommendation are extracted and presented.

As was the case in the ARMOR project, the development of a respective CPS platform involves multiple development steps, some of which will require pre-existing or prospectively-acquired data from patients, in order to develop and test the end-to-end platform. In such cases the issue of *anonymised data* utilization which cannot be linked to personal identifiers by any member of the research team is crucial. A critical clarification concerns the distinction between the terms *data*, *personal data* and *sensitive personal data*. The terms *Data*, *Personal Data*, *Sensitive Personal Data* are defined in detail in the UK Data Protection Act 1988 [1], but similar provisions apply in other EU jurisdictions.

Conditions that must be met whenever personal data are processes include: The individual who the personal data is about, has consented to the processing. The processing is necessary in relation to a contract which the individual has entered into; or because the individual has asked for something to be done so they can enter into a contract. The processing is necessary to protect the individual's "vital interests". This condition only applies in cases of life or death, such as where an individual's medical history is disclosed to a hospital's A&E department treating them after a serious road accident. The processing is necessary for administering justice, or for exercising statutory, governmental, or other public functions. The processing is in accordance with the "legitimate interests" condition. Furthermore, it is important to identify which aspects involve "sensitive personal data", and hence participants must give consent and data must be handled with appropriate arrangements and which has no personal identifiers and no means to link the data to personal identifiers—hence falls outside the scope of legislation concerning personal data.

It is noted that in the context of the ARMOR project, the consortium consulted the Research and Development Departments of Guy's and St Thomas's NHS Foundation Trust (GST) and King's College Hospital NHS Foundation Trust (KCH), the clinical departments from which data will be collected in ARMOR, and the following critical advice was provided and respectively followed.

(1) Data collected during the normal routine of clinical investigation, which was identified as suitable for ARMOR by members of the care team (such as Profs Koutroumanidis and Richardson), were anonymised by members of the care team and contains no identifying information, could be passed to members of the ARMOR consortium to be used freely, without requiring consent from the patient or any additional regulatory approval.
(2) Data collected during previous approved research projects, which has been anonymised, could be passed to members of the ARMOR consortium to

be used freely, without requiring consent from the patient or any additional regulatory approval.

(3) Data to be collected during ARMOR platform integration and evaluation required Research Ethics Committee approval.

Note that in (1) and (2) the key issues are that the data is anonymised and that the researchers cannot link the data back to any personal identifiers; and that no additional research-specific procedures are needed as part of the ARMOR project.

Another invaluable source of aspects concerning ethical and data privacy issues in the context of research in the context of highly sensitive medical conditions is the FP7 Ethical Guidelines manuscript [2]. Taking into consideration respective indications it is the purpose of this chapter to satisfactorily identify the central ethical considerations as well as indicating the measures required to prevent unnecessary exposure to risk for the participants and researchers in relative research efforts. A critical statement indicates that, "researchers have a duty to alert public authorities to the ethical and practical implications of their ICT research outcomes, as and when particular issues become apparent within the research process". With this in mind researchers must respect each volunteer's right to remain anonymous.

It is further emphasized that informed consent is required from all research participants, and the consent given must be voluntary and "based on knowledge of the purpose, procedures and outcomes of the research". In order to satisfy the knowledge requirements of a satisfactory consent, all those considering participating in the study should have a private discussion with an appropriate member of the research team, be given an information sheet that they can read and reflect upon, have the freedom to ask questions about the project (both prior to agreeing to participate and throughout their participation, as needed) as well as knowing that they are free to leave the project whenever they wish without the need to give an explanation and without this action having a detrimental effect on the standard of medical care that they receive. Furthermore, each participant must be aware of their freedom to access the data gathered (specifically about themselves) as well as the power to have this information permanently deleted should they so wish. During the time that a participant is involved in the research, they ought to be informed of any and all changes in the method, application, funding, etc. of the study to a sufficient level in order to ensure a fully informed and valid consent. All information conveyed to those involved (those considering participation, those participating, and those that leave the study) must be communicated in the clearest way possible, and every effort should be made to ensure that all information has been adequately understood; this will avoid uninformed consent being given due to excessively technical language being used, or other such complications. The autonomy of the participant will be respected throughout the entire process of the study.

Additionally, aforementioned guidelines require that the personal privacy of participants be respected, especially as ICT in the context of healthcare is "likely to raise privacy issues". For this reason, the reporting of research outcomes must be conducted in a manner which protects participant privacy and complies with all relevant data protection requirements. All data gathered should, therefore, be anonymised wherever possible and, furthermore, access to any sensitive data should be restricted. Access to such data will be granted to those that have a legitimate need to process the information in a way relevant to the research conducted and certain safeguards will be in place.

Specific aspects and issues of paramount important in any Cyberphysical systems design in highly sensitive medical application include:

Human dignity: Human dignity is inviolable. It must be respected and protected. The dignity of each participant will be respected, particularly by properly following consent procedures, by appreciating the voluntary nature of their participation, and by ensuring the security of any personal data gathered.

Right to the integrity of the person: 1. Everyone has the right to respect for his or her physical and mental integrity. 2. In the fields of medicine and biology, the following must be respected in particular: the free and informed consent of the person concerned, according to the procedures laid down by law, [...] the prohibition on making the human body and its parts as such a source of financial gain. Any wearable device will be designed with a respect for the integrity of the participant in mind. No unauthorized information will be gathered, nor will the device interfere with normal functioning in any way without consent. The device will pose no serious risk to long-term health. Procedures will be in place to identify any possibility of misuses or abuses of data as soon as they arise, and safeguards will be applied to prevent this occurring.

Respect for private and family life: Everyone has the right to respect for his or her private and family life, home and communications. Restrictions will be in place to ensure data sharing only with appropriate persons (see above), and wherever necessary data will be fully anonymised. Any wearable device will be fitted with an off function to guarantee the wearer privacy whenever they wish.

Protection of personal data: Everyone has the right to the protection of personal data concerning him or her. Such data must be processed fairly for specified purposes and on the basis of the consent of the person concerned or some other legitimate basis laid down by law. Everyone has the right of access to data which has been collected concerning him or her, and the right to have it rectified. Compliance with these rules shall be subject to control by an independent authority.

Right to respect for private and family life: Everyone has the right to respect for his private and family life, his home and his correspondence. There shall be no interference by a public authority with the exercise of this right except such as is in accordance with the law and is necessary in a democratic society in the interests of national security, public safety or the economic well-being of the country, for the prevention of disorder or crime, for the protection of health or morals, or for the protection of the rights and freedoms of others.

List of Recommendations

- Provide all participants with an appropriate information form and ensure informed and on-going (continuing) consent.
- For consent to be full and valid data subjects must agree (in principle) to the transfer of their personal data amongst involved countries (all are EU member states and so are bound by Directive 95/46/EC etc.)

- All data that is to be collected must be relevant (and not excessive) to the research project and to the stated purpose.
- All data must be anonymised wherever possible. If data is not sufficiently anonymised, extra safeguards must be in place and researchers will need to justify why anonymisation has not been possible.
- Assign responsibility amongst 'research teams' for periodic review and assessment of ethical and practical implications of ICT research outcomes, and communication of such information to relevant authorities.
- Applicants/researchers must identify the appropriate data protection authority in order that they can provide the relevant authorisations.
- Ensure that all those that will be dealing with data gathered from volunteers are familiar with the Data Protection requirements of all involved member states.
- Responsibility must be assigned to specific individuals for the timely assessment and reporting of any arising personal privacy implications of the intended or potential use of the research outcomes.
- Researchers must consider possible dual use, misuse, or malevolent use of any information gathered about research participants. Appropriate safeguards must be applied in order to protect and control data flow.
- Interface of wearable device ought to be designed with consideration of distinct access rights e.g.: the wearer (participant), a doctor, a researcher, a caregiver, a family member—as well as safeguards against unapproved access (possibly malevolent, including theft).
- Any wearable device should be fitted with an 'off' switch to allow any participant the option of immediately and autonomously leaving the study. A pause switch would allow participants privacy when desired.
- Those involved in the development of data managements processes must familiarise themselves with the requirements of the Ethical Review in FP7 document Data protection and privacy ethical guidelines (outlined below).

6.2 Highly Critical Data Management Procedures

In this section critical data management processes are presented as they have been identified and delineated in the context of the ARMOR project [3].

6.2.1 Software Related Procedures

6.2.1.1 Password Policy

Passwords are an important aspect of computer security. A poorly chosen password may result in unauthorized access and/or exploitation of CPS platform's resources. All users, including contractors and vendors with access to ARMOR Project specific systems, are responsible for taking the appropriate steps, as outlined below, to select and secure their passwords. The purpose of this policy is to establish a standard for creation of strong passwords, the protection of those passwords, and the frequency of change.

The scope of this policy includes all personnel who have or are responsible for an account (or any form of access that supports or requires a password) on any system of the CPS platform, has access to respective network, or stores any non-public information. Additionally, this policy applies to all ICT modules that human operator will or may have/request access to. These include but are not limited to User Interfaces towards Application and Information Server (and consequently to the EHR), access to the Events and Alerts Server at local site, access to the aggregator module as well as if required access to the sensors.

General Guidelines

- All system-level passwords (e.g., root, enable, Windows Administrator, application administration accounts, etc.) must be changed on at least a quarterly basis.
- All production system-level passwords must be part of ARMOR Project Leader administered global password management database.
- All user-level passwords (e.g., email, web, desktop computer, etc.) must be changed at least every 6 months.
- User accounts that have system-level privileges granted through group memberships or programs such as "sudo" must have a unique password from all other accounts held by that user.
- Where SNMP is used, the community strings must be defined as something other than the standard defaults of "public," "private" and "system" and must be different from the passwords used to log in interactively. A keyed hash must be used where available (e.g., SNMPv2).
- All user-level and system-level passwords must conform to the guidelines described below.

Technical Guidelines

- All users at ARMOR Project should be aware of how to select strong passwords.
- Avoid selecting common passwords for different purposes (e.g., personal ISP account, option trading, benefits, etc.)
- Do not shear selected passwords with anyone and in any case
- Application developers must ensure their programs contain adequate security precautions (e.g. support of individual users authentication, manager role support, TACACS+, RADIUS and/or X.509 with LDAP security retrieval support)
- Remote access is to be controlled using either a one-time password authentication or a public/private key system with a strong passphrase.
- Passphrase support as a safer measure compared to passport

6.2.1.2 Acceptable Encryption Policy

Encryption aims to protect the confidentiality, authenticity or integrity of information by cryptographic means. A policy on the use of cryptographic controls is necessary to maximize the benefits and minimize the risks of using cryptographic techniques, and to avoid inappropriate or incorrect use.

The purpose of this policy is to provide guidance that limits the use of encryption to those algorithms that have received substantial public review and have been proven to work effectively. Additionally, this policy provides direction to ensure that European regulations are followed, and legal authority is granted for the dissemination and use of encryption technologies outside of European Community.

This policy applies to all users of the CPS in question. Encryption algorithms will or may be used in various ICT components comprising the end to end architecture and thus a common policy must be applied. Such components include but are not limited to, wireless transmission from sensors towards the aggregation point, local storage at aggregation as well as (and possibly primarily) backhaul data transmission from the local site to the offline data processing and management center typically located in a remote site (e.g. hospital).

Guidelines

All CPS encryption functionality must be done based on well-known and approved cryptographic modules. Common and recommended ciphers include AES 256, Triple DES and RSA. Symmetric cryptosystem key lengths must be at least 128 bits. Asymmetric crypto-system keys must be of a length that yields equivalent strength. It is also suggested that key length requirements should be reviewed annually as part of the yearly security review and upgraded as technology allows. The use of proprietary encryption algorithms is not allowed for any purpose, unless reviewed by qualified experts outside of the vendor in question and approved by.

6.2.1.3 Anti Virus Policy Guidelines

This policy covers all ARMOR equipments capable of running relative software. These include, the local site aggregation point and of course the offline data processing and management center.

Recommended processes to prevent virus problems:

- Always run the Corporate standard, supported anti-virus software that is available from the corporate download site. Download and run the current version; download and install anti-virus software updates as they become available.
- NEVER open any files or macros attached to an email from an unknown, suspicious or untrustworthy source. Delete these attachments immediately, then "double delete" them by emptying your Trash.
- Never download files from unknown or suspicious sources.

- Avoid direct disk sharing with read/write access unless there is absolutely a business requirement to do so.
- Always scan a floppy diskette from an unknown source for viruses before using it.
- Back-up critical data and system configurations on a regular basis and store the data in a safe place.
- If lab testing conflicts with anti-virus software, run the anti-virus utility to ensure a clean machine, disable the software, then run the lab test. After the lab test, enable the anti-virus software. When the anti-virus software is disabled, do not run any applications that could transfer a virus, e.g., email or file sharing.

6.2.1.4 Database Credentials Policy

Overview

Database utilization is quite common in CPS aiming towards highly sensitive medical applications, therefore security controls aim to restrict access to database information only to authorized users. Respective policy states the requirements for securely storing and retrieving database usernames and passwords (i.e., database credentials) for use by a program that will access a database. Computer programs running on a CPS's networks often require the use of one of the many internal database servers. In order to access one of these databases, a program must authenticate to the database by presenting acceptable credentials. The database privileges that the credentials are meant to restrict can be compromised when the credentials are improperly stored. This policy applies to all software that will access CPS's, multi-user production database.

General Guideline

In order to maintain the security of ARMOR's internal databases, access by software programs must be granted only after authentication with credentials. The credentials used for this authentication must not reside in the main, executing body of the program's source code in clear text. Database credentials must not be stored in a location that can be accessed through a web server.

Technical Guidelines

- Database user names and passwords may be stored in a file separate from the executing body of the program's code. This file must not be world readable.
- Database credentials may reside on the database server. In this case, a hash number identifying the credentials may be stored in the executing body of the program's code.
- Database credentials may be stored as part of an authentication server (i.e., an entitlement directory), such as an LDAP server used for user authentication. Database authentication may occur on behalf of a program as part of the user authentication process at the authentication server. In this case, there is no need for programmatic use of database credentials.
- Database credentials may not reside in the documents tree of a web server.

- Pass through authentication (i.e., Oracle OPS$ authentication) must not allow access to the database based solely upon a remote user's authentication on the remote host.
- Passwords or pass phrases used to access a database must adhere to the Password Policy.
- Every program or every collection of programs implementing a single business function must have unique database credentials. Sharing of credentials between programs is not allowed.
- Database passwords used by programs are system-level passwords as defined by the Password Policy.
- Developer groups must have a process in place to ensure that database passwords are controlled and changed in accordance with the Password Policy. This process must include a method for restricting knowledge of database passwords to a need-to-know basis.

6.2.1.5 User Encryption Key Protection Policy

Overview

End user encryption methods aim to safeguard information transfer between two or more persons that collaborate and exchange information specific to ARMOR project. Each end user encryption method can mandate a different way to maintain or issue cryptographic keys.

This section describes Information Security's required protections for encryption keys that are under the control of end users. These protections are designed to prevent unauthorized disclosure and subsequent fraudulent use. The protection methods will include operational and technical controls, such as key backup procedures, encryption under a separate key and use of tamper-resistant hardware.

This policy applies to any encryption keys listed below and to the person responsible for any encryption key listed below. The encryption keys covered by this policy are:

- encryption keys issued by CPS specific systems; or
- encryption keys used to protect data owned by the CPS.

The public keys contained in digital certificates are specifically exempted from this policy.

The policy apply to the same CPS ICT components as indicated for adequate encryption policy. Such components include but are not limited to, wireless transmission from sensors towards the aggregation point, local storage at aggregation as well as (and possibly primarily) backhaul data transmission from the local site to the offline data processing and management center typically located in a remote site (e.g. hospital).

All encryption keys covered by this policy must be protected to prevent their unauthorized disclosure and subsequent fraudulent use.

6.2.2 Equipment Related Procedures

6.2.2.1 Removable Media Policy

Removable media is a well-known source of malware infections and has been directly tied to the loss of sensitive information in many organizations. Consequently the purpose of this policy is to effectively minimize the risk of loss or exposure of sensitive information maintained by the CPS platform and to reduce the risk of acquiring malware infections on computers operated by respective personnel. This policy covers all computers and servers operating in CPS context.

Policy

All users may only use CPS removable media in their work computers. Specific removable media may not be connected to or used in computers that are not owned or leased by the specific CPS without explicit permission of respective personnel or Committee (e.g. a respective Ethics Committee). Sensitive information should be stored on removable media only when required in the performance of any assigned duties or when providing information required by other state or federal agencies. When sensitive information is stored on removable media, it must be encrypted in accordance with the Acceptable Encryption Policy. Exceptions to this policy may be requested only on a case-by-case basis.

6.2.2.2 Equipment Disposal Policy

Technology equipment often contains parts which cannot simply be thrown away. Proper disposal of equipment is both environmentally responsible and often required by law. In addition, hard drives, USB drives, CD-ROMs and other storage media contain various kinds of data, some of which is considered sensitive. In order to protect our constituent's data, all storage mediums must be properly erased before being disposed of. However, simply deleting or even formatting data is not considered sufficient. When deleting files or formatting a device, data is marked for deletion, but is still accessible until being overwritten by a new file. Therefore, special tools must be used to securely erase data prior to equipment disposal. This policy has been developed to define the requirements for proper disposal of technology equipment in the context of a CPS platform used for highly sensitive medical purposes. This policy applies to all technology equipment in the context of the CPS in question.

Policy

When technology assets have reached the end of their useful life they should be sent be properly disposed. Information owners should securely erase all storage mediums in accordance with current industry best practices.

6.2.2.3 Information Backup Policy

Overview

A backup policy dictates how important data and systems within the context of a CPS platform are managed to ensure that there is a tolerance for various forms of data loss. The main goal is to ensure that if a specific information system suffers some kind of data loss, the whole CPS platform will not be affected and can continue to function effectively with a minimal additional overhead during data recovery. The purpose of this Information Backup Policy is to ensure data held on information systems is backed up in accordance with business needs. Such back ups are critical to the processes of ensuring the integrity and availability of information assets. This Information Backup Policy shall apply to all information processed within ARMOR Project.

Policy

Information Back Up

Regular backups of essential business information must be taken to ensure that CPS information systems can recover from a disaster, media failure or error. An appropriate backup cycle must be used and fully documented. To ensure all essential business information is backed up all employees must store their information on the network drives and not on local drives e.g. C: drive. All users of portable devices for example laptops, PDA's, smart phones and USB memory sticks must ensure the information is also stored on the network drives.

Critical paper files must be identified and backed up with either a scanned digital copy or complete photocopies stored at a remote location.

Information Restore

Full documentation of the recovery procedure must be created and stored. Regular restores of information from back up media must be tested to ensure the reliability of the backup media and restore process. The retention period for business information (in particular legal requirements) must be defined and applied to the backup data. Long term backup and restore solutions may need to be identified for certain business information.

6.2.3 Incident Handling Procedure

This section provides some general guidelines and procedures for dealing with computer security incidents on what to do if they discover a security incident. The term incident in this document is defined as any irregular or adverse event that occurs on any part of a Cyberphysical system. Some examples of possible incident categories include: compromise of system integrity; denial of system resources; illegal access to a system (either a penetration or an intrusion); malicious use of system resources, or any kind of damage to a system. Some possible scenarios for security incidents are:

- You see a strange process running and accumulating a lot of CPU time.
- You have discovered an intruder logged into your system.
- You have discovered a virus has infected your system.
- You have determined that someone from a remote site is trying to penetrate the system.

The steps involved in handling a security incident are categorized into five stages: protection of the system; identification of the problem; containment of the problem; eradication of the problem; recovering from the incident and the follow-up analysis.

Respective policy should cover all information systems that are able to be affected by malware software and mainly what human operators must do in case of indenting an incident that could be indicative of security breach. Therefore the main components related to this policy are the aggregation point and the offline data processing and management center.

6.2.3.1 Policy

Incident Specific Procedures

This section discusses the procedure for handling virus, worm and hacker/cracker incidents in a CPS system.

Virus and Worm Incidents

Although virus and worm incidents are very different, the procedures for handling each are very similar aside from the initial isolation of the system and the time criticality. Viruses are not self-replicating and, thus, incidents of this nature are not as time critical as worm or hacker incidents. Worms are self-replicating and can spread to hundreds of machines in a matter of minutes, thus, time is a critical factor when dealing with a worm attack. If you are not sure of the type of the attack, then proceed as if the attack was worm related.

- Isolate infected system(s) from the remaining network as soon as possible.
- Notify appropriate personel (e.g. the Ethics Committee) as soon as possible. If unable to reach him/her, contact the backup person.
- Try to identify and isolate the suspected virus or worm-related files and processes. Prior to removing any files or killing any processes, a snapshot of the system should be taken and saved including maintaining a copy of all system log files.

- All suspicious processes should now be halted and removed from the system. Make a full dump of the system and store in a safe place. The tapes should be carefully labeled so they will not be used by unsuspecting people in the future. Then remove all suspected infected files or worm code. In the case of a worm attack, it may be necessary to keep the system(s) isolated from the outside world until remaining systems have been inoculated and/or the other sites have been cleaned up and inoculated.
- Implement fixes and/or patches to inoculate the system(s) against further attack. Prior to implementing any fixes, it may be necessary to assess the level of damage to the system. If the virus or worm code has been analyzed, then the tasks of assessing the damage are not very difficult. However, if the offending code has not been analyzed, then it may be necessary to restore the system from backup tapes.
- Prior to bringing the systems back into full operation mode, you should notify the same group of people who were notified in stage one. The users should also be notified that the systems are returning to a fully operational state. It may be wise to request all users to change their passwords. Before restoring connectivity to the outside world, verify that all affected parties have successfully eradicated the problem and inoculated their systems. Log all actions.

Hacker/Cracker Incidents

Responding to hacker/cracker incidents is somewhat different than responding to a worm or virus incident. Some hackers are very sophisticated and will go to great depths to avoid detection. Others are naive young students looking for a thrill. Hacker incidents can be divided into three types: attempts to gain access to a system, an active session on a system, or events which have been discovered after the fact. Of the three, an active hacker/cracker session is the most severe and must be dealt with as soon as possible. There are two methods for dealing with an active hacker/cracker incident. The first method is to immediately lock the person out of the system and restore the system to a safe state. The second method is to allow the hacker/cracker to continue his probe/attack and attempt to gather information that will lead to an identification and possible criminal conviction. The method used to handle a cracker/hacker incident will be determined by the level of understanding of the risks involved.

Active Hacker/Cracker Activity

Incidents of this type would include any active session or command by an unauthorized person. Some examples would include an active rlogin or telnet session, an active ftp session, or a successful dial-back attempt. In the case of active hacker/cracker activity, a decision must be made whether to allow the activity to continue while you gather evidence or to get the hacker/cracker off the system and then lock the person out.

The decision will be based on the availability of qualified personnel to monitor and observe the hacker/cracker and the level of risk involved.

In the process of Hacker/Cracker removal from the System, the following actions are suggested:

- Snap-shot the System including all audit trail information as well as process status information. Any suspicious files should be moved to a safe place or archived to tape and then removed from the system. Also, get a listing of all active network connections. A control room analyst can provide assistance in obtaining snap-shot information on the system. Log all actions.
- Lock Out the Hacker by terminating all active processes for the hacker/cracker and remove any files or programs that he/she may have left on the system. Change passwords for any accounts that were accessed by the hacker/cracker. At this stage, the hacker/cracker should be locked out of the system. Log all actions.
- Restore the System to a normal state. Restore any data or files that the hacker/cracker may have modified. Install patches or fixes to close any security vulnerabilities that the hacker/cracker may have exploited. Inform the appropriate people. All actions taken to restore the system to a normal state should be documented in the log book for this incident. Log all actions.
- Follow-up is advisable including a short report describing the incident and actions that were taken should be and distributed to the appropriate people.

6.3 Conclusions

This chapter engages security provision from a critical yet often overlooked perspective that of adequate data management processes definition. It has been identified in many cases that the human factor comprises the weakest link in a CPS platform compromising any sophisticated and high level security technological countermeasure. Another critical observation is that security breaches often are not intentional but rather due to simple human mistakes. Such mistakes can be drastically minimized when adequate data management processes are defined and followed without deviations by all personnel handling any part of a sensitive CPS such as one that is related to brain deceases monitoring. However, a critical step in order to reach to the right data management processes is to understand the requirements of the specific system and identify the actual degree of necessity of such processes' definition. The importance of this initial evaluation is critical since inadequate results may lead, on one hand, to excessively strict data management processes degrading the usability of respective system or, on the other hand, may lead to too loose processes which effectively will leave room for error.

Driven by the aforementioned principles the first part of this chapter is devoted in offering all required information which enables a system designer to identify and evaluate the need of data management processes with respect to the specific system. The second part of the chapter is devoted in presenting specific data management processes identified in the context of the ARMOR project concerning three aspects, Software Related Procedures, Equipment Related Procedures and finally Incident Handling Procedures.

References

1. UK Data Protection Act 1988. http://www.legislation.gov.uk/ukpga/1998/29/contents
2. FP7 Ethical guidelines, "Data protection and privacy ethical guidelines" (2009) Experts Working Group on Data Protection and Privacy, Chaired by: Caroline Gans-Combe
3. Deliverable D3.2 of the ARMOR FP7 research project "Data management processes"

Chapter 7
System Architecture

Stefan Hey, Panagiota Anastasopoulou, André Bideaux, Christos P. Antonopoulos, Nikolaos S. Voros, Alberto Fernandez, Vasilis Megalooikonomou, and Arthur Krukowski

Abstract A cyberphysical system for multiparametric monitoring and analysis of patients with epilepsy needs the possibility to assess a variety of different sensor signals, transmit these signals into a central online system (electronic health record, database) and analyze the data to give a feedback to the patient or the clinician. The architecture of such a system is described in this chapter. Based on a description of the application the overall architecture and the different hardware and software modules of the systems with their interfaces are defined.

7.1 State of the Art System Architectures

Recent sensor systems permit the monitoring of neurophysiological signals, ranging from multichannel EEG monitoring systems to more complex ones including EEG and different physiological parameters for real time monitoring of multiparametric signals. In epilepsy research, several multiparametric systems for monitoring epilepsy or other brain related disorders have been used.

S. Hey (✉) • P. Anastasopoulou • A. Bideaux
KIT Karlsruhe Institute of Technology, Institute for Information Technology, Karlsruhe, Germany
e-mail: stefan.hey@kit.edu

C.P. Antonopoulos • N.S. Voros
Embedded System Design and Application Laboratory, Computer and Informatics Engineering Department, Technological Educational Institute of Western Greece, Antirio, Greece

A. Fernandez
Sensing and Control Systems S.L., Barcelona, Spain

V. Megalooikonomou
Multidimensional Data Analysis and Knowledge Discovery Laboratory, Department of Computer Engineering and Informatics, University of Patras, Patras, Greece

A. Krukowski
Intracom S. A. Telecom Solutions, Athens, Greece

N.S. Voros, C.P. Antonopoulos (eds.), *Cyberphysical Systems for Epilepsy and Related Brain Disorders*, DOI 10.1007/978-3-319-20049-1_7

A wireless physiological multiparametric monitoring system based on mobile communication networks made of front-end ambulatory monitoring equipment, ambulatory monitoring center and hospital central management system has been suggested by [1]. It has been designed to continuously acquire patient's ECG, blood pressure, SpO2, body temperature, and respiration. The front-end ambulatory monitoring equipment contains the wireless data transmission function that is girded onto the patient's waist to acquire and transmit physiological data. This provides a multi-function interface, with ECG electrodes, SpO2, thermometer and a blood pressure measurement interface. The ambulatory monitoring center is based on a Personal Digital Assistant (PDA), which is provided to authorized clinicians to access the mobile diagnosis platform. The central management system is located in the hospital center. When the physiological data appears abnormal, the front-end ambulatory monitoring equipment would automatically trigger an alarm signal.

The data is transmitted in one of two ways, either directly to the monitoring center or it is uploaded to the hospital central management system via the existing internet connection. The front ambulatory monitoring equipment has 32 MB of flash memory with an ability to store 24 h of continuous data.

The patient also carries a PDA. This is the center of the ambulatory monitoring unit as it performs a comprehensive analysis of physiological data. It is mainly used to receive and process data and display waveforms and parameters.

The Mercury system, developed at Harvard University, is a sensor network platform designed to support data-intensive applications that can adapt to fluctuations in resource availability and load [2]. This system has a varied number of sensors but is mainly based on Intel's Shimmer Sensor. Typically, it consists of three sensors placed on the arms, two sensors on the legs and a base station installed in the patient's home. Each sensor samples multiple signals from an accelerometer, a gyroscope, and other physiological data, and stores there raw into a local flash memory. Data from the accelerometer and gyroscope can be used to assess abnormal movements during myoclonic seizures or possible falls. An extra EMG sensor is placed on one arm and samples muscle activity at 500 Hz. This system is based on a round-robin driver, pocking sensors on the body in a predefined order. When an event is suspected, identified simply on threshold values of the incoming sensor signals, the driver retrieves instantly data from all sensors and alarms the system. This scheme saves power and data storage. Nevertheless, this is a simplified design, in terms of event detection-discrimination. The Mercury system has been so far used to capture up to 12 h of accelerometer and EMG data per day for a 5-day period for epileptic patients.

Gouravajhala and Khuon [3] developed a multimodality wireless sensor platform for event detection using a microcontroller. They improved the previous system by adding sensors to detect physical changes. This was done to improve the accuracy of embedded implementations of their event detection system based solely on EEG signals. The platform also included a GSR sensor, a temperature sensor, and a rotational sensor. The sensor data were successfully captured, transmitted by Bluetooth and received via the wireless settings [3].

A new design of an ambulatory epilepsy warning system for medical application was described by [4]. This device was disguised as a wig/hat wired with EEG sensors and was invisible for inexperienced observers. It consisted of a collector used for converting signals to data, Global Positioning System (GPS), PDA with Global System for Mobile (GSM) module and executed Artificial Neural Network (ANN) software to test current data with pre-learned data, and a calling center for patient assistance.

7.2 System Requirements for Multi-parametric Monitoring of Epileptic Patients

7.2.1 General Functional and Nonfunctional Requirements

Since different types of epilepsy require monitoring of different brain and body parameters, ARMORs goal was to develop a personalized and reconfigurable system that assists in diagnosis, prognosis and treatment of the disease. Some of the main requirements of the system are that it should be non-invasive, mobile, continuous, unobtrusive, reconfigurable and able to perform online analysis. The system can be divided into three main components: the sensors, the Home Gateway containing the middleware and the Personal Health Record (PHR) server. Figure 7.1 shows an overview of the system and its components.

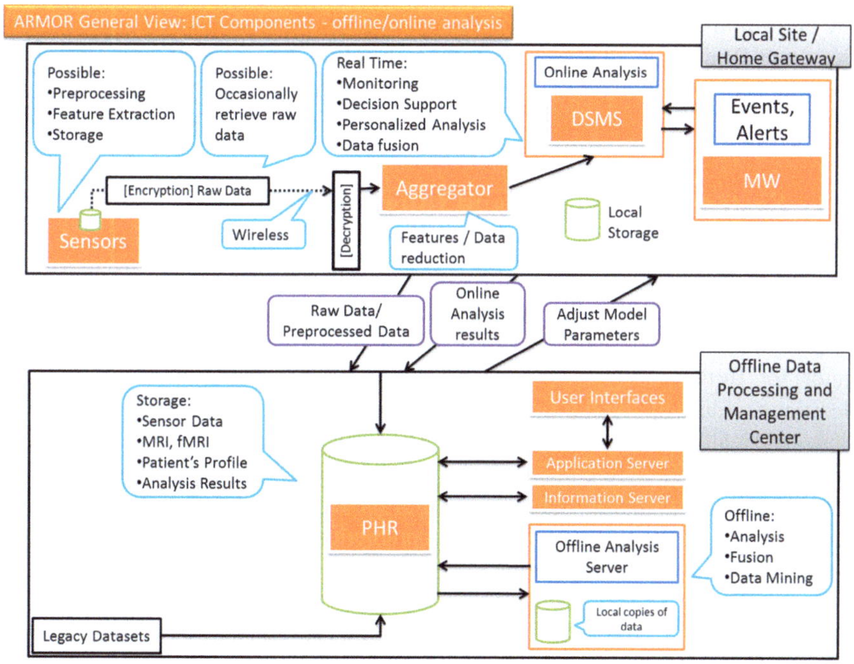

Fig. 7.1 Architecture of the ARMOR system

The PHR server contains all the information related to the patients. When a new measurement shall be started the Home Gateway downloads a profile from the PHR. This profile contains the sensor configuration. Having this information, the middleware automatically configures the sensors and starts the measurement.

The sensors are streaming data via Bluetooth to the Home Gateway. The middleware fuses the data together, stores it, uploads it to the PHR server and also streams it to the Data Stream Management System (DSMS) where different algorithms can be applied.

The different monitoring parameters used for multipara-metric analysis include ECG, GSR, multichannel EEG and physical activity sensor (acceleration sensors) [5]. Furthermore there are some additional sensors providing context information about the measurement conditions and the patient's environment. Temperature, air pressure, body position and a marker button need to be acquired and synchronized with the physiological data and then stored in a central database. The fact that for each patient some parameters are more important than others makes it necessary to adapt the system to the patient's needs. Therefor the system was designed reconfigurable regarding the sensors. This means an individually selected set of sensors is attached to each patient profile in the PHR and the System uses only these required sensors to monitor the patient.

The DSMS runs different online algorithms to detect special events. Some examples are push button, alpha rhythm and seizure detection. These algorithms are also defined in the profile and can therefore be adapted to each patient individually.

The Event Manager handles the different events of special interest found by the algorithms in the DSMS. When an event is detected the event manager notifies the PHR directly, that can then inform the clinicians via SMS or e-mail about the patients state.

7.3 System Architecture and Interface Definition

Figure 7.1 shows an overview of the whole system architecture.

The sensor platform performs the following actions: data logging, data preprocessing, data encryption and wireless transmission of the assessed data to an aggregator for further analysis. A laptop used as an aggregator performs the data decryption (needed for the online analysis), processing and transmission to the Personal Health Record (PHR) database. The transmission to the PHR is realized by a standardized SSL secure internet connection. PHR includes all the legacy datasets of the patient as well. The offline analysis of the data is realized on the information server. Based on the offline analysis, the prediction models may be adjusted, thus creating a personalized monitoring system.

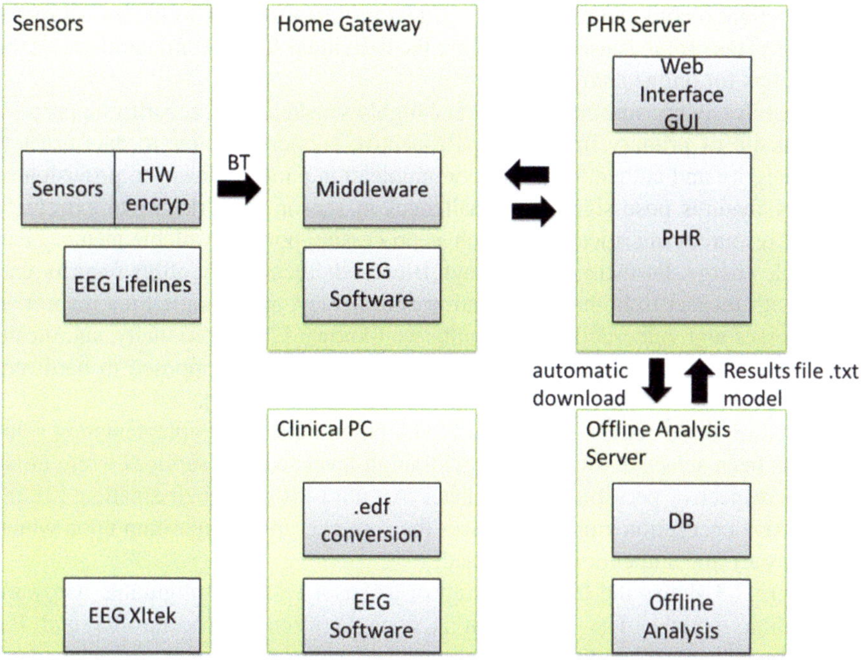

Fig. 7.2 Hardware architecture of the cyberphysical system

7.3.1 Hardware Components

The hardware of the systems can be divided in different modules that are shown in Fig. 7.2. Because of the modular architecture of the system, the sensors have to be configured regarding the individual requirements of each patient. Different types of sensors and from different manufacturers can be integrated in the system to be used on the patient. The connection to the system is done either by a wired interface or by wireless transmission of the data. The connection between the sensors and the central data base on PHR is realized by the home gateway, containing the system middleware and modules to handle sensors for offline recordings. The clinical PC is required for already existing medical equipment, such as the Xltek EEG system [6]. These clinical systems are often physically separated from the rest of other components, due to the IT-security restrictions of the hospitals. Hence for measurements with clinical devices and sensors located on the home gateway special attention has to be paid in terms of synchronization.

The central server of the overall system consists or two part, the PHR server and the offline analysis server. PHR server offers the possibility to store all the data from the patient in individual patient records and to access these data via a web based graphical user interface. To analyze the stored data, the offline analysis server downloads the data from PHR to perform the required analysis, restores the results in the patient record and modifies the patient profile according to the results of the analysis. This profile is used to configure the individual sensor configuration and the parameters for online analysis.

In wirelessly transmitting medical and highly sensitive data, security support is a prerequisite of primary importance. Respective support pertains to data privacy, data integrity and authentication of communication parties. However, provision of security features pose significant challenges in sensor network due to extremely limited resources in critical areas such as processing power, available memory and available energy. Furthermore, although Bluetooth Technology offers security features with respect to both authentication and privacy, at the same time respective weaknesses and vulnerabilities are quite well-known [7]. Additionally, significant overhead imposed by software security implementations as opposed to hardware implemented counterparts must be taken into consideration [8].

Based on these considerations, an FPGA based hardware implementation solution has been selected, aiming to provide high level security services while minimizing respective performance overhead. An ultra-low power dissipation 128 bit block AES encryption cipher comprises the core encryption algorithm upon which data privacy and authentication is based.

Figure 7.3 shows the block diagram of a sensor with reconfigurable hardware encryption module. The encryption is done between microcontroller and BT interface.

7.3.1.1 Sensor Platform

The sensor platform is based on a sensor platform developed by movisens (movisens GmbH, Karlsruhe, Germany) and consists of an ultra-low-power microcontroller (MSP430F1611, Texas Instruments) with an AD/DA converter, 2 UART interfaces, a 48 kB flash and a 10 kB RAM. The assessed data could be transmitted wirelessly by using the Bluetooth interface. To avoid data loss in case the data transmission is not possible, the raw data is in parallel stored in a 2 GB micro SD card. The stored data can then be saved on the computer by using the USB 2.0 interface that is available. The charging of the power supply of the sensors is recognized by the USB interface as well (Fig. 7.4).

The ECG module is a single channel ECG recorder with a 12 bit resolution and a sampling rate from 256 to 1024 Hz. The electrodes are integrated into a wearable chest strap, which is light, small and comfortable. The electrodes are dry, allowing the everyday use of the chest strap. The activity monitoring module consists of a triaxial acceleration sensor (adxl345, Analog Devices Inc.) with a range of ±8 g and a sampling frequency of 64 Hz. The measuring unit has an additional air pressure sensor (BMP085, Bosch GmbH) with a sampling frequency of 8 Hz and a resolution of 0.03 hPa (corresponding to 15 cm).

The galvanic skin response (GSR) module measures the skin conductance with a sampling rate of 32 Hz. The measurement range of the GSR module is 2–100 μS and its resolution is 14 bit (Fig. 7.5).

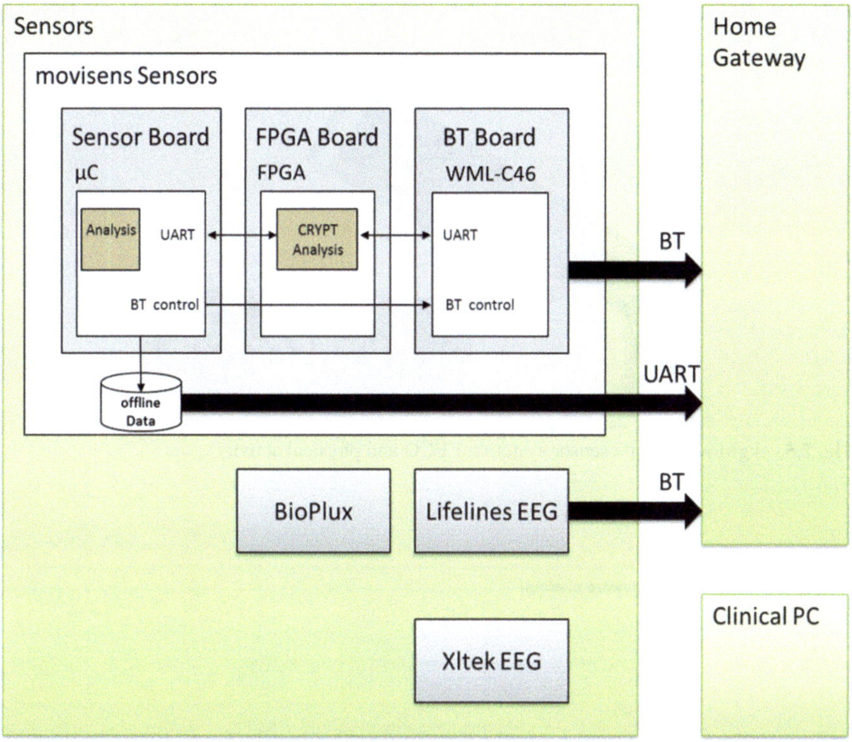

Fig. 7.3 Block diagram of hardware encryption unit

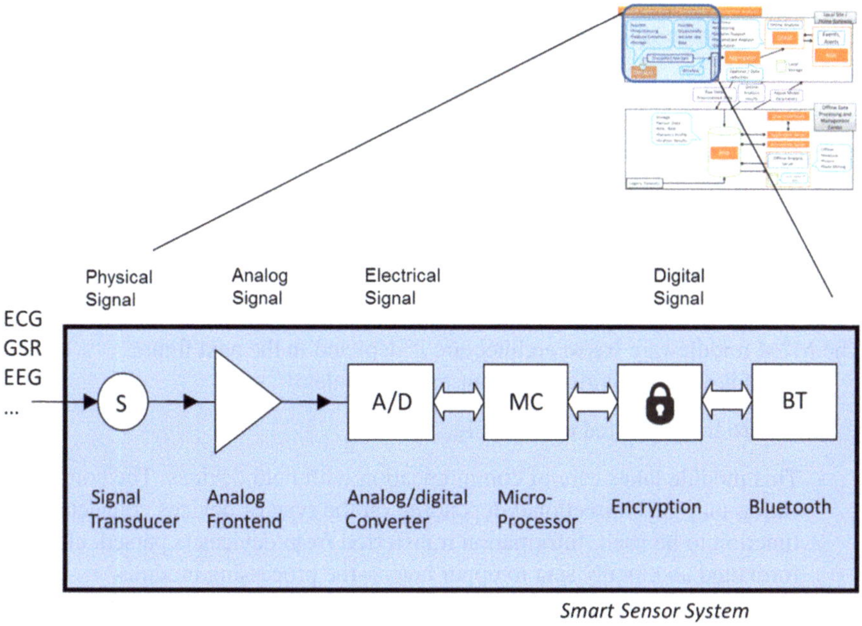

Fig. 7.4 Architecture of the sensor platform

Fig. 7.5 ekgMove, mobile sensor system for ECG and physical activity

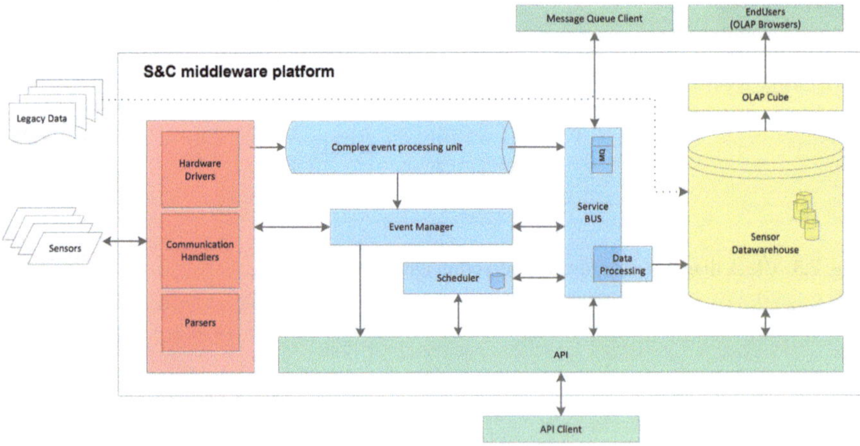

Fig. 7.6 S&C middleware architecture

7.3.2 *Software Components*

The M2M middleware based architecture is depicted in the next figure.

The middleware is divided into four main modules:

- Hardware Interface (red color in Fig. 7.6)
  - This module takes care of communication with field devices. The communication may be bidirectional depending on the type of devices connected and function to be used. Information transferred from devices is parsed, cleaned, formatted and finally sent to upper layer—the processing modules.

- Processing module (blue color in Fig. 7.6)
  - The central module handles all the processing/intelligence of the software. It has a built in event manager, that handles communication between hardware interfaces and application service bus, automation routines and alarm/events handling and auto generation based on rules.
  - Internal Service bus takes care of the poisoned and transactional elements of messaging as well as offering out of the box Pub/Sub style messaging.
- Database (yellow color in Fig. 7.6)
  - The core of the platform where all the data from sensors is stored and retrieved in the most efficient way.
  - Allows to connect from different OLAP (Online Analytical Processing) browsers to perform analysis of the data
- Interoperability Interfaces (green color in Fig. 7.6)
  - Application programming interfaces have been developed to maintain interoperability across multiple platforms, applications and systems. Web services are included in this layer and can be consumed within standard web client.

7.4 Conclusions

Driven by the architectural description presented in this chapter, the following chapters offer an in depth analysis of all respective components of the ARMOR platform. Specifically, Chap. 8 focuses on the mobile sensors used for acquiring, storing and transmitting all required multimodal signals emphasizing on critical requirements and respective characteristics.

Chapter 9 then emphasizes on designing and implementing a secure and efficient communication infrastructure. With respect to Fig. 7.1 main communication aspects concern on one hand the data exchange between the sensors and the aggregation point and on the other hand the communication link between the aggregators (i.e. ARMOR Home Gateway) and the PHR infrastructure.

As it can be easily extracted from this chapter the ARMOR middleware comprises one of the cornerstones of the ARMOR platform. Residing on the Home Gateway, the ARMOR middleware effectively host all other components such as the xAffect and the Online algorithms while at the same time conveys all measurements data, configuration data and/or command from the PHR to the sensors and vice versa. Consequently a separate chapter is dedicated to such critical entity in Chap. 10 of this book.

Without a doubt PHR represents the main component at the back-office remote site. Therefore, respective design and implementation details as well as adopted standards and prototypes are analyzed in Chap. 11.

Advanced sophisticated data processing algorithms comprise a critical added value of the ARMOR project. Both online algorithms on real time streaming data as well as offline algorithms can offer invaluable information, indications and conclusions able to advance epilepsy study and lead to increase effectiveness of respective treatments and medications. In the context of the ARMOR architecture online algorithms are part of the ARMOR home gateway and are analyzed in Chap. 14. Offline analysis as depicted in Fig. 7.1 takes place at the back-office remote site and respective algorithms and server are presented in Chap. 12.

Finally, although not corresponding to a specific component of the ARMOR architecture, Chap. 13 offers significant insight on a significant research aspect of ARMOR concerning the combinatory efficient use of MEG analysis and smart use of EEG measurements so as to deliver personalized management of epilepsy monitoring.

References

1. Dai S, Zhang Y (2006) A wireless physiological multi-parameter monitoring system based on mobile communication networks. In: Proceedings of the 19th IEEE symposium on computer based medical systems based on mobile communication networks. IEEE Computer Society, Washington, DC, pp 473–478
2. Lorincz K, Chen B, Challen G, Chowdhury A, Patel S, Bonato P, Welsh M (2009) Mercury: a wearable sensor network platform for high-fidelity motion analysis. In: Proceedings of the 7th ACM conference on embedded networked sensor systems (SenSys'09), pp 1–14
3. Gouravajhala SR, Khuon L (2012) A multi-modality sensor platform approach to detect epileptic seizure activity. In: Proceedings of the 2012 38th annual Northeast bioengineering conference (NEBEC), pp 233–234
4. Alkan A, Sahin Y, Karlik B (2006) A novel mobile epilepsy warning system. In: AI 2006 advances in artificial intelligence, 2006
5. Poh M-Z, Loddenkemper T, Swenson NC, Goyal S, Madsen JR, Picard RW (2010) Continuous monitoring of electrodermal activity during epileptic seizures using a wearable sensor. In: Proceedings of the annual international conference of the IEEE Engineering in Medicine and Biology Society, 2010, pp 4415–4418
6. http://www.natus.com/index.cfm?page=products_1&crid=223
7. Hall J (2003) Brush up on Bluetooth. GSEC practical assignment version
8. Lee J, Kapitanova K, Son SH (2010) The price of security in wireless sensor networks. Comput Netw 54:2967–2978

Chapter 8
Mobile Sensors for Multiparametric Monitoring in Epileptic Patients

Stefan Hey, Panagiota Anastasopoulou, André Bideaux, Christos P. Antonopoulos, Nikolaos S. Voros, and Mark Richardson

Abstract Multiparametric monitoring represents the assessment of physiological, behavioral and/or subjective, data with mobile sensor systems. Recent technological developments like low power microcontrollers, low power wireless standards such as Bluetooth Low Energy or Smartphones with an increasing computing capacity enhance the development of small and distributed multiparametric monitoring systems (Handbook of biomedical telemetry, Wiley, Piscataway, 2014).

Continuous monitoring of patients in the clinical setting is expensive due to the limited number and availability of assessment lots. Tele-medical applications con overcome this problem and permit the monitoring of disease symptoms at home. This may help to avoid hospitalization and to maintain therapy effects in everyday life. In particular, ambulatory monitoring technologies may support the diagnosis and treatment of neurological disorders such as epilepsy.

Depending on the type of epilepsy and due to its multifactorial causes, different brain and body parameters need to be assessed continuously over a long period to allow clinicians to have a better understanding of the patient's state of health and to be able to continuously adjust and change the medical treatment accordingly. Beside this, multi-parametric monitoring could be used for other purposes such as accurate diagnosis, detection of seizures, alerting and prevention and pre-surgical evaluation.

S. Hey (✉) • P. Anastasopoulou • A. Bideaux
KIT Karlsruhe Institute of Technology, Institute for Information Technology, Karlsruhe, Germany
e-mail: stefan.hey@kit.edu

C.P. Antonopoulos • N.S. Voros
Embedded System Design and Application Laboratory, Computer and Informatics Engineering Department, Technological Educational Institute of Western Greece, Antirio, Greece

M. Richardson
King's College London, London, UK

N.S. Voros, C.P. Antonopoulos (eds.), *Cyberphysical Systems for Epilepsy and Related Brain Disorders*, DOI 10.1007/978-3-319-20049-1_8

Abbreviations

AC	Alternating current
AES	Advanced encryption standard
ANS	Autonomic nervous system
BPF	Band pass filter
BT	Bluetooth
CPS	Cyber-physical system
DC	Direct current
ECG	Electrocardiogram
EEG	Electroencephalogram
EMG	Electro-myogram
EMU	Epilepsy monitoring unit
EOG	Electro-oculogram
GSM	Global systems for mobile communications
GSR	Galvanic Skin Response
GUI	Graphical user interface
HF	High frequency
HR	Heart rate
HRV	Heart rate variability
IED	Interictal epileptiform discharges
IGE	Idiopathic generalized epilepsy
IPS	Intermittent Photic Stimulation
IVT	Home video-EEG telemetry
LF	Low frequency
LPF	Low pass filter
LTM	Long term monitoring
MEG	Magneto-encephalogram
MRI	Magnetic resonance imaging
NEPE	Non-epileptic paroxysmal events
PSE	Photosensitive epilepsy
RAM	Random access memory
RMSSD	Root mean square of successive differences
SDNN	Standard deviation of normal-to-normal interval
SR	Sampling rate
SUDEP	Sudden Unexpected Death in Epilepsy
TLC	Temporal loss of consciousness
UART	Universal asynchronous receiver/transmitter
USB	Universal serial bus
WiFi	Wireless fidelity

8.1 Introduction

The principal objective of this chapter is to present the general requirements that should be taken into account when selecting and developing the sensor technology to be used for the mobile multiparametric patient monitoring [1].

The document is divided into three parts. In the first part some state of the art technology for monitoring epilepsy patients is presented. In addition a review of some mobile systems for epilepsy monitoring is done. In the second part the sensor requirements are listed in the form of general requirements such as data format and housing and in sensor type specific requirements such as sampling rate and measurement interval. Finally the developed sensor system architecture is described. Here both the hardware and software architecture is described.

8.2 State of the Art in Mobile Epilepsy Monitoring

8.2.1 Ambulatory EEG: State of the Art

Ambulatory electroencephalogram (EEG) is a clinic-based service, in which the patient is fitted with a miniaturized EEG recording device during a visit to the hospital, and then sent home for on-going EEG recording. Most EEG device manufacturers offer one or more ambulatory systems. Typically ambulatory systems have extreme limitations:

- Conventional clinical electrodes quickly lose signal and become detached, requiring the patient to visit the hospital for a system check every 24–48 h;
- Usually no other parameters of data are obtained, particularly video is usually not available, limiting the diagnostic accuracy of the system;
- No remote monitoring is possible;
- The system has limited data capacity;
- Limited geographical availability and limited availability of clinical experts, as for in-hospital services.

Many healthcare providers offer ambulatory EEG. The cost of ambulatory EEG is lower than inpatient long-term monitoring (LTM) because a hospital bed is not required.

The first ambulatory EEG systems were being evaluated in the 1970s and 1980s [2–4], although technology was limited to systems with very few channels.

Conventional ambulatory EEG has established benefits. An early study, comparing LTM in an epilepsy monitoring unit (EMU) against ambulatory EEG in a test-retest design in the same subjects showed concurrence of diagnostic findings in 77 % [5]. In comparison with sleep-deprived outpatient conventional EEG in 46 patients, ambulatory EEG showed similar rate of detection of interictal epileptiform discharges (IEDs) but a superior rate of detection of seizures (seven during ambulatory EEG versus zero during conventional sleep-deprived outpatient EEG) [6]. In some settings, ambulatory EEG can have a very high rate of detection of seizures, for example in 31 of a series of 54 children during 48 h of recording [7]. The role of ambulatory EEG has been extended to substitute for EMU-based LTM in presurgical evaluation, with good outcomes reported in patients with temporal lobe epilepsy [8]. Ambulatory EEG has been combined with other modalities of data collection, for example allowing the differential diagnosis between syncope and epileptic seizure by combining electrocardiogram (ECG) and EEG [9].

The absence of simultaneous video data is a serious shortcoming of conventional ambulatory EEG [10]. Given the high diagnostic utility of inpatient video-EEG telemetry (IVT), and the low rate of capture of relevant events with relatively brief conventional outpatient EEG, in some centres there has been a trend to increasing the duration of outpatient EEG and adding simultaneous video, somewhat blurring the distinction between IVT and outpatient EEG. In 100 patients with frequent events, up to 2 h of video-EEG captured events in 66, 24 of which would have been missed without video [11]. Similarly, added value of video was demonstrated in 43 children studied with video-EEG for up to 3 h, with events captured in 36 [12]. In a more recent study of 179 patients, 4 h of video-EEG was added to a routine outpatient EEG, doubling the diagnostic yield from 27 to 50 % [13]. Therefore, to maximize the value of extended monitoring in the patient's own environment, ambulatory EEG alone is likely to be suboptimal, and simultaneous video should be incorporated, through the use of home video telemetry or other analogous services.

8.2.2 Long Term Monitoring (LTM) in the Patient's Home: State of the Art

Continuous multimodal data collection outside the hospital environment has many challenges. Extended periods of monitoring in the patient's home for epilepsy have been surprisingly little reported. An innovative study 30 years ago used radiotelemetry to collect EEG in the patient's home and transmission of data over a conventional telephone line to the epilepsy center, showing the future potential of remote monitoring [14, 15]. In another system, simple improvisation was used to combine a conventional ambulatory EEG system with a portable video camera set up in the patient's home, with offline combining of the EEG and video for simultaneous viewing [16]. Probably the best-established system combines EEG and video into a single portable unit taken by the patient to their home, after EEG connection in the clinic [17]; this system also includes automated IED [18] and seizure detection [19]. This system is offered as a service by a commercial provider in the USA, on a nationwide basis (DigiTrace Ambulatory EEG, see more below).

More recently, a number of new approaches have been proposed, making use of current technologies to collect multiple parameters of data, track the patient's location and provide warning signals to carers and professionals, such as a system design which would collect EEG, automatically detect seizures, locate the patient using global systems for mobile communications, (GSM systems), and send a seizure alert to carers or professionals using SMS messaging [20]. These appear to be only proposals for systems/services, without evidence (yet) for development or commercialization.

There are many challenges in implementing LTM at home especially over extended periods. The volume of data collected may be very considerable, and in the absence of professional staff constantly monitoring the data output as in hospital, these data need to be reviewed efficiently. Automated event detection is an important innovation in this context, such as automated detection of IEDs [18, 21] and seizures [19]. IED detection has been coupled to automated data reduction, to reduce data volume and minimise the amount of data required for review by a clinician [22]. The performance of automated seizure detection methods is critical, since inadequate sensitivity will lead to seizures being undetected, and inadequate specificity will lead to non-seizure events being incorrectly detected as seizures. One method reported 73 % sensitivity with 0.15 false detections per hour [23]; although this might sound promising, an imaginary patient who has 4 real seizures in a week would have 28 seizures detected by this system during the week, of which only three would be real seizures, and one real seizure would be missed. Hence, from the patient perspective, this seizure detection method is useless. Another method reported higher sensitivity at 83 % but at the cost of an even higher false detection rate of 0.3 per hour [24]. The particular feature of absence seizures, 3 Hz generalised spike-wave, is much easier to detect in an automated system, achieving 99.1 % sensitivity, but still with a high false detection rate of 0.5 per hour [25]. A recent study has compared in detail the range of published methods for automated seizure detection, concluding that online measurement of line length and relative power in the 12.5–25 Hz band is the most efficient available method [26].

A challenge in ambulatory data, which may be extreme, is the presence of many different artefacts related to movement and other sources of physiological and non-physiological noise. Methods to remove muscle artefact can considerably aid automated event detection [27, 28].

Very long-term home monitoring will require miniaturised wearable devices which minimally hinder the patient going about their daily life. Such devices are likely to require specially-designed hardware with low power consumption [29, 30]. The most important barrier to very long-term EEG monitoring outside the clinic is electrode technology. Electrodes in current use require skilled attachment and regular inspection to maximise signal quality, and without this the signal rapidly becomes degraded or lost. In unconscious patients, subdermal wire electrodes have been used successfully for up to 60 days of continuous recording [31], but such electrodes would not be suitable for an awake active patient. New technology such as gel-free "skin-grabbing" electrodes [32] may prove to be an important step forward.

Other modalities of continuous data collection in addition to or instead of EEG and video may be relevant in certain settings. Using ECG, spectral analysis of heart rate variability may contribute to seizure detection [33]. Changes in respiration in association with seizures may also be monitored [34]. Many seizures are associated with motor activity, which may be detected using electromyogram or accelerometry.

Automated EMG-based detection of generalised tonic-clonic seizures in a very small sample detected 4 of 7 seizures in five patients, with a low false detection rate of 0.003 false detections per hour [35]. Accelerometers are relatively cheap and wearable, and may provide a means to detect and discriminate different types of motor activity, including seizures [36]. In a small study, wrist-worn accelerometers detected 7 out of 8 generalised tonic-clonic seizures in six patients, but created 204 false detections [37]. A very recent study examined the performance of a wrist-worn accelerometer seizure detection system in detecting generalised tonic-clonic seizures in 73 patients, with an average of 67 h of recording per subject; 39 generalised tonic-clonic seizures occurred, which were detected with a sensitivity of 89.7 % and false detections were low at 0.008 per hour [38]. An alternative approach to detecting motion due to seizures occurring at night is to use motion detectors placed beneath the patient's mattress. A study of such a device found only 62.5 % sensitivity and en extremely high false detection rate of 0.18 per hour [39].

8.3 Sensor Requirements for Multiparametric Monitoring of Epileptic Patients

In general, sensors used in the ARMOR project which are described in this chapter are integrated intelligent sensor systems for mobile monitoring of physiological signals and environmental or context information. These sensors consist of a signal transducer (electrode, acceleration sensor, light sensor…) with specific analog frontend. After the digitalization of the signal in the analog to digital (A/D) converter, a mobile processor (e.g. microcontroller) controls the preprocessing, mobile online analysis, event handling and communication to the middleware (Fig. 8.1).

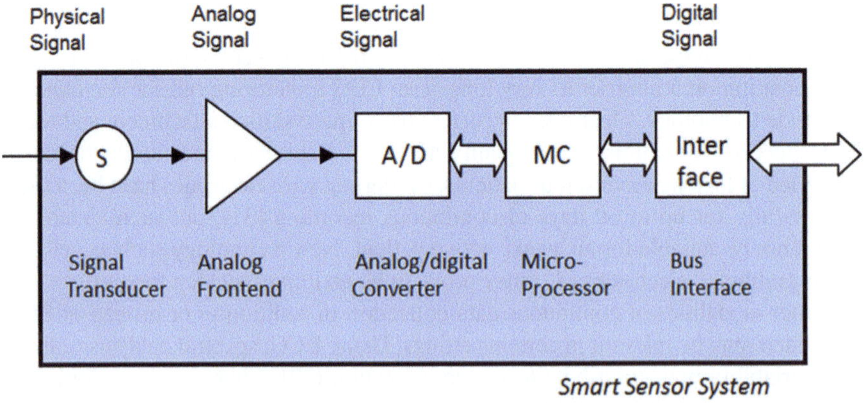

Fig. 8.1 Diagram of mobile sensor system

The mobile sensor system may have different signal transducers or a multi-channel input (in parallel or multiplexed) connected to one microprocessor. The signal transducers could be physically integrated in the same housing as the information processing unit or could be wired to this unit.

In case of multi-channel and/or multi-signal sensor system, the sensor serves as a hardware aggregator for the different channels/signals. In case of multi sensor usage connected to one middleware platform, this platform serves as a software aggregator

8.3.1 Application of Mobile Sensors in Epilepsy Monitoring

A major challenge for epilepsy monitoring is that by definition respective seizures are paroxysmal, i.e. unpredicted in onset and often also in clinical expression. Depending on the part of the brain which primarily or secondarily will sustain the hypersynchronous hyperactivity of its neurons and the role this brain area (or better brain circuit/system) normally plays in controlling body functions, a seizure may consist of a variable set of involuntary movements (i.e. convulsions), expressions from the autonomic nervous system (changes in heart rate, respiration, perspiration etc.) or cognitive (i.e. loss of consciousness) and other symptoms. Additionally some of these same symptoms may present as seizures but in fact result from causes not primarily originating in the brain (non-epileptic paroxysmal event NEPE, i.e. a syncope) and therefore need radically different treatment.

It follows that any monitoring platform has to be very versatile in order to distinguish which body or brain system is primarily causing the seizures, accurately define the onset, localize it, if possible, and all these in a quantified way appropriate for helping the patient cope with the particular disease. The demands of such a monitoring system start with the definition of appropriate clinical scenarios based not so much on their statistical prevalence (although most of them are common), but mainly aiming to exemplify the wide spectrum and variety of diagnostic problems and collectively cover all the different cases.

A critical outcome of the effort devoted in the context of the ARMOR project concerns the identification and specification of such application scenarios as a distillation of what cases must be handled by a fully grown respective cyber physical platform.

8.3.1.1 Application Scenario 1: Distinction Between Epilepsy or Non-epileptic Paroxysmal Events (NEPE)

As a scenario overview the patient is in home and he suddenly loses his consciousness. The patient has similar episodes frequently. These episodes cannot be properly defined, whether they are epilepsy or a NEPE, by simply testing his clinical history and conventional EEG studies. The patient may present frequent episodes of loss of consciousness.

The nature of these episodes cannot be ascertained by the clinical history and conventional EEG studies. In such a scenario respective cyber-physical system (CPS) should offer both clinic and home setups.

In the former case offline analysis could significantly help leading to a decision on whether the seizure belongs to Epilepsy or NEPE and possibly also on what type of Epilepsy or which NEPE. If possible online analysis may be useful in terms of rapid assistance provided either by family members or caregivers. In that context selected EEG electrodes with their number and position defined according to the specific epilepsy/seizure type that has been hypothesized from the clinical history. The minimum non-EEG electrodes sufficient for confident differentiation between epileptic seizures and NEPE must be used based on previous research on different patients groups (i.e. epileptic, vasovagal, psychogenic).

Considering corresponding risks in this scenario data acquisition and aggregation may be performed either by wireless transmission or local storage. In both cases, the former presenting the most significant challenges, data privacy and authentication must be assured. In order for private medical data to be acquired, analyzed, processed and in any way utilized, adequate conditions must be met concerning technical issues as well as soft issues including, but not limited to, which entities have access to which data, under what conditions, how data are acquired, how consent is acquired etc.

8.3.1.2 Application Scenario 2: Delineation of the Clinical EEG Expression of Different Types of Epilepsy

This scenario focuses on cases where during the day, the patient may present seizures which are defined as epileptic but the available clinic and EEG evidence is insufficient to delineate the specific type of epilepsy. This insufficiency may lead to incorrect selection of antiepileptic drug or other management. The patient presents seizures which have been documented as epileptic, but the available clinical and EEG evidence is insufficient to delineate the particular type of epilepsy/epilepsy syndrome. As the choice of antiepileptic drugs depends on the type of epilepsy, long term monitoring using novel CPS platforms at home before initiation of treatment, could provide invaluable insights using a configuration dictated by the patient's profile including video as mandatory.

In this scenario it is expected that novel CPS systems will provide full information on clinical semiology (video) and all constituents of the particular seizure(s) (including EEG, various autonomic changes, and muscle activity and tone) and their timing/sequence. Furthermore, in such scenarios offline analysis will lead to diagnosis of the type of seizures and type of epilepsy the patient is suffering from, and selection of the appropriate antiepileptic drug/other management. To achieve this insight selected EEG electrodes with their number and position defined according to the specific epilepsy/seizure type that has been hypothesized from the clinical history. Additionally, the minimum non-EEG electrodes sufficient for full identification of the most pertinent autonomic functions that will have been identified by the previous research on patients with epileptic seizures.

8.3.1.3 Application Scenario 3: Follow Up—Medication Evaluation

Following the process of diagnosis, a patient who has been diagnosed with a particular type of epilepsy may indicate some nuisances and complains about his drugs to be insufficient or that types of seizures have also become different. After the appropriate follow up and the medication evaluation, it is decided to change the treatment accordingly. This may be caused for reasons such as, the existing medication has become less effective but seizures have not changed in type, or the existing medication has become less effective because seizures may have changed in type.

In order to decide whether or not is needed to change the treatment according to the condition, long term monitoring using novel CPS's at home based on a configuration dictated by the patient's personalized profile could be extremely helpful. Respective offline analysis will lead to a decision on whether a change in drug is needed and which one that may be.

8.3.1.4 Application Scenario 4: Protection from Seizures

In the case of a patient who is diagnosed with a potentially life-threatening type of epilepsy (mainly status epilepticus) monitoring will help detect (mainly EEG) early signs of these seizure and through feedback allow efforts to protect the patient from serious consequences of such seizure. The patient or his attending person is alerted to prevent serious consequences of the episode. Assume a case where a patient is diagnosed with a type of epilepsy which either

- is most often preceded by an aura or
- is threatening to his life (status epilepticus) or
- is stereotypically triggered by specific stimuli (extreme somatic stress, reflex epilepsies triggered by specific stimuli including light or sound, or activities such as watching TV, playing video games, reading etc). In such a case long term monitoring using respective CPS can be extremely beneficial both considering a clinic setup as well as a home setup.

In the former case, a CPS system can offer online data analysis aiming to prevent and protect against serious consequences of seizures (mainly status epilepticus). Selected EEG sensors as well as non-EEG electrodes will be dictated by the patient's personalized profile. In the latter setup case the number of sensors is expected to be small. Once again online analysis support is crucial to prevent and protect against serious consequences of seizures (mainly status epilepticus).

Finally in this scenario for a respective CPS to be truly useful and efficient, it is critical to take into consideration significant risk factors such as: False Negative Seizure Identification, False Positive Seizure Identification, Delayed Seizure Identification, and Insecure handling of sensitive and private data.

8.3.1.5 Application Scenario 5: Research on Local Signs of Idiopathic Generalized Epilepsy (IGE)

This scenario aims to address the goal of treatment efficacy improvement through the use of CPS assuming a patient who is diagnosed with idiopathic generalized epilepsy (IGE) but responds poorly to the indicated treatment. About a third of patients with IGE are resistant to the appropriate antiepileptic medication. Novel CPS aim to extend and complement respective studies by collecting a multitude of local EEG signs and their correlates from the autonomic nervous system.

A clinic setup would require offline data analysis support of selected EEG electrodes with their number and position according to IGE, based on magnetoencephalographic (MEG)/EEG/magnetic resonance imaging (MRI) analysis. Additionally, minimum non-EEG electrodes that pick up changes of autonomic function and muscle activity, relevant to IGE. In a home setup case, selected EEG electrodes with their number and position according to the specific epilepsy/seizure type, and based on MEG/EEG/MRI analysis while non-EEG electrodes that pick up autonomic changes, relevant to the particular patient's profile.

8.3.1.6 Application Scenario 6: Pre-surgical Evaluation

Besides diagnosis and medication epilepsy monitoring plays a critical role in the case of positional surgical treatment. In such a context a CPS system is expected to be able to provide the wealth of clinical and neurophysiology data from many seizure events needed in order to decide (along with brain imaging, neuropsychology etc.) on whether to operate or not and with which surgical approach. For example, focal epilepsy is intractable to all antiepileptic medication and usually a surgical approach is adopted. One of the essential aims of pre-surgical evaluation is to record all habitual seizure types of the particular patient and identify the localization of their onset in the brain.

In this scenario either a clinic or a home setup can envisioned based on a wide range of sensors as well as on the personalized profile of the patient. Respective offline support is required according to the particular focal epilepsy type (lobe of origin, i.e. temporal, frontal or parietal occipital), and based on MEG/EEG/MRI analysis. At the same time non-EEG sensors are required so as to pick up autonomic changes, relevant to the particular lobe of onset, ECG vital.

8.3.1.7 Application Scenario 7: Nocturnal Seizures

In this scenario a respective CPS system is expected to offer significant advantages in monitoring a patient who presents seizures mainly or solely at sleep or to particular levels so as to delineate between several types of seizures which demand different types of treatment. It is also expected that it will enable differentiation between seizures and non-epileptic paroxysmal events that also result in electrographic or clinical arousals. Such non-epileptic paroxysmal events may include periodic leg movement disorder or obstructive sleep apnea, and can occur in patients whose clinical history gave no hint of the problem. Some people have numerous spontaneous arousals from sleep with no clear physiological precipitants. Data will be collected to understand the nature of seizures and their association to sleep macro- and microstructure. EEG electrodes configuration should comply with comfortable wearing at night.

In this case both clinic setup and home setup are to be supported based on offline line data analysis algorithms.

8.3.2 General Functional and Nonfunctional Requirements

For the realization of an ambulatory assessment system, a number of technical and non-technical requirements could be listed to describe the functionality of the system.

This section summarizes all the functional and non-functional requirements of the sensors that an efficient CPS for epilepsy monitoring must support.

- **Sampling Rate**: The sampling rate for the physiological signals will vary depending on the physiological parameter that will be assessed and will be based on the state of the art research.
- **Resolution**: The output of the sensors will have at least 12 bit resolution.
- **System architecture**: The system architecture should be modular to allow the sensor modules to work with different types and numbers of transducers and to operate the system with different sensor modules.
- **Interfaces**: Communication between the sensors for physiological data and the other components of the systems could be performed both wireless (e.g. Bluetooth, GSR, Wi-Fi, etc.) and via wired interfaces (e.g. USB). The sensors will communicate with the aggregator wirelessly by using the Bluetooth interface in order to transmit the pre-processed data for the online analysis and via USB in order to store the assessed raw data.
- **Data format**: A suitable (open, platform independent, exchangeable, etc.) data format for recording, streaming and archiving sensor data from various recording systems and with various sampling frequencies should be chosen. All data should be stored as raw data to be able to check the data for artefacts and to reanalyze the data in the case that new algorithms are available.
- **Storage capabilities**: The raw data should be stored on an internal memory (e.g. micro SD card) beside the possibility of streaming the data via wireless interface to a remote computer. Internal storage of data allows the assessment of physiological signals independently to remote systems and provides a save procedure even if the connection for data transmission is not available.
- **Time Stamp**: Time Stamp should be added to the signal on sensor level in order to be able to synchronize signals from various sensors.

- **Embedded Software**: The software of the sensors should be able to perform online data pre-processing. Online processing is a prerequisite for one type of interactive ambulatory assessment, where the actual values of the recorded physiological data triggers a synchronous acquisition of subjective data, e.g. by experience sampling methods.
- **Online Analysis**: The online analysis will be partially performed on the sensor side, where the data will be pre-processed and on the aggregator where the DSMS will perform the further analysis that will need extra computation power.
- **User interface**: Configuration software should run on a PC or Laptop to provide a broad range of configuration possibilities in order to adapt the functionality of the sensors to the specific requirements of the study.
- **Battery lifetime**: Ideally, the whole assessment period should be performed without recharging the batteries of the system.
- **Charging**: Should be done easily either by the USB interface or power plug and should be adapted to common charging cycles as they are known from everyday life systems like smartphones.
- **Security**: The wireless communication between the sensors platform and the aggregator will be performed by using encryption algorithms such as advanced encryption standard (AES).

In addition to these requirements, the systems used for the assessment of physiological data should be approved as medical devices and validated for the desired application.

In wireless transmission of personal and highly sensitive data, security support is a prerequisite of primary importance. Respective support pertains to data privacy, data integrity and authentication of communication parties. However, provision of security features pose significant challenges in the sensor network due to the extremely limited resources in critical areas such as processing power, available memory and available energy. Furthermore, although Bluetooth Technology offers security features with respect to both authentication and privacy, at the same time respective weaknesses and vulnerabilities are well-known [40]. Finally, significant overhead imposed by software security implementations must be taken into consideration [41].

Beside those general requirements some sensor specific requirements have to be taken into account. The main parameters describing the functionality of a sensor are the sampling rate, the resolution and measurement range, the measurement characteristics and the number of channels. These parameters have to be adapted to the specific requirements in ambulatory assessment and to the specific needs of a certain study. Depending on the research question to be answered, the number of channels and the sampling rate of each channel (e.g. for the measurement of EEG signals) can differ significantly.

Beside the technical requirements, there are a large number of non-technical requirements that have to be fulfilled by a system for ambulatory assessment.

- **Weight, Dimensions, Housing**: The sensor for physiological ambulatory assessment should be as unobtrusive as possible to be convenient for the user for use in everyday life, even during sleep. Of course some technical limitations such as the battery consumption/ dimension might lead to bigger housings. A trade-off between the comfort and the system lifetime will have to be made.

- **Application**: The application of sensors on the body should be easy in order to save time during preparation and robust to avoid loss of data during registration.
- **Obtrusiveness and compliance**: These parameters are directly influenced by the size and the applicability of the sensors.
- **Usability**: Usability is a very important aspect that will be taken into account, as the end-users (Doctor, Patient, Healthcare Professional, Family Member or Caregiver) might be persons with limited technical knowledge. The software for starting/stopping the sensors will have a graphical user interface (GUI) and special attentions to user friendliness will be paid.
- **Calibration**: There should be a possibility to calibrate the sensors and to check the calibration of specific units during the preparation phase.
- **Time of use**: The sensors should be able to measure at least for 24 h.
- **Costs**: To allow studies with a large number of participants, the cost of each senor unit should be as low as possible.
- **Comparability with other researchers/studies**: To be able to compare the results of a measurement with other researchers, standardized measures should be used instead of proprietary ones.
- **Interoperability to other systems**: Open data formats and standardized interfaces should be used.
- **Distribution to the end-users**: The patient will be wired-up at the hospital before he goes off to spend his/her day. The electrodes will be firmly fixed by the caregivers, whereas the sensors mostly needed for monitoring during nocturnal sleep will be easily placed/replaced by the patients before he/she goes to bed and unplugged in the morning after awakening.
- **Price**: The price of the final system will depend on the number of units that will have to be used.
- **Privacy and Security**: Effectively this requirements comprises an umbrella of required features support such as: Data protection and privacy of personal information, Protection of patient specific database records, Information security policy document, Correct processing in applications, Technical vulnerability management processes, User & Device identification and authentication, Audit logging, Activity Monitoring.

8.3.3 Sensors Selection

In the following significant types of measurements are presented and their role in epilepsy monitoring is depicted.

As ARMOR's main target are epileptic patients, it is vital to have EEG sensors present in the system, as it is an essential component in the evaluation of epilepsy.

It has been shown that ambulatory long-term EEG recordings with intensive monitoring have led to better classifications of seizures and treatment results [42]. Since 30–60 % of patients are unaware of their seizures, having these sensors present can lead to new results with optimal treatment. Without EEG recordings, false diagnosis may be made as various phenomena are similar to the resulting behavior of a seizure.

Electrocardiogram (ECG) is used to record the electrical activity of the heart. It is an effective means to help to rule out a seizure being caused by the way the heart is working. It has been noted that in some seizures, especially those located in the temporal lobe, experience a change in heart rate (HR) prior to or at the onset of the seizure [43]. A study showed an increase in heart rate of at least 10 beats/min in 73 % of seizures (93 % of patients) and this occurred most often around seizure onset. In 23 % of seizures (49 % of patients) the rate increase preceded both the electrographic and the clinical onset [44]. Such changes may clarify the timing of seizure onset and can be useful for seizure diagnosis and for automatic seizure detection.

The opposed effect of sympathetic and parasympathetic on the heart leads to an irregularity in the heart rate: the Inter-Beat-Interval (IBI) is not perfectly regular. This variation of the heart rate from beat to beat is called heart rate variability (HRV). In general a normal or increased HRV shows the flexibility of the organism to adapt itself to the changes in the environment. However a low HRV shows a limited ability of adaptation of the organism and leads to serious heart diseases. Lotufo et al. [45] demonstrated sympathovagal imbalance in epilepsy, as showed by lower high-frequency (HF) power spectrum, standard deviation of normal-to-normal interval (SDNN) and the root mean square of successive differences (RMSSD) values when compared to controls. In addition, there was a trend for higher LF (lower frequency power spectrum) values in patients receiving pharmacotherapy. As lower vagal (HF) and higher sympathetic (LF) tone are predictors of morbidity and mortality in cardiovascular samples, these findings highlight the importance of investigating autonomic nervous system (ANS) function in patients with epilepsy. Assessing HRV might also be useful when planning therapeutic interventions, as some antiepileptic drugs can show hazardous effects in cardiac excitability, potentially leading to cardiac arrhythmia.

On the other hand HRV measurements have been shown to help seizure detection in some cases [33]. Di Genaro et al. (2004) [46] observed a high incidence (92 %) of ictal HR increase in temporal lobe epilepsy seizures (preceding the EEG onset by 5 s). This suggests that the HR changes may be coupled to the functional impairment of neural circuits involved in sympathetic cardiovascular regulation, in the mesial temporal lobe structures.

Finally HRV changes have been observed in particular sleep stages in epileptic patients [47] and have been associated with sleep microstructure EEG elements of importance to epilepsy like K-complexes [48].

The autonomic nervous system obviously reflects brain activities and is strongly influenced by them. HRV as a recognized marker of ANS balance and dynamic changes appears as an excellent marker for seizure detection and follow-up especially if studied along with other ANS expressions.

Many epileptic seizures incorporate motor phenomena during their course. Ictal motor activity can be captured and analyzed with video and motion sensors (based on three dimensional acceleration sensors) in order to characterize a seizure [49]. Body sensors are important as they can be used to discriminate changes in on-going activity due to seizures or physical activity, assisted by ECG [43]. In epileptic patients, activity sensor focus has been on the distinction between seizure movements and regular nocturnal movements [50]. In a recent presentation, a group used a motion classifier to extract and cluster information about patient motion by processing body motion signals [50]. Unfortunately, they tested against only two seizure motion templates, and the project had no further presentations or publications since.

Galvanic Skin Response (GSR) acts like an ohmmeter, measuring the electrical conductance of the skin. Due to being closely related to electrodermal activity, it has been tested as a potential sensor for seizure detection [51]. In a recent presentation [51], a prototype for GSR signal integration has been described, although missing an algorithm for GSR evaluation in terms of its contribution in seizure detection. Yet, it has been reported that spontaneous epileptic seizures may be correlated with large increases in electrodermal activity [52]. These GSR changes appear to be significant in generalized tonic-clonic seizures and reflect massive sympathetic discharges, often continuing postictally [52].

Temporal loss of consciousness (TLC), which is a cardinal feature in many types of seizures, may also have alternative causes. Syncope is the commonest cause of TLC, due to cerebral hypoperfusion. Neurally mediated (vasovagal, neurocardiogenic or reflex) syncope can happen in up to 40 % of the general population and can be misdiagnosed as epilepsy, particularly when it results to cerebral hypoxia and to a reflex anoxic seizure. Blood oxygenation level along with HR are vital signs to syncope diagnosis, which are therefore very useful in differentiating between epileptic and non-seizures.

In some cases of epileptic seizures, patients have a noted drop of oxygen level in their blood. Recent studies have suggested that the cases of Sudden Unexpected Death in Epilepsy (SUDEP) may have occurred due to the brain not instructing patients to continue breathing during seizures [53]. The finds demonstrated that complex partial seizures commonly lead to significant and prolonged oxygen desaturation due to hypoventilation, with saturation levels dropping below 80 % in about one-third of seizures and below 70 % in one-eighth [54]. These pauses in breathing can last from a few seconds to minutes, and can turn deadly as the oxygen level on the blood begins to drop.

In reflex epilepsies certain stimulus may bring on seizures. One is photosensitive epilepsy (PSE), where seizures are triggered by visual stimuli that forms patterns in time or space [55]. Diagnosis can be made by combining an EEG with a device producing Intermittent Photic Stimulation (IPS). This device produces specific types of stimuli that can be controlled and adjusted [56]. To have a sensor aware of flashing/flicker lights or regular moving patterns would enable the patients to avoid the provoking stimuli and in turn prevent a seizure.

In this section a twofold selection and categorization of the most critical sensors, a novel CPS for epilepsy and related brain disorders, is presented. On one hand the sensor are presented with respect to the physiological parameter they monitor while on the other hand sensors are categorized with respect to the application scenario presented in Sect. 8.2.1.

The first approach is presented in the following table including comments of the number of sensors required as well on specialized requirements.

Parameter studied	Method	Minimum # sensors	Requirements
Brain electrical activity	Electroencephalogram (EEG)	Depends on the scenario	Positioned in specific points on head. Signal is small in microVolts and noisy
		As few as possible for each scenario, based on offline analysis of EEG/ MEG signals and MRI of the patient, and clinical evaluation	Needs reference, ground, impedance matching, amplification, filtering (low pass filter (LPF) > 50 Hz), digitization (sampling rate (SR) > 250 Hz)
		R < 5 KΩ	Mandatory for transmission: Fidelity with no cross talk between the channels and absolute synchrony of all channels
Brain electrical activity	Magnetoencephalogram (MEG)		Only for research not in final ARMOR. MEG will be used if available for one-off modeling of individual brain activity for use in ARMOR
Muscle tone	Electromyogram (EMG)	As indicated in the individual patient/ scenario (2–4) R < 5 KΩ	As in EEG, but signal is larger in few milliVolts and frequency higher around 100 Hz. Movement artifacts around 10 Hz
Eye movements	Electrooculogram (EOG)	2 R < 5 KΩ	As in EEG but signal is larger 0.1–10 mV and needs band pass filter = 0.1–10 Hz
Heart rate	Electrocardiogram (ECG)	2	As in EEG but signal is larger 0.05–3 mV and frequency 0.01–300 Hz
Respiration	Respiratory Inductance Plethysmography	1–2 (thorax and abdomen)	Signal around 6 mV. It is DC but should be recorded as AC to avoid drifts, sampling rates = 20–50 Hz
Nasal flow	Differential pressure transducer	1	SR = 200 Hz

(continued)

Parameter studied	Method	Minimum # sensors	Requirements
Oxygenation	Saturation pressure (SpO2) and pulse plethysmography	1	
Body position and movements	Accelerometers	2	To assist recognition of vasovagal trigger
			To correlate the timing of epileptic convulsions or myoclonic movements to that of EEG and other polygraphic events
Sympathetic tone changes	Galvanic skin response (GSR)	2 on fingers	Extremely slow signal (DC)
Aura	Event button	1	Pressed by patient or bystander
Seizure features	Video recording		Will be time-stamped and stored independently for the final evaluation by the MD

According to the second approach sensors are categorized according to the most critical application scenarios as follows:

8.3.3.1 Application Scenario 1: Distinction Between Epilepsy or Non-epileptic Paroxysmal Events (NEPE)

Required sensors and medical equipment: EEG, EMG, EOG, ECG, Respiration, SpO2, Activity, Nasal Flow, Light, Sound.

8.3.3.2 Application Scenario 2: Delineation of the Clinical EEG Expression of Different Types of Epilepsy

Required sensors and medical equipment: EEG, MEG.

8.3.3.3 Application Scenario 3: Follow Up—Medication Evaluation

Required sensors and medical equipment: EEG, MEG, MRI.

8.3.3.4 Application Scenario 4: Protection from Seizures

Required sensors and medical equipment: EEG, Light, Sound, Event button.

8.3.3.5 Application Scenario 5: Research on Local Signs of Idiopathic Generalized Epilepsy (IGE)

Required sensors and medical equipment: EEG, MEG, EMG, EOG, ECG, Respiration, SpO2, Activity, Nasal Flow, event button.

8.3.3.6 Application Scenario 6: Pre-surgical Evaluation

Required sensors and medical equipment: EEG, EMG, EOG, ECG, Respiration, SpO2, Activity, Nasal Flow.

8.3.3.7 Application Scenario 7: Nocturnal Seizures

Required sensors and medical equipment: EEG, ECG.

8.4 Mobile Sensor Systems

Depending on the type of epilepsy, different brain and body parameters need to be assessed in order to have a better understanding of the patient's state of health and to adapt the medical treatment accordingly. Therefore the goal is to develop a personalized system that assists in diagnosis, prognosis and treatment of the disease. Such system should fulfill the following criteria; it should be non-invasive, mobile, continuous and unobtrusive and all possible security and privacy aspects should be taken into account. Figure 8.2 shows an overview of the system.

This section describes the design of the sensors for mobile epilepsy monitoring.

8.4.1 Sensor System Architecture

In the general architecture the sensors are part of the whole system and are connected to the sensor interface of the middleware. Figure 8.3 shows the sensor architecture within the general view of ICT components (small diagram on top right of Fig. 8.3).

In general, sensors used for multiparametric monitoring in epileptic patients are integrated intelligent sensor systems for mobile monitoring of physiological signals and environmental or context information. These sensors consist of a signal transducer (electrode, acceleration sensor, light sensor,…) with specific analog frontend. After digitalization of the signal in A/D converter, a mobile processor (e.g. microcontroller) controls the preprocessing, mobile online analysis, event handling and communication to the middleware.

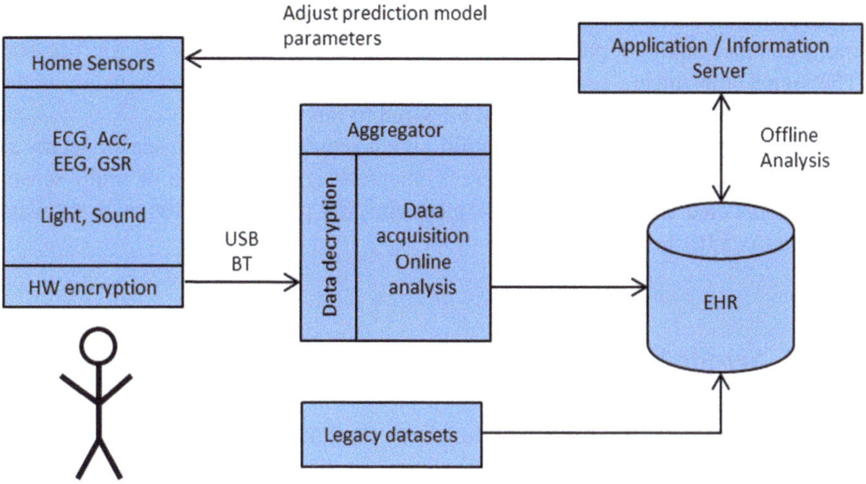

Fig. 8.2 System overview

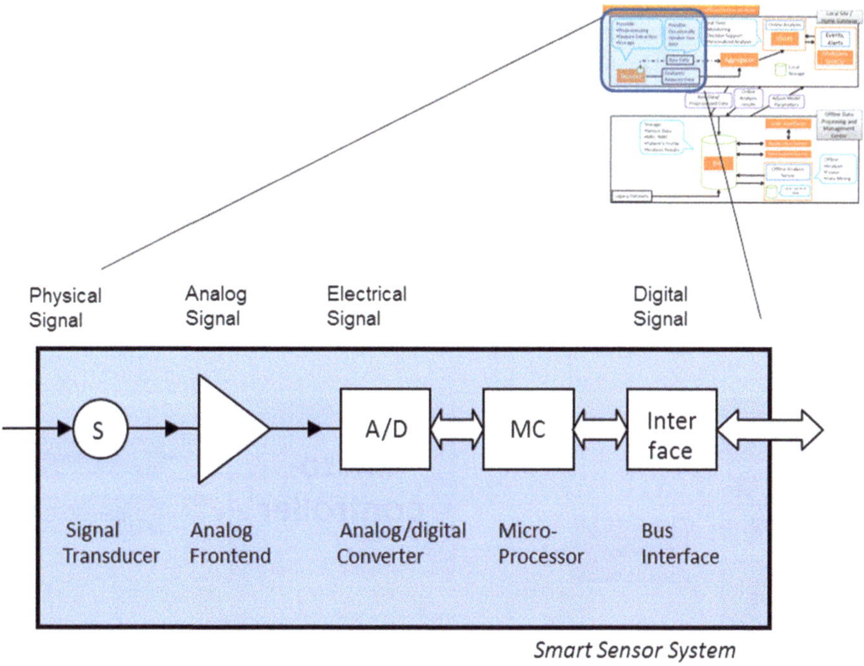

Fig. 8.3 Diagram of mobile sensor system

The mobile sensor system may have different signal transducers or a multi-channel input (in parallel or multiplexed) connected to one microprocessor. The signal transducers could be physically integrated in the same housing as the information processing unit or could be wired to this unit.

In case of multi-channel and/or multi-signal sensor system, the sensor serves as a hardware aggregator for the different channels/signals. In case of multi sensor usage connected to one middleware platform, this platform serves as a software aggregator [57].

8.4.2 Sensor Systems for Mobile Epilepsy Monitoring

The measurement of the autonomic functions should be done by light-weight, portable, low-power sensors, specially developed for the assessment in everyday life.

Our main platform is based on a sensor platform developed by movisens (movisens GmbH, Karlsruhe, Germany) and consists of an ultra-low-power microcontroller (MSP430F1611, Texas Instruments) with an AD/DA converter, 2 UART interfaces, a 48 kB flash and a 10 kB RAM. The assessed data is encrypted and then transmitted wirelessly to the aggregator by using the Bluetooth interface. To avoid data loss in case the data transmission is not possible, the data is in parallel stored in a 2 GB micro SD card. The stored data can then be saved on the computer by using the USB 2.0 interface that is available. The charging of the power supply of the sensors is recognized by the USB interface as well (Fig. 8.4).

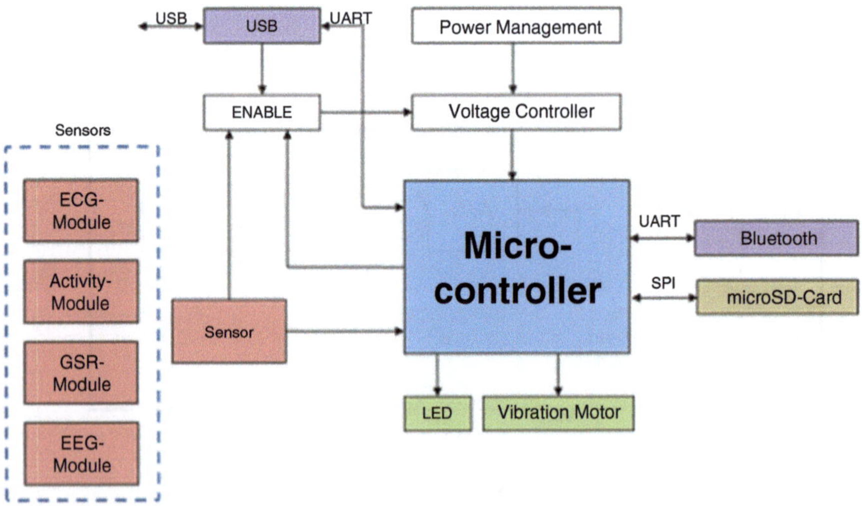

Fig. 8.4 Architecture of the sensor platform

Fig. 8.5 ECG-Sensor with activity monitoring module

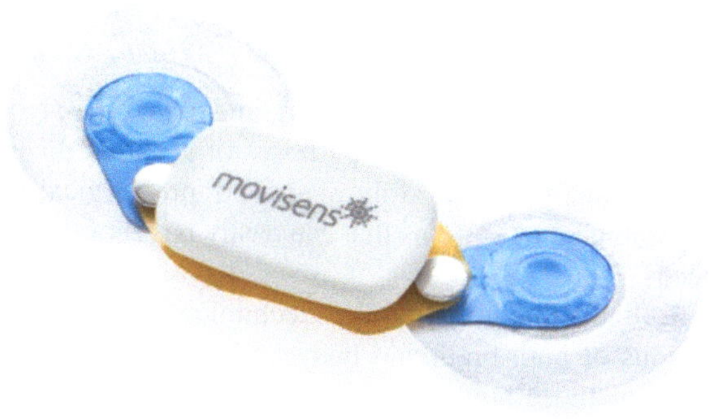

Fig. 8.6 EDA sensor

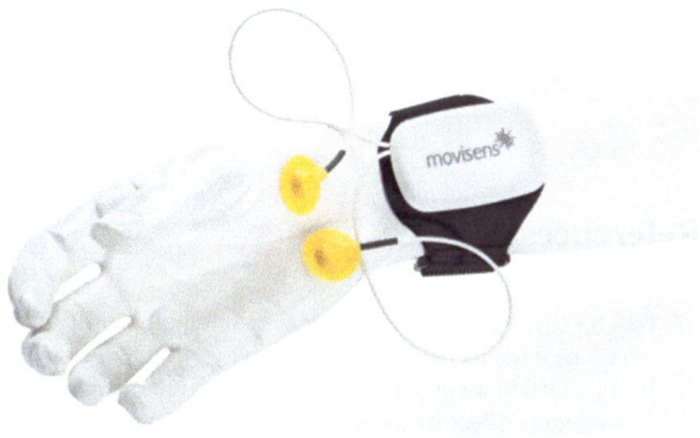

The ECG module is a single channel ECG recorder with a 12 bit resolution and a sampling rate from 256 to 1024 Hz. The electrodes are integrated into a wearable chest strap, which is light, small and comfortable. The electrodes are dry, allowing for everyday use of the chest strap (Fig. 8.5).

The activity monitoring module consists of a triaxial acceleration sensor (adxl345, Analog Devices Inc.) with a range of ±8 g and a sampling frequency of 64 Hz. The measuring unit has an additional air pressure sensor (BMP085, Bosch GmbH) with a sampling frequency of 8 Hz and a resolution of 0.03 hPa (corresponding to 15 cm).

The galvanic skin response (GSR) module measures the skin conductance with a sampling rate of 32 Hz. The measurement range of the GSR module is 2–100 µS and its resolution is 14 bit (Fig. 8.6).

Beside the above mentioned modules, the sensor platform will include an additional EEG module, which will record the brain parameters.

To assess additional context parameters like sound, light or geographic position, a smartphone can be used. Connection between Smartphone, Sensor and Aggregator is also realized by Bluetooth interface.

8.5 Conclusions

Due to its multifactorial causes and paroxysmal nature, epilepsy needs multi-parametric monitoring for purposes of accurate diagnosis, prediction, alerting and prevention, treatment follow-up and pre-surgical evaluation. This requires the development of systems that can easily be reconfigured and adapted to the patient's needs.

Reliable diagnosis requires technologies providing real-time, accurate and continuous of both brain and body multi-parametric data measurements. The assessment of those signals should be ambulatory and should not affect the patient's everyday life. Furthermore extra attention on security aspects should be paid. Within this chapter some of the main fundamental functional and non-functional requirements of such systems were presented and discussed. Possible use case scenarios were shown and the requirements in terms of sensor types and number of electrodes were underlined. Finally the architecture of a possible sensor platform was shown.

References

1. Hey S (2014) Context-aware sensing and multisensor fusion. In: Nikita KS (ed) Handbook of biomedical telemetry. Wiley, Piscataway
2. Ives JR (1975) 4-Channel 24 hour cassette recorder for long-term EEG monitoring of ambulatory patients. Electroencephalogr Clin Neurophysiol 39(1):88–92
3. Bruens JH, Kniff W (1980) Ambulatory EEG monitoring with a 24 hour cassette recorder in epileptic patients. Monogr Neural Sci 5:295–297
4. Green J, Scales D, Nealis J, Schott G, Driber T (1980) Clinical utility of ambulatory EEG monitoring. Clin Electroencephalogr 11(4):173–179
5. Ebersole JS, Leroy RF (1983) Evaluation of ambulatory cassette EEG monitoring: III. Diagnostic accuracy compared to intensive inpatient EEG monitoring. Neurology 33(7): 853–860
6. Liporace J, Tatum W IV, Lee Morris G, French J (1998) Clinical utility of sleep-deprived versus computer-assisted ambulatory 16-channel EEG in epilepsy patients: a multi-center study. Epilepsy Res 32(3):357–362
7. Saravanan K, Acomb B, Beirne M, Appleton R (2001) An audit of ambulatory cassette EEG monitoring in children. Seizure 10(8):579–582
8. Chang BS, Ives JR, Schomer DL, Drislane FW (2002) Outpatient EEG monitoring in the presurgical evaluation of patients with refractory temporal lobe epilepsy. J Clin Neurophysiol 19(2):152–156
9. Lai CW, Ziegler DK (1981) Syncope problem solved by continuous ambulatory simultaneous EEG/ECG recording. Neurology 31:1152–1154
10. Bridgers SL, Ebersole JS (1985) The clinical utility of ambulatory cassette EEG. Neurology 35:166–173
11. Chayasirisobhon S, Griggs L, Westmoreland S, Kim CS (1993) The usefulness of one to two hour video EEG monitoring in patients with refractory seizures. Clin Electroencephalogr 24:78–84
12. Connolly MB, Wong PKH, Karim Y, Smith S, Farrell K (1994) Outpatient video-EEG monitoring in children. Epilepsia 35(3):477–481

13. Modur PN, Rigdon B (2008) Diagnostic yield of sequential routine EEG and extended outpatient video-EEG monitoring. Clin Neurophysiol 119(1):190–196
14. Kamp A (1984) Long-term supervised domiciliary EEG monitoring in epileptic patients employing radio telemetry and telephone telemetry. II. Radio telemetry system. Electroencephalogr Clin Neurophysiol 57(6):584–587
15. van der Weide H, Kamp A (1984) Long-term supervised domiciliary EEG monitoring in epileptic patients employing radio telemetry and telephone telemetry. I. Telephone telemetry system. Electroencephalogr Clin Neurophysiol 57(6):581–583
16. Kerr AM, Amos PM, Etchells AH, Irwin AW, Holmes T, Stephenson JB (1988) A low-cost method for simultaneous video-recording of ambulant subject and electroencephalograph: the Quarrier's system. J Ment Defic Res 32(Pt 6):497–500
17. Schomer DL, Ives JR, Schachter SC (1999) The role of ambulatory EEG in the evaluation of patients for epilepsy surgery. J Clin Neurophysiol 16:116–129
18. Gotman J, Ives JR, Gloor P (1979) Automatic recognition of inter-ictal epileptic activity in prolonged EEG recordings. Electroencephalogr Clin Neurophysiol 46(5):510–520
19. Gotman J (1982) Automatic recognition of epileptic seizures in the EEG. Electroencephalogr Clin Neurophysiol 54(5):530–540
20. Giansanti D, Ricci G, Maccioni G (2008) Toward the design of a wearable system for the remote monitoring of epileptic crisis. Telemed J E Health 14(10):1130–1135
21. Casson AJ, Rodriguez-Villegas E (2008) On data reduction in EEG monitoring: comparison between ambulatory and non-ambulatory recordings. Conf Proc IEEE Eng Med Biol Soc 2008:5885–5888
22. Casson AJ, Yates DC, Patel S, Rodriguez-Villegas E (2007) Algorithm for AEEG data selection leading to wireless and long term epilepsy monitoring. Conf Proc IEEE Eng Med Biol Soc 2007:2456–2459
23. Schad A, Schindler K, Schelter B, Maiwald T, Brandt A, Timmer J, Schulze-Bonhage A (2008) Application of a multivariate seizure detection and prediction method to non-invasive and intracranial long-term EEG recordings. Clin Neurophysiol 119(1):197–211
24. Hartmann MM, Furbass F, Perko H, Skupch A, Lackmayer K, Baumgartner C, Kluge T (2011) EpiScan: online seizure detection for epilepsy monitoring units. Conf Proc IEEE Eng Med Biol Soc 2011:6096–6099
25. Petersen EB, Duun-Henriksen J, Mazzaretto A, Kjaer TW, Thomsen CE, Sorensen HB (2011) Generic single-channel detection of absence seizures. Conf Proc IEEE Eng Med Biol Soc 2011:4820–4823
26. Logesparan L, Casson AJ, Rodriguez-Villegas E (2012) Optimal features for online seizure detection. Med Biol Eng Comput 50(7):659–669
27. De Clercq W, Vergult A, Vanrumste B, Van Hees J, Palmini A, Van Paesschen W, Van Huffel S (2005) A new muscle artifact removal technique to improve the interpretation of the ictal scalp electroencephalogram. Conf Proc IEEE Eng Med Biol Soc 1:944–947
28. Logesparan L, Rodriguez-Villegas E (2011) A novel phase congruency based algorithm for online data reduction in ambulatory EEG systems. IEEE Trans Biomed Eng 58:2825–2834
29. Yates D, Lopez-Morillo E, Carvajal RG, Ramirez-Angulo J, Rodriguez-Villegas E (2007) A low-voltage low-power front-end for wearable EEG systems. Conf Proc IEEE Eng Med Biol Soc 2007:5282–5285
30. Altini M, Del Din S, Patel S, Schachter S, Penders J, Bonato P (2011) A low-power multimodal body sensor network with application to epileptic seizure monitoring. Conf Proc IEEE Eng Med Biol Soc 2011:1806–1809
31. Martz GU, Hucek C, Quigg M (2009) Sixty day continuous use of subdermal wire electrodes for EEG monitoring during treatment of status epilepticus. Neurocrit Care 11(2):223–227
32. Sun M, Jia W, Liang W, Sclabassi RJ (2012) A low-impedance, skin-grabbing, and gel-free EEG electrode. Conf Proc IEEE Eng Med Biol Soc 2012:1992–1995
33. Jeppesen J, Beniczky S, Fuglsang-Frederiksen A, Sidenius P, Jasemian Y (2010) Detection of epileptic-seizures by means of power spectrum analysis of heart rate variability: a pilot study. Technol Health Care 18(6):417–426

34. Corbishley P, Rodríguez-Villegas E (2008) Breathing detection: towards a miniaturized, wearable, battery-operated monitoring system. IEEE Trans Biomed Eng 55(1):196–204
35. Conradsen I, Beniczky S, Wolf P, Jennum P, Sorensen HB (2012) Evaluation of novel algorithm embedded in a wearable sEMG device for seizure detection. Conf Proc IEEE Eng Med Biol Soc 2012:2048–2051
36. Becq G, Bonnet S, Minotti L, Antonakios M, Guillemaud R, Kahane P (2007) Collection and exploratory analysis of attitude sensor data in an epilepsy monitoring unit. Conf Proc IEEE Eng Med Biol Soc 2007:2775–2778
37. Lockman J, Fisher RS, Olson DM (2011) Detection of seizure-like movements using a wrist accelerometer. Epilepsy Behav 20(4):638–641
38. Beniczky S, Polster T, Kjaer TW, Hjalgrim H (2013) Detection of generalized tonic-clonic seizures by a wireless wrist accelerometer: a prospective, multicenter study. Epilepsia 54: e58–e61
39. Carlson C, Arnedo V, Cahill M, Devinsky O (2009) Detecting nocturnal convulsions: efficacy of the MP5 monitor. Seizure 18(3):225–227
40. Hall J (2003) Brush up on Bluetooth. GSEC practical assignment version 1.4b
41. Lee J, Kapitanova K, Son SH (2010) The price of security in wireless sensor networks. Comput Netw 54:2967–2978
42. Stefan H, Hopfengärtner R (2009) Epilepsy monitoring for therapy: challenges and perspectives. Clin Neurophysiol 120(4):653–658
43. Jones VM, Huis In't Veld R, Tonis T, Bults RB, Van Beijnum B, Widya I, Vollenbroek-Hutten M, Hermens H (2008) Biosignal and context monitoring: distributed multimedia applications of body area networks in healthcare. In: Proceedings of the 2008 IEEE 10th workshop on multimedia signal processing, MMSP 2008, pp 820–825
44. Zijlmans M, Flanagan D, Gotman J (2002) Heart rate changes and ECG abnormalities during epileptic seizures: prevalence and definition of an objective clinical sign. Epilepsia 43(8): 847–854
45. Lotufo PA, Valiengo L, Benseñor IM, Brunoni AR (2012) A systematic review and meta-analysis of heart rate variability in epilepsy and antiepileptic drugs. Epilepsia 53(2):272–282
46. Di Gennaro G, Quarato PP, Sebastiano F, Esposito V, Onorati P, Grammaldo LG, Meldolesi GN, Mascia A, Falco C, Scoppetta C, Eusebi F, Manfredi M, Cantore G (2004) Ictal heart rate increase precedes EEG discharge in drug-resistant mesial temporal lobe seizures. Clin Neurophysiol 115(5):1169–1177
47. Ferri R, Curzi-Dascalova L, Arzimanoglou A, Bourgeois M, Beaud C, Nunes ML, Elia M, Musumeci SA, Tripodi M (2002) Heart rate variability during sleep in children with partial epilepsy. J Sleep Res 11(2):153–160
48. Colrain IM (2005) The K-complex: a 7-decade history. Sleep 28(2):255–273
49. Nijsen TME, Arends JBAM, Griep PAM, Cluitmans PJM (2005) The potential value of three-dimensional accelerometry for detection of motor seizures in severe epilepsy. Epilepsy Behav 7(1):74–84
50. Jallon P, Bonnet S, Antonakios M, Guillemaud R (2009) Detection system of motor epileptic seizures through motion analysis with 3D accelerometers. Conf Proc IEEE Eng Med Biol Soc 2009:2466–2469
51. Gouravajhala SR, Khuon L (2012) A multi-modality sensor platform approach to detect epileptic seizure activity. In: Proceedings of the 38th annual Northeast bioengineering conference (NEBEC), Philadelphia, 2012, pp 233–234
52. Poh M-Z, Loddenkemper T, Swenson NC, Goyal S, Madsen JR, Picard RW (2010) Continuous monitoring of electrodermal activity during epileptic seizures using a wearable sensor. Conf Proc IEEE Eng Med Biol Soc 2010:4415–4418
53. Devinsky O (2011) Sudden, unexpected death in epilepsy. N Engl J Med 365(19):1801–1811
54. Hirsch LJ (2009) Breathing new life into the fight against sudden death in epilepsy. Epilepsy Curr 9(5):137–139

55. Verrotti A, Trotta D, Salladini C, di Corcia G, Chiarelli F (2004) Photosensitivity and epilepsy. J Child Neurol 19(8):571–578
56. Guerrini R, Genton P (2004) Epileptic syndromes and visually induced seizures. Epilepsia 45(Suppl 1):14–18
57. Anastasopoulou P, Antonopoulos C, Shgir H, Krikis G, Voros NS, Hey S (2012) Mobile multi-parametric sensor system for diagnosis of epilepsy and brain related disorders. In: Proceedings of the 3rd international conference on wireless mobile communication and healthcare, Paris, vol 61, pp 207–214

Chapter 9
Secure and Efficient WSN Communication Infrastructure

Christos P. Antonopoulos, Nikolaos S. Voros, Stefan Hey, Panagiota Anastasolpoulou, and André Bideaux

Abstract Efficient wireless communication in conjunction with high level security provision concerning highly challenging medical applications is of paramount importance. Such requirements however comprise multifaceted challenges for CyberPhysical systems relying on state of the art WSN communication technologies. In that respect this chapter highlights two critical relative aspects as follows: The first aspect focuses on the sensor devices. Based on identified and presented requirements various security countermeasures are analyzed offering adequate security level while taking into consideration specific requirements such as low resource availability and low cost as well as the fact that sensors are miniaturized devices operating for long time periods. The second aspect addresses security concerning the wireless transmission with respect to performance and resource overhead. This chapter focuses on two different wireless communication technologies, comprising probably the most vulnerable part of a CPS platform. It is noted that the presented information effectively comprise the main outcome following an extensive evaluation based on both simulation as well as real platform measurements conducted in the context of the ARMOR FP7 project. Last but not least, a novel ultra low power hardware encryption module design is presented in order to further enhance the security level provided by the system while minimizing resource wastage.

C.P. Antonopoulos (✉) • N.S. Voros
Embedded System Design and Application Laboratory, Department of Computer and Informatics Engineering, Technological Educational Institute of Western Greece, Antirio, Greece
e-mail: cantonopoulos@teiwest.gr

S. Hey • P. Anastasolpoulou • A. Bideaux
KIT Karlsruhe Institute of Technology, Institute for Information Technology, Karlsruhe, Germany

N.S. Voros, C.P. Antonopoulos (eds.), *Cyberphysical Systems for Epilepsy and Related Brain Disorders*, DOI 10.1007/978-3-319-20049-1_9

9.1 Wireless Network Sensors Secure Data Acquisition and Local Storage

9.1.1 Requirements

The ARMOR system is appointed to develop a platform to manage and analyse acquired multimodal and advanced technology data from brain and body activities of epileptic patients emphasising privacy and security issues. It is therefore important to analyse the security risks of the system and implications of compromise of the system or data. A secure system will protect data confidentiality and integrity as well as protect its availability.

Information security enables ARMOR to meet its objectives by implementing multimodal data collection with due consideration of information technology (IT) capacity and constraints. ARMOR members meet this goal by striving to accomplish the following security related objectives.

9.1.1.1 Availability

The ongoing availability of systems addresses the processes, policies, and controls used to ensure authorized users have prompt access to information. This objective protects against intentional or accidental attempts to deny legitimate users access to information or systems.

Certain ARMOR scenarios have several time critical functions. For example in emergency life threatening seizure detection, a situation where every second is important, doctors should have feasible access to the data they need. Also, appropriately appointed caregivers should be alarmed in a timely way.

9.1.1.2 Integrity of Data or Systems

System and data integrity relate to the processes, policies, and controls used to ensure information has not been altered in an unauthorized manner and that systems are free from unauthorized manipulation that will compromise accuracy, completeness, and reliability.

ARMOR features such patient risk assessment and decision support for the professionals, rely on accurate information. Corrupted data may cause unexpected behavior on the system. With fabricated data, a malicious party may try to affect the behavior of the ARMOR system. Data can be corrupted during transmission or while stored. Attempts may be made to enter fabricated data to the system through normal ARMOR input devices e.g. touch screen or via open communication channel.

9.1.1.3 Confidentiality of Data or Systems

Confidentiality covers the processes, policies, and controls employed to protect information of external parties and ARMOR against unauthorized access or use. The ARMOR applications will access and use information about the patient that is sensitive e.g. health status. Actual or perceived risk of such information being available to unauthorized personnel will affect negatively the acceptability of the ARMOR solution.

Protection of user privacy is thus important. Confidentiality of the ARMOR data can be compromised at data storage, during data transmission or gaining access to one of the devices through which ARMOR provides data output e.g. health professional's computer.

9.1.1.4 Accountability

Clear accountability involves the processes, policies, and controls necessary to trace actions to their source. Accountability directly supports non-repudiation, deterrence, intrusion prevention, security monitoring, recovery, and legal admissibility of records.

9.1.1.5 Assurance

Assurance addresses the processes, policies, and controls used to develop confidence that technical and operational security measures work as intended. Assurance levels are part of the system design include availability, integrity, confidentiality, and accountability. Assurance highlights the notion that secure systems provide the intended functionality while preventing undesired actions.

9.1.2 Challenges

This section includes the mobile sensors that will be used for the monitoring of the vital parameters of the patients. All the interfaces to the energy supply or information exchange can be thought as possible attack points. The sensors will be equipped with a wireless communication port, a port for the configuration/programming of the sensor and a power supply.

The configuration port is used in order to load the software to the sensor and to upload the new software. In such systems there is sometimes an extra interface, which can only be used in order to update the program code. This interface allows no access to the memory and can only be used to transmit program data to the system (Bootloader).

The wireless communication port is used for the direct transmission of the data. The possible attacks in the wireless communication port will be discussed in the next section.

Below we discuss common attacks against databases and servers:

9.1.2.1 Excessive Privileges

When users (or applications) are granted database privileges that exceed their job function, these privileges may be used to gain access to confidential information. For example, a user whose job requires read-only access to records may take advantage of excessive update privileges to modify data.

9.1.2.2 Privilege Abuse

Users may abuse legitimate data access privileges for unauthorized purposes. A user with privileges to view individual patient records via a custom healthcare application client may abuse that privilege to retrieve all patient records via a MS-Excel client.

9.1.2.3 Unauthorized Privilege Elevation

Attackers may take advantage of vulnerabilities in database management software to convert low-level access privileges to high-level access privileges, e.g. an attacker might take advantage of a database buffer overflow vulnerability to gain administrative privileges.

9.1.2.4 Platform Vulnerabilities

Vulnerabilities in underlying operating systems may lead to unauthorized data access and corruption.

9.1.2.5 SQL Injection

SQL injection attacks involve a user who takes advantage of vulnerabilities in front-end web applications and stored procedures to send unauthorized database queries, often with elevated privileges. Using SQL injection, attackers could even gain unrestricted access to an entire database.

9.1.2.6 Weak Audit

Weak audit policy and technology represent risks in terms of compliance, deterrence, detection, forensics and recovery. Unfortunately, native database management system (DBMS) audit capabilities result in unacceptable performance degradation and are vulnerable to privilege-related attacks—i.e. developers or database administrators (DBAs) can turn off auditing.

9.1.2.7 Denial of Service

Denial of service (DoS) may be invoked through many techniques. Common DoS techniques include buffer overflows, data corruption, network flooding and resource consumption. The latter is unique to the database environment and frequently overlooked.

9.1.2.8 Database Protocol Vulnerabilities

Vulnerabilities in database protocols may allow unauthorized data access, corruption or availability. For example, the SQL Slammer worm took advantage of a Microsoft SQL Server protocol vulnerability to execute attack code on target database servers.

9.1.2.9 Weak Authentication

Weak authentication schemes allow attackers to assume the identity of legitimate database users. Specific attack strategies include brute force attacks, social engineering, and so on.

9.1.3 State of the Art Security Countermeasures

9.1.3.1 Attacks on the ARMOR Sensor

The ARMOR sensors used to collect the biological parameters are systems of low complexity and so the attacks points on the sensor are limited. As shown in Fig. 9.1 the ARMOR sensors dispose of a Bluetooth interface, a debugger interface, a power supply and a local storage on a SD card. The debugger is used to read or flash the sensor and debug the program.

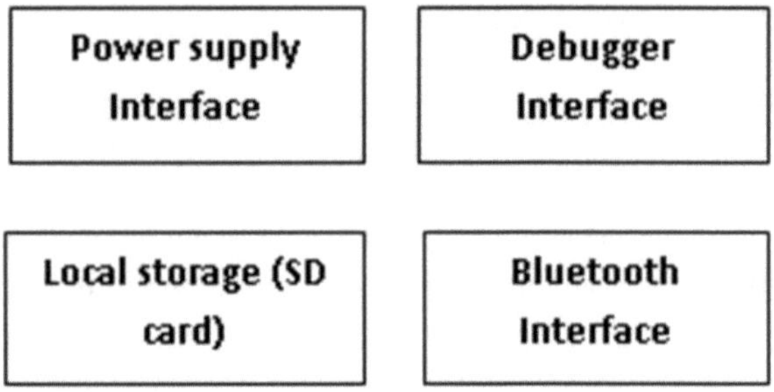

Fig. 9.1 Sensor platform

Security of the Bluetooth Interface

The Bluetooth standard technology is designed to authenticate the connection with a partner where a PIN code is needed. The encryption of the connection is also needed and will be described in the next section.

Deactivation of the Debugger Interface

To avoid the read of the data or the program code from the microcontroller, the debugger interface is destroyed. The Sensor platform disposes of a special interface to update the flash with a program code: the so called "bootloader". This interface gives no access to the storage and can only be used to transfer a program. However using a software solution for the encryption of the data on SD card will store the secret key in the microcontroller. This means that over the update interface (bootloader) a possible attack could be done to get the secret key. Therefore in the case of using an encryption of the local storage it is advisable that this interface is destroyed as well.

Attacks over the Power Supply Interface

Attacks over the power supply are normally made in laboratory to get a secret key that is not changing and need to use measurement instruments direct on the sensor. In the case of a CPS using data encryption, the secret key is changing periodically with respect to new measurements. The secret key is needed while saving the data on the local storage and so this kind of attacks will be noticed by the user of the sensor.

9.1.3.2 Design Alternatives for Mobile Cryptography

Assuming acquired data from the CPS sensors are stored on a SD card, encrypted should be used to avoid possible abuse, if the sensor has been loss and the data fall in the "wrong hand" or if the sensor was intentionally stolen from the patient. In the following possible design alternatives for mobile cryptography are described.

Application Specific Integrated Circuit (ASIC)

An application Specific Integrated Circuit (ASIC) is an integrated circuit customized for a particular-purpose use. An ASIC can include all the needed end-to-end security such as the cryptography algorithms, the key management or all the methods to avoid a hardware manipulation [1].

Hardware Based Crypto Accelerator

Hardware based crypto accelerator are special chip designed to handle cryptographic algorithm with a high transfer rate. Respective chips can be plugged into the central processor of a computer and assume the encryption and decryption of the data. In this area the most important things for the developer is to get a higher data rate as well as to minimize the power consumption [2].

Software Based Solution

All cryptographic algorithm and key handling can be implemented on a mobile system in a software form. Because of the high processing power, complex asymmetric methods are not possible to implement on limited embedded microcontrollers found in WSN platforms like the TI MSP 430 used in various respective implementations. Symmetric and asymmetric methods using a few complex mathematic operations are appropriate for implementation on an embedded microcontroller [3, 4].

For the encryption of the CPS sensor data, software solution based on a symmetric AES algorithm are presented as a commercial solution by key WSN processing chip manufactures like Texas Instruments offering the following features.

9.1.3.3 Texas Instruments AES Crypto Software for the MSP 430

Texas Instruments offers a performance optimized software implementation of the Advanced Encryption Standard (AES). This implementation is designed for the 16 bit RISC architecture of the Texas Instruments MSP 430 controller family and it is provide as a C interface. In the following the description of the TI AES crypto feature [5]:

Functionality

- AES-128 and AES-256 encryption and decryption in ECB mode
- On-the fly roundkey generation
- Automatic decryption key calculation

Performance:

- AES-128 encryption in 5432 cycles decryption in 8802 cycles
- AES-256 encryption in 7552 cycles decryption in 12,258 cycles

Codesizes:

- Codesize for AES-128 is 2536 bytes
- Codesize for AES-256 is 2830 bytes
- Codesize for AES-128 & AES-256 is 3992 bytes

9.1.3.4 Attacks on Computer and Infrastructure

Because the sensors will be configured on a computer and the data from the sensor could be extracted from the local storage on a computer: the computer itself and the infrastructure connected to it are also part of the security even if the data are encrypted. The encryption supposes the existence of a secure environment to be efficient: for example to avoid the interception of the secret key.

- Computer: The main threat in this context is the possible existence of malware and virus on the computer that is used to configure or extract the data. This can lead behind the interception of the encryption secret key to a possible lose of the data.
- LAN and Internet connection: Through the network connection on a Workgroup infrastructure or over the internet an interception or manipulation of the dataflow on the same network is possible.

To avoid those attacks the sensor configuration must be done on a high secure infrastructure (e.g. Hospital) and not at home on a private computer. Concerning the data extraction the data decryption occurs also only at the hospital on a high secure environment or on the server. This depends on where the secret key will be generated.

9.1.3.5 Evaluation and Conclusion

In general the security of the data is realized by the anonymisation of the data. That means only an Id of the person is saved on the local storage of the sensor. In the unlikely case that a physical attack takes place data is secured by the following mechanisms:

- Data is saved without file system
- Data is saved on a binary format
- No information about the data format (Sampling rate, LSB value …)
- Knowledge on the data compression algorithm is need to extract the data

Research on similar commercial products has shown that no encryption mechanism on the local storage is used [6, 7]. Based on the above risk analysis no encryption on the local storage is required.

9.2 Wireless Sensor Network Secure and Efficient Communication

9.2.1 Requirements

In this section critical concepts regarding the implementation of security control will be presented from a system wide perspective. In later sections detailed analysis will be provided following a more technical aspect regarding networking, WSN and Bluetooth based network aspects.

9.2.1.1 Access Control

The goal of access control is to allow access by authorized individuals and devices and to disallow access to all others. Access should be authorized and provided only to individuals whose identity is established, and their activities should be limited to the minimum required for respective purposes. An effective control mechanism includes numerous controls to safeguard and limits access to key information system assets at all layers in the network stack.

9.2.1.2 Access Rights Administration

System devices, programs, and data are system resources. Each system resource may need to be accessed by individuals (users) in order for work to be performed. Access beyond the minimum required for work to be performed exposes the systems and information to a loss of confidentiality, integrity, and availability. Accordingly, the goal of access rights administration is to identify and restrict access to any particular system resource to the minimum required for work to be performed.

9.2.1.3 Authentication

Authentication is the verification of identity by a system based on the presentation of unique credentials to that system. The unique credentials are in the form of something the user knows, something the user has, or something the user is. Those forms exist as shared secrets, tokens, or biometrics. More than one form can be used in any authentication process. Authentication that relies on more than one form is called multi-factor authentication and is generally stronger than any single-factor authentication method. Authentication contributes to the confidentiality of data and the accountability of actions performed on the system by verifying the unique identity of the system user.

Authentication over the WSN based CPS delivery channel presents unique challenges. That channel does not benefit from physical security and controlled computing and communications devices like internal local area networks (LANs), and is used by people whose actions cannot be controlled. It should be considered the use of single-factor authentication in that environment, as the only control mechanism, to be inadequate for high-risk transactions involving access to patient information or the movement of healthcare information to other parties. Authentication does not provide assurance that the initial identification of a system user is correct.

9.2.1.4 Encryption

Encryption is used to secure communications and data storage particularly, authentication credentials and the transmission of sensitive information. It can be used throughout technological environment, including the operating systems, middleware, applications, file systems, and communications protocols. Encryption can be used as a preventive control, a detective control, or both. As a prevention control, encryption acts to protect data from disclosure to unauthorized parties. As a detective control, encryption is used to allow discovery of unauthorized changes to data and to assign responsibility for data among authorized parties. When prevention and detection are joined, encryption is a key control in ensuring confidentiality, data integrity, and accountability.

Properly used, encryption can significantly strengthen the security of a CPS system. Encryption also has the potential, however, to weaken other security aspects. For instance, encrypted data drastically lessens the effectiveness of any security mechanism that relies on inspections of the data, such as anti-virus scanning and intrusion detection systems. When encrypted communications are used, networks may have to be reconfigured to allow for adequate detection of malicious code and system intrusions.

A sensor network is an area with unique requirements compared to a typical network. Therefore, the requirements of the WSN encompass both the typical network requirements and the unique requirements suited solely to wireless sensor networks. To ensure the security of WSNs, the following major security objectives are of paramount importance.

9.2.1.5 Privacy

Data privacy is the most important issue in network security. Every network with any security focus will typically address this problem first. Privacy is a service that is used to prevent the disclosure of information to unauthorized parties, means that certain information is only accessible to those who have been authorized to access it, and keep them secret from all other entities that do not have required privileges.

Privacy is achieved using cryptography through the use of symmetric or asymmetric cipher algorithms able render the information unintelligible except by authorized entities. The information may become intelligible again by using decryption. From a user point of view this requires the authenticated communicating parties to have the required keys.

9.2.1.6 Authentication-Integrity

Authentication is a service that is used to establish the origin of information. Essentially ensures that participants in communication are genuine and not impersonators. It is necessary for the communication participants to prove their identities as what they have claimed using some techniques so as to ensure the authenticity. If there is no such authentication mechanism, the adversary could impersonate a benign node and thus get access to confidential resources, or even propagate fake messages to disturb the normal network operations. Packet integrity pertains to the assurance that data are not modified or in any way tampered with between transmitter and receiver entities. Most commonly, authentication is provided by digital signatures or message authentication codes.

9.2.1.7 Authorization

Authorization is a process in which an entity is issued a credential, which specifies the privileges and permissions it has and cannot be falsified, by the certificate authority. Normally, authorization is granted following a process of authentication. A non-cryptographic analog of the interaction between authentication and authorization is the examination of an individual's credentials to establish their identity (authentication); upon proving identity, the individual is then provided with the key or password that will allow access to specific resources, such as a locked room (authorization). Authentication can be used to authorize a role rather than to identify an individual. Once authenticated to a role, an entity is authorized for all the privileges associated with the role.

9.2.1.8 Availability

Availability ensures that a node should maintain its ability to provide all the designed services regardless of the security state of it. This security criterion is challenged mainly during the denial-of-service attacks, in which all the nodes in the network can be the attack target and thus some selfish nodes make some of the network services unavailable.

9.2.2 Challenges

In this section, major types of attacks in WSN are depicted. Wireless Sensors Network has commonalities with a usual computer networks. Because sensors have limited capabilities like computational power, memory size, etc, they are exposed to attacks and make it easy to collapse when basic security is not provided. In this section, we will do a small reference to most popular attacks like DoS, Sybil Attack, traffic Analysis Attack, Node Replication attack and Privacy attacks that could easily affect WSN networks in the contest of CPS systems used in medical application scenarios.

9.2.2.1 Denial of Service

A DoS attack or Denial of Service [8] is the most well known attack in communication networks and services. Therefore, this type of attack can harm or destroy completely a WSN. As DoS is characterized any event that diminish or eliminate a network's capacity to perform its expected function. Some of these events are hardware failures, software bugs, resource exhaustion, environmental conditions, or any complicated interaction between these factors. Although attackers commonly use the Internet to exploit software bugs when making DoS attacks, here is considered primarily protocol- or design-level vulnerabilities.

Determining if a fault or collection of faults is the result of an intentional DoS attack presents a concern of its own—one that becomes even more difficult in large-scale deployments, which may have a higher nominal failure rate of individual nodes. An intrusion-detection system monitors a host or network for suspicious activity patterns such as those that match some pre-programmed or possibly learned rules about what constitutes normal or abnormal behavior.

9.2.2.2 Sybil Attack

Sybil Attack [9–11] is a particularly dangerous attack against sensor networks and CPS platforms. A malicious node, which is called the Sybil node, illegitimately claims multiple false identities by either fabricating new identities or impersonating existing ones. Sybil attack's goal is to gain a disproportionate amount of influence over the network via its false identities. In the worst case, an attacker may generate an arbitrary number of additional node identities, using only one physical device. The result is especially harmful, because often these attacks are the home gateway to other attacks (such as those on resource exhaustion, voting, etc.).

9.2.2.3 Traffic Analysis Attack

A sensor network has a base station and a number of nodes. Each node processes data that it received from its group of neighboring sensor nodes and sends that processed data to the base station through multiple hops. The most critical part of a sensor network is the base station as all the relevant data collected by the sensor nodes are directed towards the base station where the data is aggregated and processed. So if an adversary can detect the base station and compromise it, the entire WSN will be rendered useless [12].

Adversary is guided by anyone of the following motives (for WSN) [13]:

- **Data Benefit**: Gain access to the sensitive data being transmitted or monitored.
- **Mission interference**: Intent to damage the WSN rather than gain access to the relayed data.

If these are the motives, it is obvious that detecting and compromising the base station will be of most benefit to it. Since the centre of data aggregation is the base station, to compromise the base station it will be able to access the majority of data flowing through the network. Also by rendering the base station non-functional, the entire WSN will collapse.

9.2.2.4 Node Replication Attack

An application-independent attack unique to wireless sensor networks is the node replication attack [14–16]. An adversary prepares its own low-cost sensor nodes and induces them to the network to accept them as legitimate ones. For a successful attack, the adversary only needs to capture one node physically, reveal its secret credentials, replicate the node in large quantity, and deploy these malicious nodes back into the network so as to subvert the network with little effort.

9.2.2.5 Privacy Attacks

Wireless sensor networks and even more in CPS used in medical applications can be used to determine the activities of daily living and provide data for longitudinal studies and this poses opportunities to violate privacy [17, 18]. The importance of securing such systems will continue to rise as their adoption rate increases. In addition to policy and database query privacy violations, WSNs are susceptible to new side channel privacy attacks that gain information by observing the radio transmissions of sensors to deduce private activities, even when the transmissions are encrypted. This physical layer attack needs only the time of transmission and the fingerprint of each message, where a fingerprint is a set of features of an RF waveform that are unique to a particular transmitter. Thus, this is called the fingerprint and timing based snooping (FATS) attack. To execute a FATS attack, an adversary eavesdrops on the sensors' radio to collect the timestamps and fingerprints of all radio transmissions. The adversary then uses the fingerprints to associate each message with a unique transmitter, and uses multiple phases of inference to deduce the location and type of each sensor. Once this is known, various private user activities and health conditions can be inferred.

Furthermore, Wireless Sensor Networks are notorious for strict constraints and scarce resource availability compared to a traditional computer network comprising an additional set of challenges. Due to these constraints it is difficult to directly employ the existing security approaches to the area of wireless sensor networks. Therefore, prior to developing useful security mechanisms while borrowing the ideas from the current security techniques, it is necessary to know and understand these constraints.

9.2.2.6 Limited Resources

All security approaches require a certain amount of resources for the implementation and operation, including data memory, code space and energy to power the sensor. However, currently these resources are very limited in a typical wireless sensor node.

Limited Memory and Storage Space

A sensor is a small device with only a small amount of memory and storage space for the code. In order to build an effective security mechanism, it is necessary to limit the code size of the security algorithm as well as the run-time requirements.

Power Limitation

Energy is the biggest constraint to wireless sensor capabilities. We assume that once sensor nodes are deployed in a sensor network, they cannot be easily replaced or recharged. Therefore, the battery provided energy must be conserved to extend the life of the individual sensor node and the entire sensor network. When implementing a cryptographic function or protocol within a sensor node, the energy impact of the added security code must be considered. When adding security to a sensor node, we are interested in the impact that security has on the lifespan of a sensor. The extra power consumed by sensor nodes due to security is related to the processing required for security functions, the energy required to transmit the security related data or overhead, and the energy required to store security parameters in a secure manner.

Limited Bandwidth

An important limitation to wireless sensor nodes is the low radio bandwidth. Commonly utilized IEEE 802.15.4 based radio (e.g. such chipcon CC2420) can offer raw data rates of roughly 250 kbps. However, overheads caused by packet framing, medium access control (MAC) and multihop routing reduce the achievable data rate significantly, even in a single-hop network.

Limited Processing Power

Another aspect characterized by scarce resource availability is the processing power offered by typical Micro-Controller Units (MCU) found in respective platforms. Indeed they comprise processing units synchronized in the area of 10 MHz, based on 16 Bit RISC architecture offering very limited features such as Flash and RAM memory, hardware accelerators, timers etc. This is emphasized by the effort devoted by various groups to optimize security related algorithms' implementations in order to be efficiently executed by respective MCUs.

9.2.2.7 Unreliable Communication Medium

Certainly, unreliable communication is another threat to sensor security. The security of the network relies heavily on a defined protocol, which in turn depends on communication.

Unreliable Data Transfer

Normally the packet-based routing of the sensor network is connectionless and thus inherently unreliable. Packets may get damaged due to channel errors or dropped at highly congested nodes. The result is lost or missing packets. Higher channel error rate also forces the software developer to devote resources to error handling. More importantly, if the protocol lacks the appropriate error handling it is possible to lose critical security packets. This may include, for example, a cryptographic key.

Conflicts

Even if the channel is reliable, the communication may still be unreliable. This is due to the broadcast nature of the wireless sensor network. If packets are been concurrently transmitted, conflicts may occur at the receiver and the transfer itself will fail. In a crowded sensor network, this can be a major problem.

Latency

The multi-hop routing, network congestion and node processing are some of the reasons leading to higher latency in the network, thus making it difficult to achieve synchronization among sensor nodes. The synchronization issues can be critical to sensor security where the security mechanism relies on critical event reports and cryptographic key distribution.

9.2.2.8 Unattended Operation

Depending on the function of the particular sensor network, the sensor nodes may be left unattended for long periods of time. There are three main caveats to unattended sensor nodes:

Exposure to Physical Attacks

The sensor may be deployed in an environment open to adversaries, bad weather, and so on. The likelihood that a sensor suffers a physical attack in such an environment is therefore much higher than the typical PCs, which is located in a secure place and mainly faces attacks from a network.

Remote Management

Remote management of a sensor network makes it virtually impossible to detect physical tampering and physical maintenance issues.

No Central Management Point

A sensor network should be a distributed network without a central management point. This will increase the vitality of the sensor network. However, if designed incorrectly, it will make the network organization difficult, inefficient, and fragile. Perhaps most importantly, the longer a sensor is left unattended the more likely it is to be compromised by an adversary.

9.2.2.9 Key Management Approaches

Key management is a fundamental security issue in wireless sensor networks. It is the basis to establishing secure communication using cryptographic technologies between sensor nodes in a monitored area. Key management is the process by which cryptographic keys are generated, stored, protected, transferred, loaded, used, and destroyed.

Aiming towards a robust wireless sensor network, where there is no possibility of adding or removing nodes to or from the network, where the sensor nodes of the network will be in inside a predefined area, where anyone would not have access to it, it is adequate to use simple key management approaches which at the same time assure secure and reliable encryption key agreement between authorized communicating parties [19–21].

9.2.3 State of the Art Measures

Nowadays several WSN platforms are available targeting highly sensitive and demanding medical applications such as CPS systems for brain related disorders. In the context of the ARMOR projects a comprehensive evaluation had been undertaken concerning respective prominent technologies and focusing on networking performance but even more security provisions capabilities. In that respect in this section the main outcome of this research is presented concerning two prominent relative technologies i.e. IEEE 802.15.4 and Bluetooth communication protocols emphasizing. Furthermore, a proposed hardware based encryption module designed and implemented in the context of ARMOR is presented offering advanced performance capabilities and low resource requirements while offering high level security.

9.2.3.1 IEEE 802.15.4 Based Solutions

Respective solutions comprise prominent candidates as they are utilized in several experimental and commercial scenarios. Their popularity, gained throughout the years, is based on significant advantages when aiming towards very low power consumption, low complexity, low price and low traffic demands characteristics. Popular IEEE 802.15.4 based platforms include TelosB and MicaZ [22, 23]. The former comprises probably the most well know WSN platform upon which many projects have been based including medical oriented ones [24, 23], offering can deliver 250 Kbps data rate at 2.4 GHz frequency band. Processing is based on the (also widely utilized) 8 MHz Texas Instrument MSP430 16 bit microprocessor. Concerning memory capabilities the developer is provided with 48 KB program flash, 10 KB data ram and a 1 MB external flash. Shimmer is also a very interesting case since it offers a very versatile environment a wide range of medical sensors thus utilized extensively in this evaluation effort. MicaZ also comprises a prominent platform used in various scenarios and offering analogous characteristics as the TelosB. The main difference concerns the processing module, which is based on Atmel ATmega128L for both the radio and processing tasks offering 128 KB program and 512 KB data memory.

Applying and utilizing robust and valid encryption algorithms comprises a necessity when sensitive data are conveyed through Wireless Sensor Network. However, adopting such algorithms by a resource limited node as typically found in

WSN poses specific and important requirements. Therefore, a critical aspect is to identify and reveal the main areas where these requirements apply as well as provide quantitative insight concerning these requirements. Furthermore, this section considers the AES and RC5 algorithms as prominent candidates well known and studied by research community and even more importantly already successfully ported to WSN platforms.

Encryption Algorithms' Requirements

Identifying the imposed requirements to WSN platforms is a multifaceted challenge. The first approach focuses on requirements that are independent to specific platform implementations. In that respect one of the most limited resources is the available memory which can be distinguished between the memory allocated for the source code (ROM) and the memory allocated from the run-time requirements (RAM). In that respect significant efforts have been made to efficiently port respective algorithms to these platforms, aiming towards highly optimized implementations requiring very little memory as depicted in the following table presenting relative information from [25]. Differences observed between platforms are attributed on different MCU architecture and compiler tools (Table 9.1).

Furthermore, a critical resource to be considered is the processing power of the WSN node. It is noted, and applied to memory as well, that a WSN is expected to perform multiple tasks in parallel i.e. sampling data, processing data, assisting the communication radio etc. Concerning this aspect, relative information is presented for AES and RC5. Additionally, regarding the block/key/round both algorithms follow WSN suitable configuration which offer adequate security level while keeping the computation overhead to acceptable levels. Specifically AES follows the 128/128/10 while the RC5 the 64/128/18 configuration. In [26] these two algorithms are evaluated on the required CPU cycles to encrypt and decrypt a common 30 Byte plaintext which is a realistic packet size for WSN applications. Since the plaintext is bigger than the blocksize of each algorithm the operational mode is of considerable significance. Although, various modes are considered in [26] here results regarding only Cipher-Block-Chaining (CBC) which by supporting cipher stealing feature avoids padding. Even more, it is noted that CPU cycles measurement differences for different operational modes as shown in [26] are either very small or even negligible. Therefore, taking into consideration previous assumptions, respective requirements are shown in the following Table 9.2.

Table 9.1 Security algorithms memory requirements

	MicaZ		TelosB	
Encryption Alg.	RAM (kB)	ROM (kB)	RAM (kB)	ROM (kB)
AES	2	10	1.8	9
RC5	0.2	2.5	0.2	6

Table 9.2 MCU cycles requirements for 30 Byte plaintext

		Encryption	Decryption
AES	Key setup	1313	5034
	CBC	218	223
RC5	Key setup	40,565	40,565
	CBC	712	699

Table 9.3 CC2420 HW AES energy and delay performance

	Encryption delay (ms)	Encryption energy (μJ)
MicaZ	0.023	1.83
TelosB	0.225	14.3

Various observations can be made on this table. Without a doubt the most resource consuming operation is the key setup process. This is quite apparent on RC5 case comparing the Key Setup cycles to CBC required cycles. This is reflected on energy measurements and the delay measurements presented in [27]. Furthermore, it is indicated that CBC encryption/decryption differences are rather limited while AES requires significantly less CMU cycles. However, these operations are closely related to implementation characteristics of the MCU and to memory access so actual measurements may vary in specific cases significantly.

Another aspect that poses specific and potentially significant requirements is the overhead if message authentication code algorithms are deployed. These algorithms are applied on whole message and add an HMAC code overhead. This code in more resource rich environments is about 8 or 16 Bytes which is unacceptable for WSN cases where 4 Bytes HMAC code can be found. Still, however, both the byte overhead as well as the delay/energy imposed by the respective algorithms represents a substantial burden to an already resource limited environment. So it is argued that if possible such overhead should be avoided. Such an approach could refer to specific key management approaches leading to pair wise key distribution which could provide authentication for the payload of a packet.

However, it must be noted that, AES being one the most popular and widely utilized algorithms, is found in hardware implementation as well in one of the most prominent radio chips (i.e. CC2420 used in both MicaZ and TelosB). Therefore, it would be incomplete not to reference respective performance, which is also provided in [25] and shown in Table 9.3. In this table Encryption include both key setup and actual encryption operations.

Comparing software to hardware implementation of the same code, hardware implementation, as anticipated, clearly outperforms any software implementation in all aspects. Driven by such performance indications respective in-depth simulation based evaluation has been carried out concerning the effect of such security provision upon the communication performance of TelosB and MicaZ focusing on throughput, delay and power consumption metrics. As reported in [27, 28] the overhead imposed by software implementation is quite excessive leading to emphatic performance degradation as opposed to hardware based solutions where respective effect is drastically less significantly.

Finally, driven by respective conclusions in the context of ARMOR project a highly efficient and power conservative hardware implementation is presented aiming to minimize any performance overhead imposed.

9.2.3.2 Bluetooth Based Solutions

Bluetooth is a wireless radio specification designed to replace cables as the medium for data and voice signals between electronic devices. The specification is defined by the Bluetooth Special Interest Group (SIG) which is made up of over 1000 electronics manufacturers. Intended primarily for mobile devices, Bluetooth's design places a high priority on small size, low power consumption and low costs. The Bluetooth specification seeks to simplify communication between electronic devices by automating the connection process. Operating at 2.4 GHz frequency band Bluetooth based solutions throughput capabilities can vary considerably depending both on the version of the protocol supported and even more on the specific implementation's characteristics. Therefore, concerning data rates solutions covering a wide range from 300 Kbps up to 1.5 Mbps can be found. Indicative examples of relative solutions include Shimmer [23] and MoviSens [29] platforms. The former utilized the Roving Networks based Bluetooth modules [30].

Bluetooth Security Characteristics and Weaknesses

Bluetooth security is based on three critical services: *authentication*, *authorization*, and *encryption*. The authentication service is tasked with ensuring that a device seeking a connection is indeed who it claims to be. Authorization is the process that determines whether or not a requesting device is allowed access to specific information or services. Encryption helps to ensure confidentiality by protecting private data from being viewed by unintended recipients.

Bluetooth devices can be set in one of three different security modes. In security mode 1, no security measures are utilized. Any other Bluetooth device can access the data and services of a device in security mode 1. Security mode 2 enacts security measures based on authorization. In this mode, different trust levels can be defined for each of the services offered by the device. Security mode 3 requires both authentication and encryption. The security features of the current Bluetooth specification provide for secure communication at the link level. However, there are some weaknesses that need to be addressed. These weaknesses arise from the specification's heavy reliance on device authentication for security services as well as the level of control that the user has over Bluetooth devices and their configuration. The current Bluetooth specification does not provide any means of user authentication. The lack of any means of user authentication coupled with the reliance on device authentication leaves Bluetooth particularly vulnerable to spoofing attacks and the misuse of authenticated devices.

The pairing process is a particularly vulnerable point for Bluetooth. The Bluetooth Security Expert Group writes, "The user should be aware that when using the pairing procedure from the Bluetooth Baseband specification, the initial exchange of keys using non-encrypted channels is the weakest part of the security procedure." If an attacker could intercept the transmissions of the pairing process, he could then derive the initialization key by calculating initialization keys for every possible passkey and comparing the results to the intercepted transmissions. The initialization key could then be used to compute the link key. The link key calculated by the attacker could then be compared to the intercepted transmission to see if it is correct. It should be noted that one of the reasons for this vulnerability is the fact that Bluetooth allows for the use of short passkeys. Short passkeys significantly reduce the complexity of initialization and link keys thus making it easier for an attacker to derive these keys from intercepted pairing transmissions. The use of unit keys is another cause for concern. A device that uses a unit key is a security risk because it uses the same link key for each device that it establishes a secure connection with. As a result, any one of the devices that has the unit key can use it to eavesdrop on secured connections that use that link key. Bluetooth security does not take into account the identity or the intent of the user. As far as Bluetooth's link level security is concerned, a device with the proper unit key is secure.

In June of 2003, Bluetooth security was put under the microscope with the release of Redfang. Developed by Ollie Whitehouse of computer security firm @ Stake, Redfang is the first hacking tool to target Bluetooth devices. One of the security measures of the Bluetooth specification is a stealth mode where a device will ignore any inquiry broadcasts it receives. By ignoring these broadcasts, the device keeps its Bluetooth address a secret and communication links cannot be established. Redfang gets around this security measure by using a brute force attack to determine device ID's of any Bluetooth enabled devices within range of the attacker. Assuming Redfang could determine a target device's Bluetooth address, it could then transmit a page message and attempt to establish a connection. While the practical use of Redfang has been called into question, it has succeeded in proving a theoretical weakness in Bluetooth security and focusing the attention of the security community on Bluetooth.

Network Performance Overhead Due to Bluetooth Security

In order to evaluate the effect of enabling security features at Bluetooth (BT) communication the Shimmer [Shimmer] platform was utilized offering a highly configurable environment and a standardized BT communication over SPP (Serial Port Profile) service offering trouble-free compatibility with various BT receivers test (both PC embedded and USB dongle). Evaluation experiments critical parameters considered include the following:

- Packet creation interval: 2–50 ms
- Concurrent transmitters: 1, 2, 4, 6 (7 being the maximum by BT requirements)
- Packet payload: 8, 15, 20 Byte (Considering that in one packet multiple samples can be inserted for optimal use of the transmission medium)

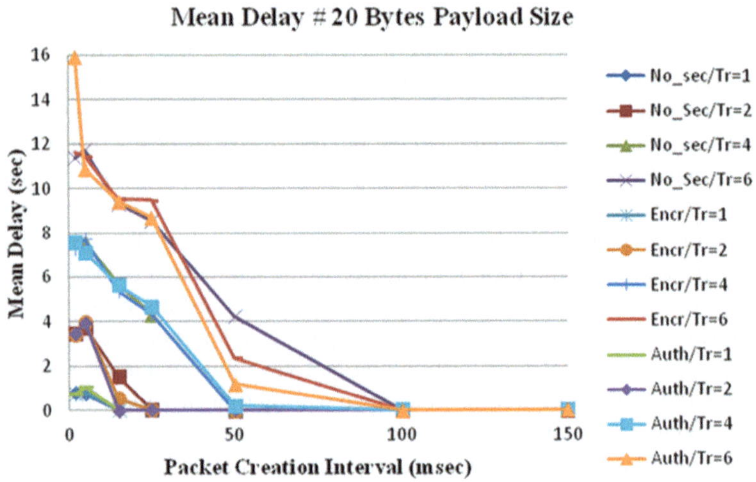

Fig. 9.2 Delay capabilities differentiation wrt to different security configurations

The first and of outmost importance observation extracted for respective measurements with respect to the document's objective concerns that in all scenarios there was a 100 % success in packet transmission. It is shown that connection-oriented communication and FHSS physical layer transmission techniques offer significant advantages as far as communication robustness is concerned leading to zero packet loss either with or without security provisions as well as while varying the packet payload. Such an observation emphatically advocates the use of such technology is demanding application scenarios where highly sensitive data are handled.

Considering Mean Delay performance Fig. 9.2 indicates the effect of different security level provisioning as well as network parameter on respective metric considering 20 Byte packet payload.

As clearly depicted security provisions does not affect the overall network performance while the parameter which affects the delay performance of a Bluetooth network is the number of competing nodes rather than anything else, at least with respect to parameters and ranges considered in the specific evaluation effort.

Last but not least experiments were conducted to estimate the effect of security provision to the node lifetime. It is very important to clarify that the absolute measurements presented here are platform dependant, contrary to the delay/throughput measurements which (since compiling with specific standard) are expected to be the same or analogous in any platform utilized. Therefore, the main purpose of these experiments was to see whether secure oriented connection had any significant effect on the system lifetime in two transmission rates, a high one (250 Hz adequate for ECG and hopefully for EEG accompanied by efficient preprocessing algorithm) and a low one (10 Hz adequate for low demanding sampling sequences such as accelerometer, gyroscope etc.). On the other hand some notion of lifetime capabilities is also provided.

As depicted when high sampling and transmission rate was considered a lifetime of ~15.5 h was measured while decreasing the transmission rate added an additional 3 h lifetime, reaching up to ~18.5–19 h possible lifetime. But in both cases having a security option enabled did not affect lifetime in a considerable way. That indicates that the CPU is always active either executing a security algorithm or not. Therefore, assuming CPU consumption is present all the time executing a security algorithm doesn't decrease expected lifetime. However, if low power states of CPU were exploited differences would be quite more apparent. From another perspective reducing the transmission rate 25 times, added only 3 more hours of life that being a 20 % lifetime increase.

9.2.3.3 ARMOR Proposed HW Solution

AES algorithm has been standardized by the National Institute of Standards and Technology (NIST) as a highly secure block ciphering method. It has replaced the old DES algorithm, whose key sizes were becoming too small. The developed encryption module (see Fig. 9.3) implements AES algorithm in FPGA technology and is highly optimized for ultra-low power dissipation rendering it an ideal solution for WSN network applications.

The encryption process transforms a 128-bit plaintext block into an unintelligible block of the same size, which is also referred to as the ciphertext. It supports 128-bit, 192-bit or 256-bit encryption keys. The longer the key length the higher the

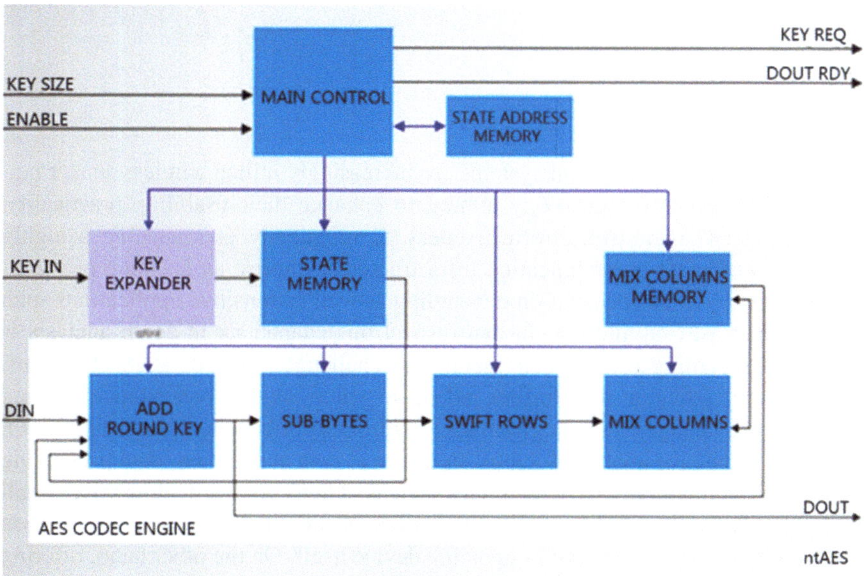

Fig. 9.3 Block diagram of the encryption module

Table 9.4 Hardware encryption module implementation performance

Impl.	Area	Process (μm)	Max freq (in MHz)	Throughput (in Mbps)	
				@ fmax	@ 50 MHz
Our	1.2k	0.18	515	137	13.33
Ref [13]	4k	0.13	300	100	16.67
Ref [14]	3.4k	0.09	152	56	18.4
Ref [15]	3.6k	0.35	N/A	N/A	6.3

security level, at the cost of increased processing load. The keys are fully programmable, thus data security level can be easily reconfigured on-the-fly depending on application throughput rate and power requirements [31].

The encryption module has been carefully designed to require minimum logic resources as well as utilizing power aware design techniques in architectural (8-bit datapath, use of sequential structures, resource reuse, optimized Galois Field Multiplier which is the structural datapath element of the cryptographic engine, pipelining, path balancing, one hot FSM encoding) as well as in FPGA implementation level (clock and data disabling). The use of these techniques has resulted in a significant performance/silicon footprint ratio. A comparison with other third-party implementations [32–34] is demonstrated in [31] as well as in Table 9.4.

As depicted, our design is clearly the most compact since it requires only 1200 gates compared to 3.4k, 3.6k and 4k gates proposed in [32–34] respectively. Our design is the fastest, since it can achieve 515 MHz operating frequency even at slower CMOS process (0.18 μm) compared to the ones presented in [32, 33] respectively.

9.3 Conclusions

State of the art CyberPhysical systems are increasingly utilize wireless sensor network communication technology aiming to enhance their usability, practicality, widespread and their overall effectiveness. Consequently guaranteeing a highly secure and efficient communication infrastructure comprise an absolute prerequisite, especially when focusing on demanding as well as sensitive applications such as brain deceases monitoring. In that respect this chapter an in-depth analysis is attempted regarding main requirements and challenges towards such objectives. Furthermore, going a step further, adequate solutions are presented analyzing respective advantages and disadvantages. Aiming to offer a complete analysis on one hand respective analysis focuses on the local data acquisition, data processing and storage components. Characteristics such as of low complexity, low price, small and operating unattended for extended periods of time make a typical WSN node quite vulnerable to direct attacks upon the device itself. On the other hand, offering ultra low power communication based very simple communication protocols with minimum control overhead constitute the wireless transmission channel as another quite vulnerable point of attach for anyone wanting to access data, control the system or even bring the whole platform down. Respective analysis of countermeasures is presented considering efficiency as opposed to resource requirements comprising probably the most challenging tradeoff. Driven by this observation in the final part of this chapter a highly effective as well as resource conservative hardware encryption module implementation is presented offering high level security being based upon the well know encryption algorithms AES and offering high level of flexibility allowing high level of re-configurability.

References

1. Kumar S, Girimondo M, Weimerskirch A, Paar C, Patel A, Wander AS (2003) Embedded end-to-end wireless security with ECDH key exchange. In: Proceedings of the 46th IEEE international Midwest symposium on circuits and systems, 2003. MWSCAS'03, vol 2, pp 786–789
2. Smyth N, McLoone M, McCanny JV (2005) Reconfigurable cryptographic RISC microprocessor. In: Proceedings of the 2005 IEEE VLSI-TSA international symposium on VLSI design, automation and test (VLSI-TSA-DAT), vol 2005, pp 29–32
3. Schneier B, Kelsey J, Whiting D, Wagner D, Hall C, Ferguson N, Kohno T, Stay M (2000) The Twofish team's final comments on AES Selection
4. Batina L, Mentens N, Sakiyama K, Preneel B, Verbauwhede I (2006) Low-cost elliptic curve cryptography for wireless sensor networks. In: Workshop on security and privacy in ad-hoc and sensor networks, vol 4357. Springer, Berlin, pp 6–17
5. Texas Instruments MSP430 microcontroller crypto software features informations. http://jce.iaik.tugraz.at/sic/Products/Crypto_Software_for_Microcontrollers/Texas_Instruments_MSP430_Microcontrollers
6. Somnomedics. http://somnomedics.eu/home.html
7. Medset telesmart. http://www.medset.com/index.php?id=45
8. Wood AD, Stankovic JA (2002) Denial of service in sensor networks. IEEE Comput 35(10):54–62
9. Hsu K, Leung M-K, Su B (2008) Security analysis on defenses against sybil attacks in wireless sensor networks. IEEE J
10. Newsome J, Shi E, Song D, Perrig A (2004) The sybil attack in sensor networks: analysis & defenses. In: Proceedings of the 3rd international symposium on information processing in sensor networks. ACM, New York, pp 259–268
11. Baig ZA (2008) Distributed denial of service attack detection in wireless sensor networks. Thesis, Monash University
12. Biswas S, Mukherjee S, Mukhopadhyaya K (2007) A countermeasure against traffic-analysis based base station detection in WSN. In: Web proceedings of the international conference on high performance computing, HiPC 2007
13. Sabbah E, Majeed A, Don K et al. (2006) An application driven perspective on wireless sensor network security. In: Proceedings of the 2nd ACM international workshop on quality of service & security for wireless and mobile networks. ACM, New York
14. Zhu WT (2011) Node replication attacks in wireless sensor networks: bypassing the neighbor-based detection scheme. In: Proceedings of the international conference on network computing and information security, pp 156–160. http://ieeexplore.ieee.org/xpl/login.jsp?tp=&arnumber=5948813&url=http%3A%2F%2Fieeexplore.ieee.org%2Fxpls%2Fabs_all.jsp%3Farnumber%3D5948813
15. Manjula V, Chellappan C (2011) Replication attack mitigations for static and mobile WSN. arXiv preprint arXiv:1103.3378

16. Kalita HK, Kar A (2009) Wireless sensor network security analysis. Int J Next-Generation Netw (IJNGN) 1(1):1–10
17. Ko J, Lu C, Srivastava MB, Stankovic JA, Terzis A, Welsh M (2010) Wireless sensor networks for healthcare. Proc IEEE 98(11):1947–1960
18. Srinivasan V, Stankovic J, Whitehouse K (2007) Protecting your daily in-home activity information from a wireless snooping attack. In: Proceedings of the 10th international conference on ubiquitous computing, Seoul, pp 202–211
19. Canniere CD, Rechberger C (2006) Find SHA-1 characteristics: general results and applications. In: Lai X, Chen K (ed) Advances in cryptology—ASIACRYPT 2006. Springer, Heidelberg
20. Bojkovic Z, Bakmaz B, Bakmaz M (2008) Security issues in wireless sensor networks. Int J Commun 2(1):106–115
21. Camtete S, Yener B (2005) Key distribution mechanisms for wireless sensor networks: a survey. Rensselaer Polytechnic Institute
22. http://www.advanticsys.com/shop/mtmcm5000msp-p-14.html
23. http://www.shimmer-research.com/
24. http://fiji.eecs.harvard.edu/CodeBlue
25. Lee J, Kapitanova K, Son SH (2010) The price of security in wireless sensor networks. Comput Netw 54(17):2967–2978
26. Law YW, Doumen J, Hartel P (2006) Survey and benchmark of block ciphers for wireless sensor networks. ACM Trans Sens Netw (TOSN) 2(1):65–93
27. Antonopoulos CP, Petropoulos C, Antonopoulos K, Voros NS (2012) The effect of symmetric block ciphers on WSN performance and behavior. In: Proceedings of the 5th IEEE international workshop on selected topics in wireless and mobile computing (STWiMob 2012), Barcelona, 8–10 Oct 2012
28. Deliverable D3.3 of the ARMOR FP7 research project "end-to-end security infrastructure"
29. http://www.movisens.com/en
30. Revoing (2011) Revoking networks "advanced user manual" version 4.77 2/3/2011
31. Krikis G, Antonopoulos CP, Voros NS (2013) Design and implementation of efficient reconfigurable cipher algorithms for wireless sensor networks. In: Proceedings of the 11th IEEE international conference on industrial informatics, Bohum, 29–31 July 2013
32. Helion technology. OVERVIEW DATASHEET–Ultra-low resource AES (Rijndael) cores for ASIC. Revision 2.0.1
33. Jagannatha KB, Bhanuprakash R (2012) ASIC core design for low power AES. World J Sci Technol 2(5):29–33. ISSN: 2231–2587
34. Feldhofer M, Dominikus S, Wolkerstorfer J (2004) Strong authentication for RFID systems using the AES algorithm. In: Proceedings of the 6th international workshop on cryptographic hardware and embedded systems (CHES 2004), Boston, 11–13 Aug 2004, pp 357–370

Chapter 10
System Middleware

André Bideaux, Stefan Hey, Panagiota Anastasolpoulou, Alberto Fernandez, Christos P. Antonopoulos, and Nikolaos S. Voros

Abstract This document defines the middleware of a cyberphysical monitoring system for epilepsy and related brain disorders. Taking into account requirements, this document provides insights about the service functionalities of the middleware and the interfaces connecting the different technologies in place within the system middleware.

The document is divided into two main parts, the first part describes the functional, non-functional and security aspects of the middleware, while the second part is all necessary information about the middleware architecture.

The middleware runs within the Home Gateway, it is the ICT part responsible to connect sensor data to upper software layers (like personal health record service and the tele-alarm and messaging manager). Furthermore it provides the necessary infrastructure (thanks to the inclusion of a data stream management system) for the development of on-line multi-parametric data processing. Upper software layers connected are the Tele-alarm and Messaging Manager and the Personal Health Record. All middleware components follow a SOA (service oriented architecture) paradigm that allows easy scalability of the system.

A. Bideaux (✉) • S. Hey • P. Anastasolpoulou
KIT Karlsruhe Institute of Technology, Institute for Information Technology, Karlsruhe, Germany
e-mail: andre.bideaux@kit.edu

A. Fernandez
Sensing and Control Systems S.L., Barcelona, Spain

C.P. Antonopoulos • N.S. Voros
Embedded System Design and Application Laboratory, Computer and Informatics Engineering Department, Technological Educational Institute of Western Greece, Antirio, Greece

N.S. Voros, C.P. Antonopoulos (eds.), *Cyberphysical Systems for Epilepsy and Related Brain Disorders*, DOI 10.1007/978-3-319-20049-1_10

10.1 Middleware Requirements

10.1.1 Introduction

This chapter defines the requirements related to the middleware platform. From sensors devices to electronic health record interface and notification applications, the middleware address the needs of the interoperability that is required among the different technologies in place within cyberphysical monitoring for epilepsy and related brain disorders and its functional goals.

The main source of information used to develop middleware requirements are user and sensor requirements, with main focus on the description of the services to be delivered to the users; patients and healthcare professionals, and the specification of the scenarios and use cases.

Specific privacy and security requirements have been devoted to security issues, extracting main objectives to be accomplished at sensor/WSN and extrapolating the implications to be developed at middleware level.

10.1.2 Functional Requirements

The main purpose of Middleware is to create an isolation layer between the physical world (sensors/field devices) and high level applications.

Due to the heterogeneous nature of the physical world, it is expected to have many different field devices (which likely will include sensors and a communication system) running different protocols and communication interfaces (serial, ip (udp or tcp), etc.).

At high level applications side, the creation of a standardised way of accessing sensor data for online/offline processing is important for the application developers. Application developers are managing the information from the service creation point of view, hence isolating data access from underlying communication and commissioning of field sensors is mandatory.

The communication between physical and application worlds is accomplished at intermediate level between the two defined interfaces (online and offline). In this middle layer several functions supported by a database will run to match the services necessities. The description of the functional requirements is divided into the following four parts: Functional Sensor Requirements, Functional Requirements of Data Stream Management System (DSMS), Functional Requirements of Application programming interfaces (API) and Functional Requirements of Notification services.

10.1.2.1 Functional Sensor Requirements

Within this section, all the functional requirements for/from sensors involved in the whole System and related to the middleware are described. The following requirements are related to acquiring, processing and storing data from sensors described in Chap. 8:

- All the data must be time stamped appropriately, at a sufficient level to perform synchronisation among separate data modalities, e.g. EEG & ECG
- Data from Sensors must be streamed to the middleware
- Middleware drivers must collect the chunks of data from Sensor and pass them to the online data stream management system (DSMS)
- Middleware can pull previously recorded data from the sensors.
- Middleware supports data input from sensors specified in Chap. 8
- Middleware is able to store the data locally on the Home Gateway
- Alarms/Warnings can be originated by sensors (i.e. push button)

10.1.2.2 Functional Requirements of Data Stream Management System (Online Analysis)

The Data Stream Management System (DSMS) is the backbone for online analysis. The following functional requirements describe the methods for online management, fusion and analysis of sensor data:

- Detection of abnormal values: The system must detect abnormal values due to improper use (e.g. motion artefacts) of sensors.
- The system need to use a variety of algorithms, to be able to deal with the streaming nature of data in order to perform the online analysis efficiently.
- The DSMS has to be configured to perform the desired analysis.
- Application errors need to be logged for efficient error handling.
- The online algorithms must be able to detect epileptic seizures and their patterns.
- Alarms/Warnings need to be originated by DSMS.
- Changing the configuration parameters must be possible.
- Detection of modalities including a preset range of normal values (i.e. excessive tachycardia and oxygen level excursions) for each patient individually.
- Use of activity sensor for diagnostic purposes involving cardiogenic triggers of seizures.

10.1.2.3 Functional Requirements of the Application Programming Interface (API)

How do we interact with the middleware? What are the different integrations between systems and middleware for user services? These functional requirements are described as follow:

- Middleware will initiate the uploading of sensor data to the PHR via PHR-API by notifying the upper layers.
- Middleware will send the raw data from sensors to the PHR-API by notifying the upper layers.
- Configuration parameters related to sensors are provided by upper layers (GUI/PHR).
- Middleware will receive Patient ID from underlayed processes and services which are controlled by the GUI.
- Alarms/Warnings can be originated by the upper layers.

10.1.2.4 Functional Requirements of Notification services

The middleware must ensure the communication between components. All the requirements needed to communicate events, intercommunication processes and send data to other managed systems are described as follow:

- Alarm notification: Middleware must be able to pass alarms to the upper layers.
- Middleware will receive the configuration parameters from the upper layers (GUI).
- Middleware will be able to send the status of operations to the upper layers (Upload progress, performing analysis, data sent, data error, etc.).
- Application errors can be send as notifications to the upper layers (GUI).
- Middleware will receive a 'send sensor data' trigger from the upper layers (GUI).
- Middleware will notify upper layers (GUI) whether the function used is in the ON or OFF mode.
- Middleware will receive a 'capture sensor data' trigger from the upper layers (GUI).
- Middleware will receive a 'pause' and 'resume' trigger from the upper layers (GUI).

10.1.3 Non-functional Requirements

Non-functional requirements for the middleware are qualities or standards that the system has, but which are not tasks or features automated by the platform.

Non-functional requirements need to be made precise and actionable. "SMART" requirements [1] have the following characteristics:

- Specific: without ambiguity, using consistent terminology, simple and at the appropriate level of detail.
- Measurable: it is possible to verify that this requirement has been met.
- Attainable: technically feasible.
- Realizable: realistic, given the resources.
- Traceable: linked from its conception through its specification to its subsequent design, implementation and test.

Regarding this set of characteristics, non-functional specifications for the middleware are described following the next areas:

10.1.3.1 Physical Technology

The following technologies are used to ensure correct operation of the middleware and they form the nucleus of all installations within the Home Gateway:

- Windows Web Server 2008 R2 64 bits
- SQL Server 2008 R2
- StreamInsigth™

10.1.3.2 Security

The security aspects of middleware follow the requirements and specifications which are documented in Chaps. 9 and 11. The following points give an overview about the most important requirements:

- Storing the activity for auditing.
- Anonymization of patient data (only the Middleware knows the patient ID).
- Using cryptographic random number generators to generate session IDs.
- Store credentials in a secure manner.
- Use Strong password policies.
- Encrypt communication channels to secure the authentication tokens.
- Authenticated API commands should be supported.

10.1.3.3 Compatibility/Interoperability

Middleware establishes an outgoing applicative connection (TCP, UDP, SOAP, and REST) towards the platform.

10.1.3.4 Reliability/Adaptivity

New types of sensor networks and dynamical (re-) configuration of data sources are supported in the middleware through xAffect and unisens format. Additionally, the middleware provides the basic building blocks for data communication between different components like PHR, DSMS and SHACU. The particularities of the information sent between those components are unknown by the middleware.

10.1.3.5 Robustness

Periodical retrieval/upload of data is automated by the drivers and all the transfers are reliable. Any failure in the system triggers automatic events to the administrator in order to be monitored and recovered.

10.1.3.6 Connectivity

The middleware needs a permanent or part time available, internet connection, to upload the data and event notifications to the PHR.

10.1.3.7 Scalability

The middleware is designed as a SOA (Service Oriented Architecture) platform. The architecture is extremely loosely coupled and well distributed to provide full scalability and efficiency.

10.1.3.8 Deployment

The middleware is tailored to be deployed locally at the Home Gateway. However, the technology used allows it to be ready to be deployed on Fully Managed Web Clusters. Therein supporting Software as a Service (SaaS) model with small modifications, so being ready to scale on the cloud (making use of the massive scalability of the cloud environment), and designed for fault tolerance, including management and monitoring software components.

10.2 Middleware Architecture

The principal objective of this Chapter is presenting an overview of the middleware package. The document will provide insights on the functional and non-functional aspects of the middleware. The architecture is based on ICT components which are described in Fig. 10.1. It highlights the middleware architecture within the overall system.

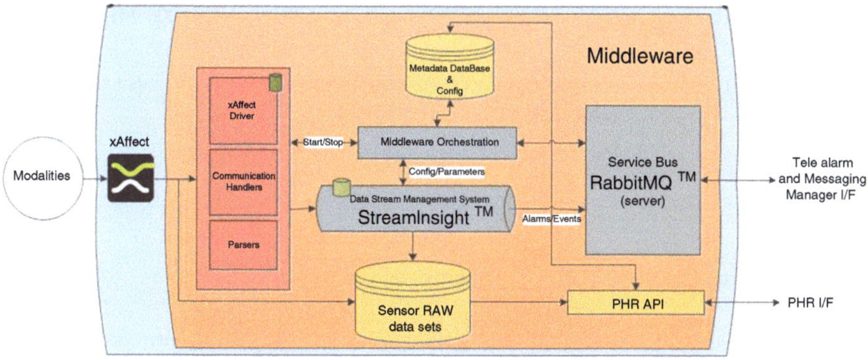

Fig. 10.1 Middleware architecture

As can be interpreted from Fig. 10.1, middleware will be installed in a local personal computer (the Home Gateway). Data from the different modalities (EEG, GSR, SPO2, etc…) will be acquired to the personal computer throughout an encrypted wireless channel. Data will be stored and uploaded to PHR server in a time configurable way (mainly in a daily basis scheme), from where data will be accessible and presented in a friendly manner to authorized persons (doctors, caregivers, etc…). When the on-line processing takes place, an event notification trigger can occur, by pressing the push button or by any event detection algorithm in the DSMS component. Depending on the severity of the event, immediate notification to a caregiver will be triggered using the MS (Management Service—SHACU) and immediate sensor data plus event information will be uploaded to PHR. Notice that to allow upload of data a broadband connection at home is required.

10.2.1 Secure Incoming Data

Security and privacy of patient data is of big concern in these cyberphysical systems which have been addressed in Chap. 9.

Regarding the communication, part of the efforts in the project is being concentrated to secure communications at RF level between the sensors and the home gateway by the provision of encryption/decryption engines.

To provide a secure communication between sensor (modalities) and home gateway, both hardware and software solutions are possible. While Chap. 9 explains the encryption phase, the following subsections describe two alternatives for the decryption of sensor data being sent through RF channels. The encrypted data reaching the middleware needs to be decrypted for online processing. There are two ways of decrypting the data, which are described within this chapter, hardware and software decryption.

Table 10.1 Decryption module processing delay and throughput rate (@200 MHz operating frequency)

Key size (bits)	Processing delay (in clock cycles)	Throughput rate (Mbps)
128	480	53.3
192	582	44.4
256	684	37.4

10.2.1.1 Hardware Decryption Engine

Hardware decryption module follows the same 8-bit architecture presented in Chap. 9. As done in encryption phase, decryption commences by executing the expansion operation in order to produce all intermediate requires keys. After the completion of the key expansion phase the decryption module is ready for the 128-bit data blocks processing. Due to the 8-bit data-path architecture that is used, 16 clock cycles are required to load/unload the 128-bit ciphertext/plaintext block. The input block is decrypted by performing four different byte-oriented transformations, which are executed sequentially and repeatedly (as rounds); the transformations are: Add Round Key, InvSubstitute Bytes, InvShift Rows and InvMix Columns. The resulted intermediate decipher result, known as state, is stored and updated at the end of each round. The number of rounds depends on the size of the key. The actual key size and the number of rounds are configured via the KEY_SIZE input as shown in Table 10.1. Note that, since the key size determines the number of rounds, it also affects the latency of the decryption module. Respective delay and throughput performance is presented in Table 10.1.

10.2.1.2 Software Decryption Engine

A decryption module can be developed within xAffect. Its deciphering process can be abstracted as much as possible providing wide algorithm customization.

This module should be designed to be flexible and adaptable which results a highly reusable ciphering/deciphering component that can be used within all the functionalities that xAffect provides, like data collection, data logging, data sending, etc.

10.2.2 Data Fusion

Data Fusion is a process dealing with the association, correlation, and combination of data and information from single and multiple sources. Multi sensor data fusion refers to the acquisition, processing and synergistic combination of information gathered by various knowledge sources and sensors to provide better understanding of a phenomenon.

The most important part of data fusion is, the data itself. Therefore, it is crucial to take care during the capturing in order to minimize uncertainties. However, there is one aspect that is sometimes left behind and in some cases it is the most relevant: adapt metadata in a common format in order to simplify the process.

The data fusion can be classified as low level fusion of data, high level variable fusion and mixture level fusion. When the data fusion is performed before analysis, it is classified as low level. When the data fusion is performed after some data analysis, it is classified as high level variable fusion. There are some situations where we can fuse data and variables.

The overall system architecture, depicted in Fig. 10.1, shows the two paths where processing of sensor data takes place. On-line multi-parametric data processing and analysis takes place at the Home Gateway, while the off-line multi-parametric data processing takes place in different server/database, by (i) accessing/acquiring sensor data from the PHR (previously captured by the system) or by (ii) accessing/acquiring sensor data from external databases not belonging to the system itself. The data flow is depicted in Fig. 10.2.

An interesting relation in Fig. 10.2 is between off and on line processing. The goal of Off-line processing is to detect relations between different modalities throughout extensive data analysis. Online processing on the other hand is used to detect events of special interest and reduce the data amount by processing the raw data.

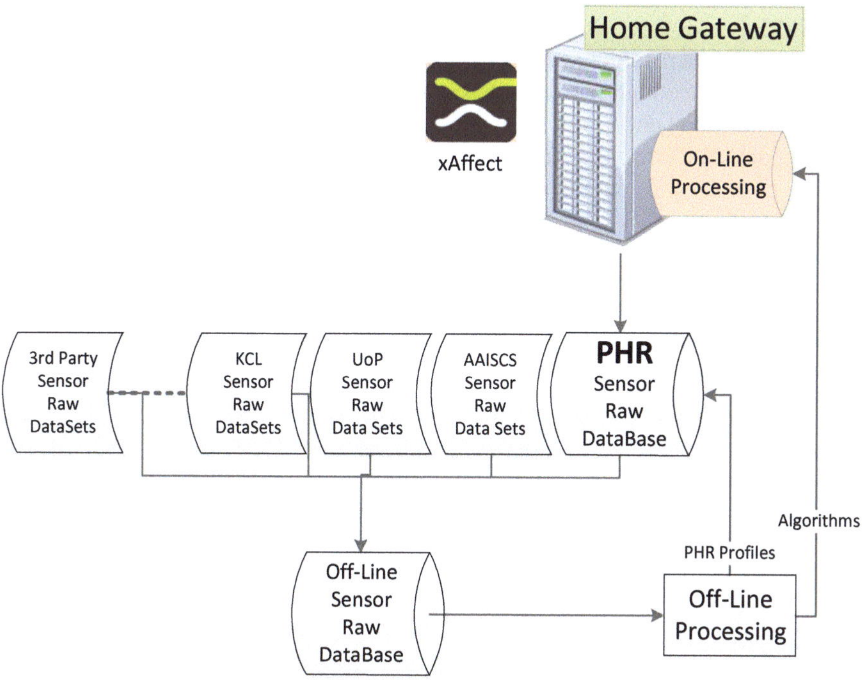

Fig. 10.2 Data fusion within the system

10.2.2.1 xAffect Framework

To connect all the data online, synchronize and handle it a lean and open framework, which can be personalized and reconfigured for each patient, is needed. xAffect is a software framework developed by the Research Center for Information Technology, Karlsruhe, Germany. It was developed in Java to fulfill real-time data processing, easy integration of different data sources, easy integration of algorithms and data logging of raw as well as derived data [2]. Libraries for some common sensors already exist in xAffect. To use a broad spectrum of bio-signals, additional libraries need to be implemented. The data format which is being used is the unisens-format. This is a universal and generic format suitable for recording and archiving sensor data from various recording systems and with various sampling frequencies [3].

The current version of xAffect can be modified in order to customize the interface for the middleware, which is necessary to achieve the performance and the required functionality for the system.

The changes that need to be made are:

- Additional libraries for sensors (To use a broad spectrum of sensors, non-existing libraries had to be written).
- Decryption module (data from sensors need to be ciphered).
- Data acquisition pause/resume to achieve the needs for the control of the sensor data acquisition.
- A custom notification module for communicating xAffect state to middleware DSMS.
- Extended data recording functionalities to provide configurable file splitting (in order to reduce high network consumption during heavy data uploads), alarm signal detection and a communication system with middleware and PHR.
- Extended data streaming functionalities to provide hot-plug client connections and custom xml output data formats (including gzip for network traffic optimization).

10.2.3 Graphical User Interface (GUI)

The Graphical User Interface handles the communication between the components and the User.

The GUI, shown in Fig. 10.3 is an example of how this part could look like. In the GUI, the user enters username and password and the patient-id. With the patient-id, the middleware will download the personalized profile from the PHR. Afterwards the user can press the configure button to initialize the communication with the sensors. Furthermore the profile contains information about the alarm settings. This enables the middleware to set up the DSMS with the customized alarm set [4].

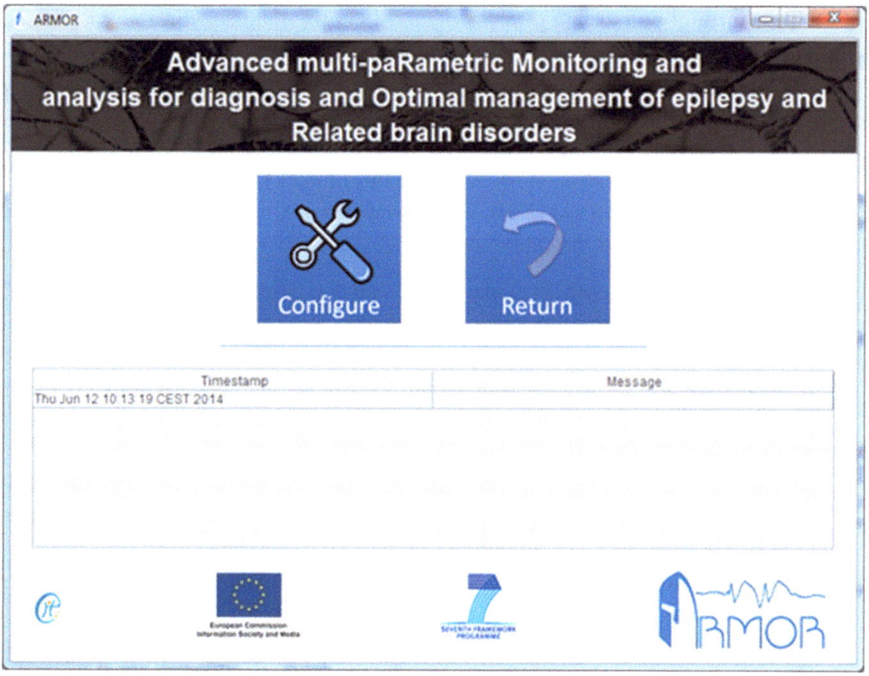

Fig. 10.3 Example of a graphical user interface for a system middleware

When the configuration process finished successfully the user is able to start the measurement by pressing the record button. During recording the data is streamed from the sensors through xAffect towards the DSMS and the storage. The middleware takes the data from the storage and uploads it in junks to the PHR. In case an Event occurs the data will be uploaded immediately. This ensures that when an alarm will be send to a clinician, the data will be ready for downloading and viewing. Furthermore, the GUI allows users to pause or resume the measurement. This allows the subject to interrupt the data acquisition and move out of the systems Bluetooth range.

10.2.4 Data Stream Management System (DSMS)

DSMS function takes place in on-line scenario, where real time processing of modalities is performed. In order to achieve it, minor adaptation of the xAffect™ stream connector of the sensor data being delivered to Middleware has been introduced in order to assure no sensor data is lost.

DSMS is based on Microsoft™ StreamInsight™ platform created for the development and deployment of complex event processing (CEP) applications.

It's a high-throughput stream processing architecture that uses .Framework-based development platform.

The development of DSMS allows sensor data to be received from xAffect™ in real time, synchronize it in a lossless way and feed it to computation algorithms. At the end, a framework is provided that enables the creation of own queries and policies over data (from now on Middleware Framework).

In order to explain data workflow within Middleware, we need to introduce some concepts related to StreamInsight™ as follow:

- **Sources**: They are data providers and can be implemented with adapters, IEnumerable or IObservable objects. They are in charge of collecting data, fit it into a payload, generate an appropriate timestamp (if needed) and redirect it to StreamInsight's core adding appropriate CTI (Current Time Increment) insertions. Sources are IObservable objects built with information coming from xAffect through TCP sockets.
- **Queries**: All data from sources goes directly to StreamInsight core (in any number of streams) where it's processed in order to satisfy one or more queries (LINQ). That will produce an output that will go to the data consumers (observers) where custom operations can be defined. It can also compute different operations depending on query such as aggregation, unions, max/min etc.
- **Consumers**: They are pieces of software which main function is processing data output from StreamInsight™ queries. They are usually implemented with the observer interface so it can be easy to notify them when new data is available.

Lossless stream between xAffect and DSMS can be achieved by using TCP channels. Then synchronized sensor data can be provided by using the built-in architecture of StreamInsight™ available through the highly flexible and configurable Insight framework which allows researchers to gather and analyze customized data structures derived from StreamInsight™ output.

10.2.4.1 Push Button Processing

One of the services provided by middleware is the processing of the marker button in the Bioplux™ device if it is configured as alarm push button.

When the user press the Bioplux™ push button, xAffect™ captures the event and records it. Additionally it sends the push button raw data through the IP connection to Middleware DSMS.

Middleware DSMS canalizes the push button raw data to StreamInsight™ core. The Middleware Framework contains a developed query to detect alarm signals, if it successfully detects an alarm condition. It notifies the event manager which builds and sends a XML message to the upper layers of the software (as can be seen in Fig. 10.1). Also if the user has defined more queries to detect alarms they would produce appropriate output that would be handled within a defined observer. Figure 10.4 depicts the complete data workflow.

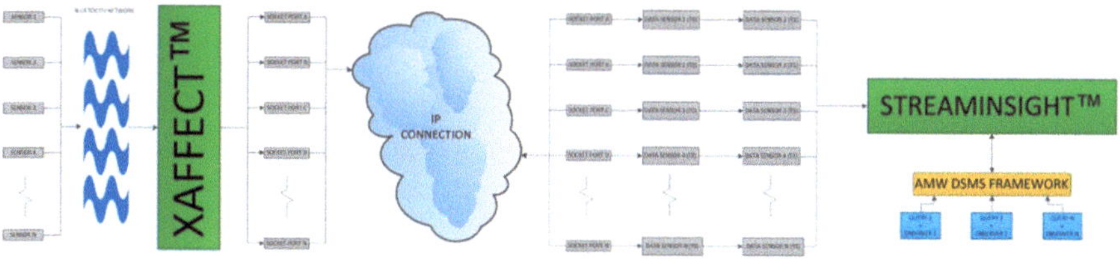

Fig. 10.4 Middlewaredata work flow

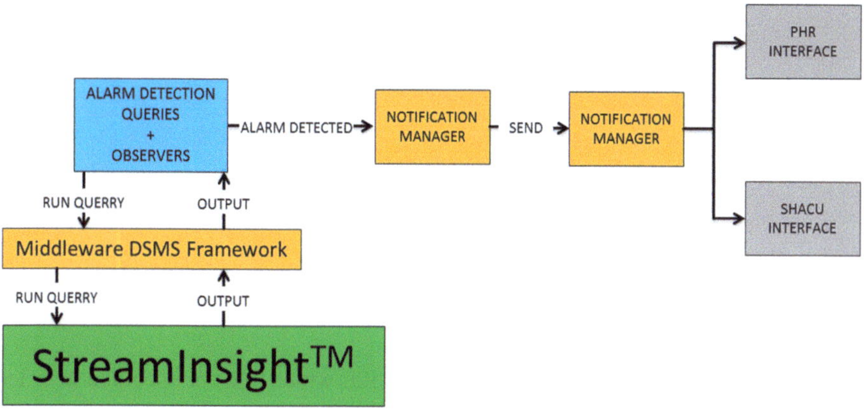

Fig. 10.5 Middleware DSMS notification manager module workflow

10.2.5 Event Notification

One of the objectives of Middleware DSMS is to be able to deliver alarm/warning events to high level applications. One of these events is the push button detection. Further events are for example alpha rhythm detection or seizure detection. The workflow of the event notification is shown in Fig. 10.5.

The module that communicates events within the middleware is called the Notification Manager which is explained in the next subsection.

10.2.5.1 Event Observer Output Format

In order to have a common way of communicating alarm/warning events from the observers to the notification manager, a guideline has been established. This guideline specifies the usage of a common Alarm data structure defined inside the Middleware Framework.

Table 10.2 Example XML notification for push-button event


```xml
<?xml version="1.0" encoding="utf-8"?>
<messagexmlns="http://armor.tesyd.teimes.gr"
xmlns:xsi="http://armor.tesyd.teimes.gr/2013/XMLSchema-instance"
xsi:schemaLocation="http://armor.tesyd.teimes.gr message.xsd">
     <header>
     </header>
     <body>
            <event>
                   <id>129</id>
                   <type>alarm</type>
                   <datetime>
                          <timestamp>1363773623</timestamp>
                          <utc_offset>+01</utc_offset>
                   </datetime>

            </event>
     </body>
</message>
```


This data structure defines three common fields which must be specified in order to generate a valid notification:

- ID: Identifies the event (for instance, 129 = push button)
- Type: Describe the kind of notification (ALARM/WARNING)
- Timestamp: Describes the event detection time

With this data the Notification Manager will be able to generate valid alert messages for each defined interface. An example can be found in Table 10.2 for the push button event.

10.2.5.2 Notification Manager

The Notification Manager module is mainly designed to detect anomalies or special situations in DSMS input-output raw data. It generates custom messages and forwards them to the specified endpoints. We can see in Fig. 10.5 how it is integrated within Middleware Data workflow.

After detecting an event that requires to be notified, the Notification Manager generates a suitable message (according to its configuration rules) and queues it. Then an automatic process takes the message and sends it through many interfaces as needed using any specific communication channel or protocol that each interface requires.

By default this module is configured to send XML messages related to alarms/events. The XML basic structure is defined to identify when the event happened, the event type and to which ID it is related. Table 10.2 shows an example of xml packet related to a notification of an alert event.

10.3 Conclusions

Sets of multiple sensors are required to acquiring the data needed to handle patient's epileptic disorders. That's why the middleware is the most important part of the whole system. The encrypted data streams will be decrypted, fused and streamed to the DSMS. The DSMS handles the live data analysis and will detect events of special interest. These events will then be reported via PHR to the clinicians, who will have direct access to the important data.

Tests under clinical conditions showed that the described system is able to handle online analysis with the sensor data. Events like a push button or the detection of alpha rhythms can be detected reliably and sent from the notification manager towards the PHR.

References

1. Mannion M, Keepence B (1995) SMART requirements. ACM SIGSOFT Software Engineering Notes 20(2): 42–47. http://www.win.tue.nl/~wstomv/edu/2ip30/references/smart-requirements.pdf
2. Schaaff K, Mueller L, Kirst M. http://www.xaffect.org
3. Kirst M, Ottenbacher J. www.unisens.org
4. Bideaux A, Anastasopoulou P, Hey S, Cañadas A, Fernandez A (2014) Mobile monitoring of epileptic patients using a reconfigurable cyberphysical system that handles multi-parametric data acquisition and analysis. In: Proceedings of the 2014 EAI international conference on wireless mobile communication and healthcare (MobiHealth), Athens, pp 377–380

Chapter 11
Personal Health Record

Artur Krukowski, Carlos Cavero Barca, Juan Mario Rodríguez, and Emmanouela Vogiatzaki

Abstract This chapter focusses on the introduction of the Electronic (EHR) and Personal Health Records (PHR) as new technological approaches aimed at standardising electronic management of medical information between the patient and its physicians, as well as among medical organisations collaborating in providing integrated medical care services. It presents combined experiences in developing e-Health platforms and services with respect of supporting medical research into the causes and relationships among physiological parameters and health problems concerning different chronic diseases, cardiovascular, stroke, epilepsy, and others. The Personal Health Records (PHR) is presented as a new technological approach aimed at standardizing electronic management of medical information between the patient and its physicians, as well as among medical organizations collaborating in providing integrated medical care services. On the examples of most common commercial as well as open-source implementations of such system we aim to describe roles and aims behind electronic health recording, follow with applicable legal and standardizations frameworks and European activities in this area, leading towards introduction to most common commercial as well as open-source implementations of such systems and concluding with indication of specific adaptations enabling the use of stored personal health data for scientific research into causes and evaluation of chronic illnesses. We describe also ethical and privacy concerns that are relevant to using and exchanging electronic health information.

A. Krukowski (✉)
Intracom S. A. Telecom Solutions, Athens, Greece
e-mail: krukowa@intracom-telecom.com

C.C. Barca • J.M. Rodríguez
ATOS Research and Innovation (ARI), ATOS, Madrid, Spain

E. Vogiatzaki
Research for Science, Art and Technology (RFSAT) Ltd., Sheffield, UK

N.S. Voros, C.P. Antonopoulos (eds.), *Cyberphysical Systems for Epilepsy and Related Brain Disorders*, DOI 10.1007/978-3-319-20049-1_11

11.1 Role and Structure of PHR

The Electronic Health Record (EHR) of a patient can be defined as digitally stored health care information about individual's lifetime with the purpose of supporting continuity of care [1], education and research, and ensuring confidentiality at all times. A patient's healthcare information may be spread out over a number of different institutes that do not interoperate. In order to provide continuity of care, clinicians should be able to capture the complete clinical history of the patient. The Personal Health Record (PHR) is the electronic part of the health-related information of a person (such as diagnoses, medications, allergies, lab test results, immunization records but also administrative tasks such as appointment or prescription renewals) that can be extracted from multiple sources, but always under the control of the consumer, patient or informal caregiver. This is the relevant difference between the PHR and the EHR or electronic medical record, which is maintained by the healthcare providers and payers [2].

The EHR standardisation aims to ensure that patient records are used to support shared care among clinicians with different specialisations, while enabling the mobility within and among countries for people who give and receive healthcare.

From the viewpoint of standardization, the single most important characteristic of the EHR is the ability to share EHR information between different authorized users. In technical terms, this requires interoperability of information in the EHR and interoperability of EHR systems that exchange and share this information. We distinguish two major levels of interoperability (info sharing) of information [3]:

- Functional interoperability: it is the ability of two or more systems to exchange information (so that it is human readable by the receiver); and
- Semantic interoperability: it is the ability for information shared by systems to be understood at the level of formally defined domain concepts (so that information is digitised by the receiving system). Semantic interoperability is not an all-or-nothing concept. The degree of semantic interoperability depends on the level of agreement on terminology and on the content of archetypes and templates used by the sender and receiver of information.

One of the key requirements for interoperability of the EHR is to break the nexus between the EHR and the EHR system (i.e. the EHR should conform to an information model independent of both the physical database schema used for local storage and the applications, which create, maintain, and retrieve EHR). This EHR information model should be independent of any particular implementation technology (i.e. it should be a logical information model). Technology independence is essential to make EHR 'future proof' to enable a lifetime EHR possibility.

In order to achieve semantic interoperability of EHR information, there are three prerequisites, with the first ones being required for functional interoperability [4]:

1. A standardized EHR reference model, i.e. the EHR information architecture, between the sender (or sharer) and receiver of the information;

2. Standardized service interface models to provide interoperability between the EHR service and other services such as demographics, terminology, access control and security services in a comprehensive clinical information system;
3. A standardized set of domain-specific concept models, i.e. archetypes and templates for clinical, demographic, and other domain-specific concepts; and Standardized terminologies, which underpin the archetypes. This does not mean that there is need to have a single standardized terminology for each health domain but rather, terminologies used should be associated with controlled vocabularies.

11.2 EHR Standardization Bodies

One of the main factors hindering the widespread adoption of integrated PHRs is the lack of technical standards for interoperability, which is the ability of systems to exchange information using the same mechanisms. "The immaturity and slow diffusion of standards for interoperability and data portability are key barriers to the integration and exchange of structured data among PHRs and the range of relevant entities that provide and finance health care" [5]. The success and final adoption of the PHR systems depends on the capability of interacting with Electronic Health Records (EHRs) and other sources of personal health data e.g. Personal Health Records and Personal Health Record Systems published by the U.S. Department of Health and Human Services in 2006. Currently after years of pursuing the interoperability, the EHR standards continue lacking of the public adoption and immaturity mentioned.

A number of standardization efforts are progressing to provide the interoperability of EHRs such as the CEN/TC 251 [6], ENV 13606 HER Communications standard (http://www.centc251.org) [7], openEHR (http://www.openehr.org) and HL7 Clinical Document Architecture 2.0 (http://xml.coverpages.org/CDA-20040830v3.pdf). These standards aim to structure and mark-up the clinical content for the purpose of exchange. A complementary initiative addressing the issue of how to exchange EHR complying with different content standards is the Integrating the Healthcare Enterprise IHE (http://www.ihe.net) Cross-Enterprise Document Sharing (XDS) integration profile detailed in its Technical Framework: http://www.ihe.net/Technical_Framework.

11.2.1 ISO/TC215

CEN/TC 251 [6] is the technical committee on Health Informatics of the European Committee for Standardization. Its mission is to achieve compatibility and interoperability between independent health systems and to enable modularity by means of standardization. This includes requirements on health information structure to support clinical and administrative procedures, technical methods to support interoperable systems as well as requirements regarding safety, security and quality [8].

The CEN pre-standard ENV 13606:2000 "Electronic Healthcare Record Communication" is a message-based standard for the exchange of EHR content [7]. This standard defines an EHR information model, called the "extended architecture" since it is an extension of the earlier pre-standard ENV 12265.

It also defines a list of machine-readable domain terms that can be used to structure EHR content, a method of specifying "distribution rules", that is, rules under which certain EHR content may be shared with other systems and, finally, request and response messages that allow systems to exchange subsets of an EHR. ENV 13606 does not attempt to specify a complete EHR system; instead, it focuses on the interfaces relevant for a communication between EHR systems.

ENV 13606 [7] was intended to be the first fully implementable EHR standard, and subsets of it were implemented in a number of EHR projects in the UK, Denmark, the Netherlands, Sweden, and Norway. However, none of these projects used the complete ENV 13606 specification; moreover, the implementation experience showed a number of weaknesses in the standard that limited its usefulness and market uptake: the single-level modelling approach made the information model extremely complex, with lot of optionality and a level of abstraction that made quite difficult to comprehend and implement the model.

In 2001, CEN/TC 251 [6] decided to revise ENV 13606 into a full European Standard, taking into account the existing implementation experience and to adopt the openEHR archetype methodology. ENV 13606 is a standard that is now gradually being approved, and consists of five parts:

- Reference Model: it defines the hierarchy of generic building blocks of the EHR through a set of classes. It represents the stable characteristics of the EHR entries, how they are aggregated, and the context information required to meet ethical, legal and provenance requirements;
- Archetype Interchange Specification: each archetype defines legal combinations of the building block classes defined in the Reference Model for particular clinical domains, organizations. The archetype model is syntactically equivalent to those of the Good Electronic Health Record project, and the openEHR standard;
- Reference Archetypes and Term Lists: it includes the vocabularies for attributes, and archetypes to represent HL7 specialized Acts and openEHR specialised ENTRYs;
- Security Features: it defines an interoperable specification for EHR disclosure consent, and an interoperable disclosure log;
- Exchange Models: this part is still under discussion.

ENV 13606 standard was not intended to specify the internal architecture or database design of EHR systems or components. Nor is it intended to prescribe the kinds of clinical applications that might request or contribute EHR data in particular settings, domains or specialties. For this reason, the information model proposed there was called the *EHR Extract*, and might be used to define a message, an XML document or schema, or an object interface. The information model in this European Standard is an ISO RM-ODP Information Viewpoint of the EHR Extract. This European Standard considers the EHR to be the persistent longitudinal and potentially multi-enterprise or multi-national record of health and care provision relating to a single subject of care (the patient), created and stored in one or more physical systems in order to inform the subject's future health care and to provide a medico-legal record of care that has been provided.

11.2.2 ISO/EN EN13606

ISO/EN EN13606 [7] is a norm designed to achieve semantic interoperability in EHR-related data communication among different Health Information Systems (HIS). Its main goal is to define a stable and reliable information structure in order to communicate EHR parts of the same patient (CEN/TC251–ISO/TC215 2010).

The first version of the 13606 four-part pre-standard was published in 1999–2000 but attempts to implement this pre-standard in software proved to be difficult and those implementations which were undertaken suffered from the "HL7 v2 problem" of too much optionality. In 2002 CEN made a decision to revise the 13606 pre-standard and upgrade it to a full normative European standard (EN 13606, also called EHRcom). The ISO/EN13606 standard were completed and ratified after Part 5 by ISO and CEN in February 2010. ISO/EN 13606 architecture provides a framework to communicate EHR data using the dual model approach (reference model and archetypes) to provide the semantic interoperability. The ISO/EN 13606 consists of five parts:

1. **Part 1**: CEN 2007: ISO 2008: The Reference Model, the generic common information model. The global characteristics of health record components.
2. **Part 2**: CEN 2007: ISO 2008: Archetype Interchange Specification, information model of the metadata to represent the domain-specific characteristics of electronic health record entries. This chapter defines how to share archetypes, and not how to exchange them within particular systems.
3. **Part 3**: CEN 2008: ISO 2009: Reference Archetypes and Term Lists, establishing the normative terminologies and controlled vocabularies.
4. **Part 4**: CEN 2007: ISO 2009: Security Features, covering security mechanisms and methodology.
5. **Part 5**: CEN/ISO 2010: Exchange Models, interface designed to request specific extracts, archetypes or audit log.

The relevant components of the generic reference model are (Fig. 11.1): an "HER Extract" is the root node of the EHR and contains "Compositions" which can be organised using "Folders" (in the same way as Microsoft Windows explorer folders). The "Entry" can be an observation, medication order, diagnoses and can be organized within "Sections". The leaf nodes (data) are "Elements" which are included in the "Entry" and optionally organized within "Clusters".

The ISO/EN 13606 is a subset of the full openEHR specification [9]. Within the shared classes the main difference between with the openEHR reference model is that "Entries" are broken down in the corresponding kind of information stored (Munoz et al. 2011).

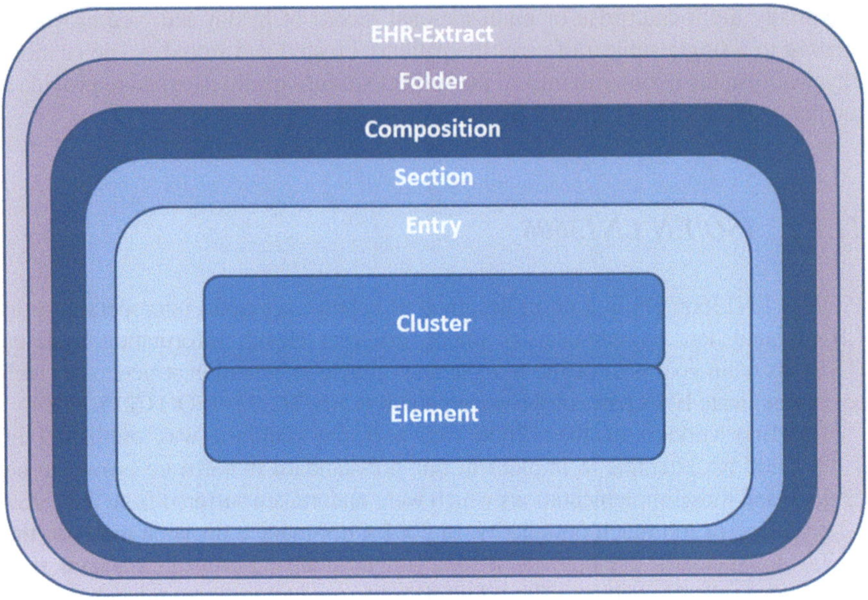

Fig. 11.1 Components of the ISO/EN 13606 association (2009)

11.2.3 GEHR/OpenEHR

The GEHR/openEHR initiative was started in 1992 as an EU research project, called "Good European Health Record", in the 3rd Framework Program. The initiative was later continued under the name "Good Electronic Health Record" with strong participation from Australia. Currently it is maintained by the openEHR Foundation, a non-profit organization defining itself as "*an international, on-line community whose aim is to promote and facilitate progress towards EHRs of high quality, to support the needs of patients and clinicians everywhere*". The openEHR is a foundation that supports the development of an open and semantic-connected platform for eHealth systems (www.openehr.org). It is based on 15 years research, focused engineering design and real-world implementation experience, rather than being created as a formal consensus standard. However, over the last years it has had a significant influence over the development of EHR standards by the three main international eHealth standards organisations: CEN, HL7 and ISO. The information model covers the EHR architecture and describes classes such as Folder, Composition, Section and Entry, and of the basic data structure and types. The Care Entry class "define the semantics of all the 'hard' information in the record" [10], the Admin Entry represents information recorded during administrative issues. Figure 11.2 shows the ontology leading to the Entry model.

The most noteworthy concept introduced by GEHR/openEHR is the "archetype" concept. This approach uses a two-level methodology to model the EHR structure.

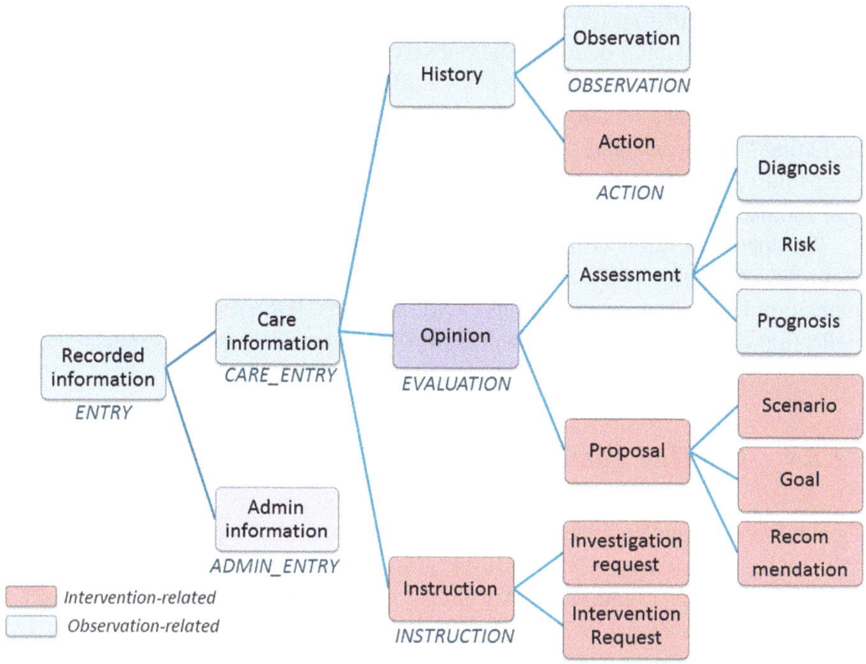

Fig. 11.2 OpenEHR ontology of recorded information [28]

In the first level, a generic reference model that is specific to the healthcare domain but still very general is developed. This model typically contains only a few classes (e.g. role, act, entity, participation) and must be stable over time. In the second level, healthcare and application specific concepts such as blood pressure, lab results etc. are modelled as archetypes, that is, constraint rules that specialize the generic data structures that can be implemented using the reference model. As an example, a constraint may restrict a generic "Observation" class to, e.g., "Blood Pressure" archetype.

An archetype definition consists of three parts: descriptive data, constraint rules and ontological definitions. The descriptive data contains a unique identifier for the archetype, a machine-readable code describing the clinical concept modelled by the archetype and various metadata such as author, version, and purpose. It also states whether an archetype is a specialization of another archetype. The constraint rules are the core of the archetype and define restrictions on the valid structure, cardinality and content of EHR record component instances complying with the archetype. The ontological part defines the controlled vocabulary (that is, machine-readable codes) that may be used in specific places in instances of the archetype. It may contain language translations of code meanings and bindings from the local code values used within the archetype to external vocabularies such as SNOMED or LOINC. It may also define additional constraints on the relationship between coded entries in the archetype based on the code value. As mentioned above the "Care Entry" concept covers the most common and medically relevant information.

In the openEHR four types of entries can be distinguished: observations, evaluations, instructions, and actions. The "observation" and "action" classes represent statements about the past events of the individual subject of record. The "evaluation" classes represent current assessment by the attending health professional, including "diagnosis" and "prognosis", as well as the representation of the imagined future, like "goals" and "scenarios". "Instructions" represents future events that should take place as prescribed by the health professional.

The openEHR framework includes a reference information model, the Archetype Definition Language (ADL) for expressing archetypes, an archetype library, implementation technology specifications (XML schemas, IDL specifications etc.) and a collection of open source implementations of the openEHR specifications.

11.2.4 The HL7 Family of Standards

Health Level Seven HL7 (www.hl7.org) is an international organization founded in 1987 and supported by ANSI with the goal of develops global standards related with eHealth. This organization has already defined a set of standards for clinical information interchange, whose name is HL7 standards. Among them HL7 CDA (Clinical Document Architecture) defines the Architecture of electronic documents used within Health domain and it is HL7's current main strategy for EHR interoperability. Besides, HL7 supports RIM (Reference Information Model), a model of healthcare information as viewed within the scope of HL7 standards, which provides a static view of information needs along with use case models, interaction models, data type models, terminology models, and other types of models to provide a complete view of the requirements and HL7 standards design, thus giving a valid starting point for any HL7-compliant Architecture Design [11].

Meta-classes can be identified in RIM [12], as observed in Fig. 11.3:

- **Act**: actions in the healthcare management
- **Participation**: context for an act: Who? For whom? Where?
- **Entity**: physical things, subjects or targets taking part in healthcare act.
- **Role**: establishes roles that entities play in its participation in healthcare acts.
- **Act Relationship**: represents a relationship between two acts.
- **Role Link**: represents a dependency between roles.

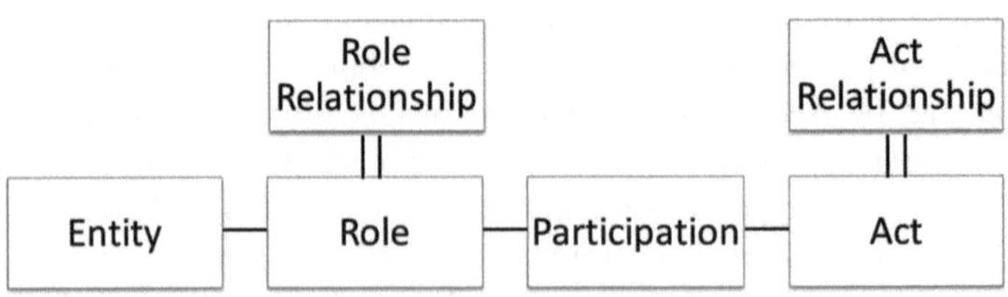

Fig. 11.3 RIM structure

Finally, the HL7 v3 Template is "an expression of a set of constraints on the RIM which is used to apply additional constraints to a portion of an instance of data which is expressed in terms of some other Static Model. Templates are used to further define and refine these existing models within a narrower and focused scope".

11.2.5 Relationship Between Standards

The openEHR supports the creation, storage, maintenance, and querying of complete EHRs. ISO/EN 13606 is a subset of the full openEHR implementation and it is an appropriate standard for exchange of EHR extracts. At the same time, ISO/EN 13606 offers a partial alignment with HL7 Clinical Document Architecture (CDA) (Fig. 11.4).

Release 2.0, which complies with hl7 RIM. e.g., HL7 CDA is based on the HL7 RIM, but it was designed to represent patient summaries, not thinking on providing decision support capabilities. All the aforementioned reference models are archetype-based. HL7v2.x messaging is an appropriate standard, at least in short/medium term, for transmission of information from clinical information systems to EHR systems.

11.3 Data Structuring Algorithms in PHR

According to European Committee for Standardisation in standards CEN/TC251–ISO/TC215 2010 [6], the EHR is "*the persistent longitudinal and potentially multi-enterprise or multinational record of health and care provision relating to a single*

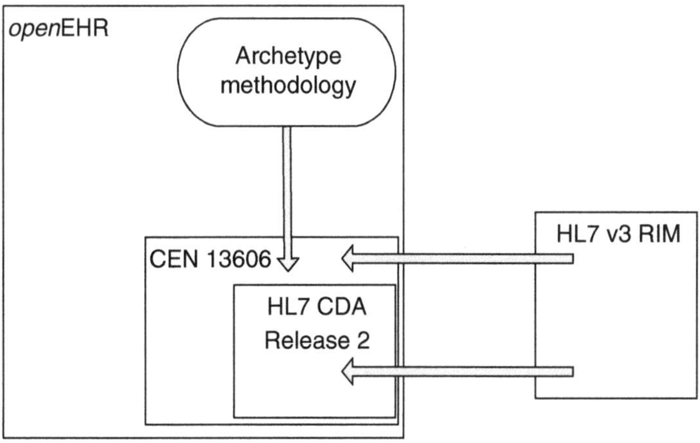

Fig. 11.4 Relationships between standards (openEHR, EN13606 and HL7 CDA)

subject of care (the patient), created and stored in one or more physical systems in order to inform the subject's future health care and to provide a medico-legal record of care that has been provided".

Patient data is managed, in current Hospital Information Systems (HIS), as a digital and well-structured record, which contains all individual-related health data, such as, demographic, diseases, allergies, history and activity during illness periods, etc.

The semantic interoperability of health data can be achieved only with the standardization of EHR. During the last 10 years, the international organizations have been driving great efforts to define the architecture for exchanging properly the health information between EHR coming from diverse systems (Munoz et al. 2011).

11.3.1 The Dual Model Approach

The huge amount of clinical concepts and volatility of the information are the main drawbacks to deploy long-term EHR systems due to the single model approach. The ad-hoc solutions are implemented by technical stuff gathering the user requirements from the clinicians providing tools probably perfect at design time but a predictable out-of-date system requiring new releases in the near future if additional information is needed. The "dual model approach", also called the two-level modelling [13–15] separates out the clinical knowledge (volatile) and the reference model (static).

The medical concepts are modelled using archetypes based on a stable reference model [16]. This approach is attractive for the stake-holders due to its stability, the EHR products are installed once and additional clinical functionalities could be extended using archetypes, on the contrary the ever changing clinical practice jeopardises return on EHR investment in ad-hoc systems.

The dual model approach defines a generic information model, the reference model with domain-invariant classes to be instantiated as well as specific clinical models, which support semantic interoperability, which are called archetypes, containing specific clinical information designed using a common language as it is shown in Fig. 11.5. This approach has been supported by many working groups within different international initiatives.

11.3.2 Detailed Clinical Models (DCM)

The Detailed Clinical Models are actually "a new way to structure medical information. It combines expert knowledge, data specification and terminology and enables various technical applications". They specify the information models and the way the data is exchanged: user interfaces, data management, decision support systems and so on. It could be considered equivalent to Archetypes, Templates or Clinical Statements. The Clinical Information Modelling Initiative (CIMI) is working to establish those common models from different standards, initially modelling these DCMs in both ADL (archetype definition language) and UML (www.uml.org).

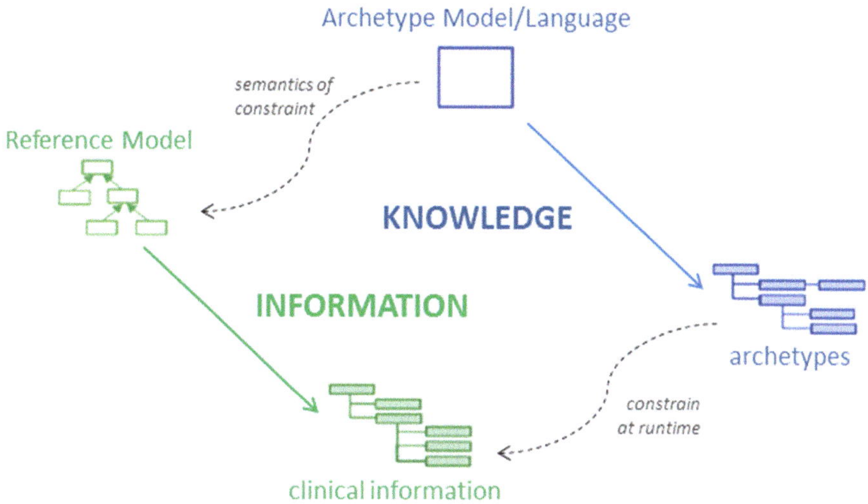

Fig. 11.5 Dual model approach

11.4 Standardisation of User Interfaces to PHR

11.4.1 Continuity of Care Record (CCR)

The American Society for Testing and Materials (ASTM International) Continuity of Care Record (CCR) is a clinical framework that was first developed by health care practitioners to meet the information exchange needs of primary care providers. ASTM defines the CCR as a "summary of the patient's health status (e.g., problems, medications, allergies) and basic information about insurance, advance directives, care documentation, and care plan recommendations" [17].

11.4.2 CCD (Continuity of Care Document)

A CCD is a joint effort between HL7 International and ASTM approved as an ANSI standard in 2007 in order to use HL7 CDA for sharing the CCR (Continuity of Care Record) patient summary. It represents a complete implementation of CCR, combining the best of HL7 technologies with the richness of CCRs clinical data representation, and does not disrupt the existing data flows in payer, provider of pharmacy organizations.

The Continuity of Care Document (CCD) establishes a detailed set of constraints and templates, covering the main sections of the summary record to be represented as CDA elements, according to the HL7/ASTM Implementation Guide for CDA Release 2-Continuity of Care Document (CCD) Release 1 (2007).

11.4.3 Integrating the Healthcare Enterprise (IHE)

International is a global association of healthcare IT vendors, user organisations, clinical professional societies, and advocacy groups that promotes interoperability through the co-ordinated use of established standards such as Digital Imaging and Communications in Medicine (DICOM) and HL7. IHE, more than any other single organisation, paved a way for practical medical interoperability [8].

11.5 Terminologies

Coded elements are used in the healthcare environment to precisely define the clinical concepts language-independently. The use of medical terminology is one of the bases to provide semantic interoperability.

- **Systematised Nomenclature of Medicine Clinical Terms** (**SNOMED-CT**) http://www.ihtsdo.org/snomed-ct consists of controlled medical vocabularies (CMVs),—accumulated medical concepts updated in a rigorous fashion. It has been gaining momentum as the primary coding method for clinical concepts. SNOMED has become the presumptive source of clinical codes and concepts within its member countries.
- **Logical Observation Identifiers Names and Codes** (**LOINC**) http://loinc.org is a database and universal standard for identifying medical laboratory observations. It was developed in 1994 and maintained by Regenstrief Institute, a US non-profit medical research organization. It was created in response to the demand for an electronic database for clinical care and management and is freely available.
- **International Classification of Diseases** (**ICD**) http://www.who.int/classifications/icd/en is the standard diagnostic tool for epidemiology, health management and clinical purposes. This includes the analysis of the general health situation of population groups. It is used to monitor the incidence and prevalence of diseases and other health problems. It is used to classify diseases and other health problems recorded on many types of health and vital records including death certificates and health records. In addition to enabling the storage and retrieval of diagnostic information for clinical, epidemiological and quality purposes, these records also provide the basis for the compilation of national mortality and morbidity statistics by WHO Member States. It is used for reimbursement and resource allocation decision-making by countries.

The crucial issue regarding the deployment of an EHR product in the medical environment is the ease of mapping to existing local data stores, as well as to national specifications (i.e. as an interface specification, for instance from hl7 v3 to archetypes). In order to provide the semantic interoperability the connection with external health terminology bindings is mandatory to be language independent [18].

The term binding is supported for manual or semi-automatic creation between archetypes and the concepts in terminology systems [19, 20] but the main drawbacks still reside in,

- The big amount of clinical concepts and their relationships, which makes necessary to filter out the terms using a powerful and intelligent tool.
- Mapping and translating concepts across vocabularies.

11.6 European R&D Projects Related to EHR Standardization

According to an eHealth ERA report related to eHealth priorities and strategies in European countries [21], the achievement of a European EHR is not yet an overarching goal, but collaboration on developing individual countries' EHR or even basic patient summaries is a first step. Electronic Health Record is a rather fuzzy term, which has various definitions. A long-term objective of most European countries is a system of regional or nationwide summaries, or sometimes even full (occasionally life-long) document-based or deeply structured records for each citizen. Such a summary or record may be viewed by any of the following:

- Either all the necessary persons concerned
- Only by those who need access in order to ensure safe health services
- Only by those who have been directly authorised by the patient.

The eventual development of EHR is evident in 25 out of the 32 countries reviewed during the preparation of the above-mentioned report. Six countries report that they currently have widespread local EHR in hospitals and other health provider organisations, which, however, are not yet fully interconnected. Three countries have a national EHR, although they are yet restricted in scope. Luxembourg, for example, maintains radiology records for its citizens; and in Sweden, citizens have a medication record. Germany, Sweden and Turkey are currently developing the structures of a patient summary or minimal data set. Consistent with its regions-based healthcare system, Spain is developing this work on a regional level. Only one country has a fully implemented EHR system of a countrywide scope—the Czech Republic. The Danish MedCom infrastructure supporting the electronic exchange of various healthcare related messages between healthcare and other service providers is expanded towards a countrywide EHR system as well.

Interoperability seems not to be as high on most countries' agendas as one might expect, given that it is one of the key issues in the EU eHealth Action Plan, it is a core element of current discussions among European Member States, and is also vividly present in international discussions. Only about one-third of the countries' fact sheets mention interoperability explicitly. With the exception of Italy, Romania and Spain, which have made technical and semantic interoperability priority issues, interoperability is seen as a challenge that needs to be addressed as part of a larger

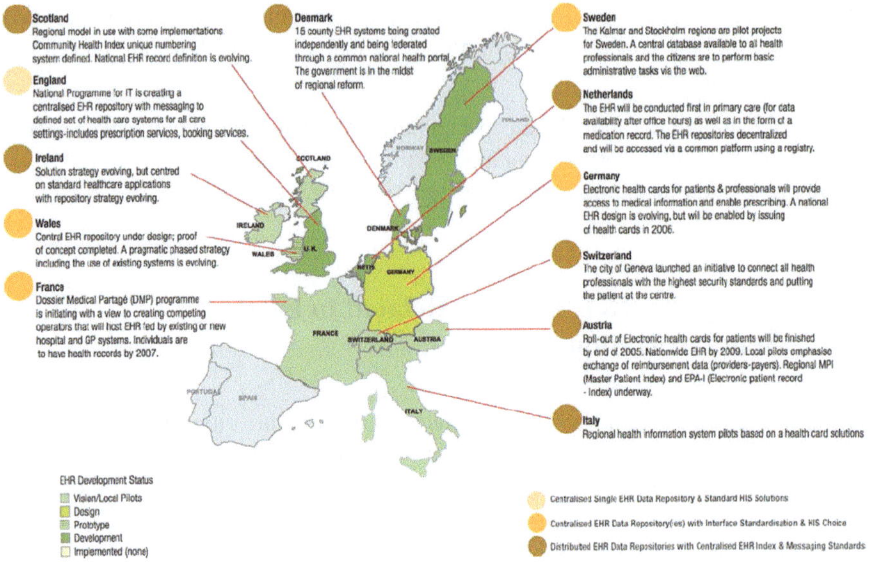

Fig. 11.6 Overview of status of implementation and uptake of hospital information systems and electronic health records throughout Europe (www.capgemini.com)

initiative. In Denmark, for example, MedCom has already developed a platform for technical standards and interoperability for eMessages—the Danish Health Data Network—, and SNOMED CT (Systematised Nomenclature of Medicine Clinical Terms) is currently being translated to provide semantic interoperability. Figure 11.6 summarizes the status in several European countries regarding existence of EHR.

11.6.1 European R&D Projects Related to EHR Standardization

The objectives of the EU funded (2007–2010) EHR-IMPLEMENT project (http://www.ehr-implement.eu) are to collect, analyse and compare national initiatives of broad scale EHR implementations among European countries focusing on socio-organisational issues and to provide best practice, policy and strategic recommendations to facilitate EHR implementation throughout Europe. The project aimed to:

- Analyse selected national policies, strategies and initiatives for broad scale EHR implementation taking into account cultural and organizational diversities of health systems in six European Member States (Belgium, Denmark, France, Ireland, Slovenia and United Kingdom);

- Carry out a survey of national policy and action plans for broad scale EHR implementation across European Member States;
- Identify the best practices towards broad scale EHR implementations in European countries;
- Raise the awareness of decision and policy makers regarding socio-cultural and organizational issues of broad scale EHR implementation; and
- Support the creation of a multidisciplinary community of scientific experts, technical personnel and National Health System representatives to promote information sharing and mutual learning

11.7 PHR/EHR Implementations

11.7.1 Commercial Implementations

Since the beginning major companies have invested on developing proprietary though freely accessible, own brands of Personal and Electronic Health systems, most well-known ones are listed below.

11.7.1.1 Google Health

Google Health was a personal health information centralization service, introduced by Google in 2008 and cancelled in 2011. Google Health was under development from mid-2006. In 2008, the service underwent a 2-month pilot test with 1600 patients of The Cleveland Clinic. Starting on May 20, 2008, Google Health was released to the public as a service in beta test stage. On 15 of September, 2010 Google updated Google Health with a new look and feel. On 24 of June, 2011 Google announced it was retiring Google Health on 1st of January 2012. The reason Google gave for abandoning the project was a lack of widespread adoption.

The service allowed Google users to volunteer their health records—either manually or by logging into their accounts at partnered health services providers—into the Google Health system, thereby merging potentially separate health records into one centralized Google Health profile.

Volunteered information could include "health conditions, medications, allergies, and lab results". Once entered, Google Health used the information to provide the user with a merged health record, information on conditions, and possible interactions between drugs, conditions, and allergies. Google Health's API was based on a subset of the Continuity of Care Record.

Google Health was an opt-in service, meaning it could only access medical information volunteered by individuals. It did not retrieve any part of a person's medical records without his or her explicit consent and action. However, it did encourage users to set up profiles for other individuals.

11.7.1.2 Microsoft Health Vault

Microsoft HealthVault (http://www.healthvault.com) is a WEB-based platform from Microsoft to store and maintain health and fitness information. It started in October 2007 in the US. As of 2013, the website addresses both individuals and healthcare professionals in the UK and Germany and the list of national deployments constantly grows.

A HealthVault record stores an individual's health information. Access to a record is through a HealthVault account, which may be authorized to access records for multiple individuals, so that a mother may manage records for each of her children or a son may have access to his father's record to help the father deal with medical issues. Authorization of the account can be through Windows Live ID, Facebook or a limited set of OpenID providers.

An individual interacts with their HealthVault record through the HealthVault site, or, more typically, through an application that talks to the HealthVault platform. When an individual first uses a HealthVault application, they are asked to authorize the application to access a specific set of data types, and those data types are the only ones the application can use. An individual can also share a part (some data types) or the whole of their health record with another interested individual such as a doctor, a spouse, a parent, etc.

HealthVault Connection Centre allows health and fitness data to be transferred from devices (such as heart rate watches, blood pressure monitors, Withings Wi-Fi body scales etc.) into an individual's HealthVault record. User can find and download drivers for medical devices. A dedicated Device Driver Development Package from Microsoft allows also device manufacturers to develop the software support for their devices such that they can communicate with the Health Vault.

HealthVault supports storage of DICOM (http://dicom.nema.org) based medical imaging. Consumers can upload and download medical imaging DVD through HealthVault connection centre. Third parties can also upload and download medical imaging to/from HealthVault. In addition, there has been plethora of HealthVault medical imaging viewers released by the third party to connect to HealthVault even on mobile phones.

HealthVault supports a number of exchange formats including industry standards such as the Continuity of Care Document (CCD), Continuity of Care Record (CCR) and Clinical Document Architecture (CDA). Support for industry standards makes it possible to integrate with diverse personal health record solutions.

A list of WEB applications from 3rd-party providers is available at the HealthVault website. Health service providers can develop their own support for MS Health Vaults via a HealthVault .NET Software Development Kit. Examples include:

- InstantPHR released by Get Real Health (www.getrealhealth.com)
- HealthUnity PHR Gateway (www.healthvault.com) by HealthUnity
- PassportMD developed by PassportMD (www.passportmd.com)
- ActivePHR released by ActiveHealth (www.activehealth.com)

11.7.1.3 World Medical Card

World Medical Card (www.wmc-card.com) is a product and registered trademark belonging to World Medical Centre, a Norwegian company headquartered in Bergen, Norway. The company's business is Health information technology, more specifically a supplier of Personal health records.

The World Medical Centre was established in 1998 for creating a system for improving the safety of people in situations requiring immediate medical treatment by a doctor, not familiar with the person's medical history. The international, personal medical card (World Medical Card) system was developed in cooperation with specialists in acute medicine and the University of Bergen, to allow individuals to carry essential medical information with them at all times and everywhere in the world.

Early versions included smart card and data matrix versions, but were abandoned as they required specific infrastructure to be installed at the facility receiving the patient in order to be useful. The latest generation of cards include the information in printed letters, but is only accessible by physically cutting open the card. This also reveals an emergency code that allows the medical professional access to a read-only web profile describing the cardholder.

Later the product portfolio was extended to include a WEB-based Personal health record, allowing the user to manage personal information while also serving as the source of information for producing the card, and a mobile application. Currently it is offered today in a form of three main elements:

1. Online ("onWeb") health profile for adding and editing personal health data
2. Multilingual ("onMobile") WAP phone application for accessing same data
3. Sealed ("onCard") physical card containing compact holder's health data

The company has been a pioneer in promoting the use of the International Classification of Diseases (ICD) ICD-10 document "Classification of Mental and Behavioural Disorders" and Anatomical Therapeutic Chemical (ATC) codes as defined by World Health Organization (www.who.int) for processing all medical data.

11.7.2 FREE and Open Source Implementations

The popularity of EHR/PHR systems have given raise to development of Open-Source platform too. Their capabilities and features can closely compete with commercial and proprietary implementation, except when it comes to interoperability and flexibility in developing of add-on services and applications. The list of Open Source EHR/PHR solutions (http://www.goomedic.com/open-source-emr-list#) suitable for serving as a base for the development of the project solution include:

- **INDIVO Health** (indivohealth.org) original personally controlled health record (PCHR) system. A PCHR enables an individual to own and manage a complete, secure, digital copy of her health and wellness information. INDIVO integrates health information across sites of care and over time. INDIVO is free and open-source, uses open, unencumbered standards, and is actively deployed in diverse settings, in particular our own Children's Hospital Boston and the Dossia Consortium.

- **TOLVEN Patient/Clinician HR** (www.tolven.org) focusing on delivering:
  - Personal Health Record (ePHR) that enables patients to record and selectively share healthcare information about themselves and their loved ones in a secure manner.
  - Clinician Health Record (eCHR) enables healthcare actors to securely access healthcare information collated from any number of trusted sources relating to individual patients in a structured and accessible way.
  - Healthcare Informatics Platform enables all healthcare data to be stored and accessed via ePHR and eCHR solutions. It uses industry standard technologies and data models.
  - Health Analytics solution that enables all data stored in the TOLVEN Platform to be extracted or analysed for statistical purposes
- **HealtheMe** from KRM Associates Inc. (http://www.krminc.com), an open source PHR system developed as part of a Medicaid eHealth transformation initiative for use in West Virginia, known also as HealtheMountaineer.
- **OpenEMR** (http://www.open-emr.org) is a Free and Open Source electronic health records and medical practice management application that can run on Windows, Linux, Mac OS X, and many other platforms. OpenEMR is ONC Complete Ambulatory EHR certified and is one of the most popular open source electronic medical records in use today. OpenEMR is supported by a strong community of volunteers and professionals. The OpenEMR community maintains OpenEMR's as a free, software solution for medical practices.
- **OpenMRS** (http://www.open-emr.org) is both software and a community. OpenMRS is a Java-based, web-based electronic medical record. It started from a simple data model, wrapped into an API, and then built a web-based application that uses the API. The OpenMRS API works like a "black box," hiding the complexities of the data model beneath it and ensuring that applications and modules using the API work with a similar set of business rules for managing the electronic medical record system data.

  At the heart of OpenMRS is a concept dictionary. It defines all of the unique concepts used throughout the system. Using combinations of questions and answers, observations (observable data) can be defined as well as forms that gather multiple observations within a single encounter.

  OpenMRS is constructed to support modules. Using modules, implementations are able to modify the behaviour of the system to meet their local needs without everyone having to agree on a single approach. Modules have full access to the system, so they can add tables in the database, alter behaviour of the API, and/or add or change web pages in the web application as needed to meet their needs

11.7.3 Implementations from European Research Projects

Many EHR platforms have been developed through European activities, funded by the European Commission through programs such as ICT, ICT-PSP, AAL, ARTEMIS and other ones. The list of most commonly known is provided below.

11.7.3.1 LinkCare Platform

It has been developed by the LinkCare Alliance (www.linkcarealliance.org) as part of the eTEN Project. It aimed to deliver proven information systems for chronic care, linking hospital care, primary care and home care. This concept fills a critical unmet need in current healthcare systems that are challenged by the need of different actors to cooperate in specific scenarios, notably those corresponding to the management of chronic patients.

From the end-user standpoint, LinkCare provides a single vendor integrated solution, a computer supported cooperative work environment, targeting core business areas where costs are greatest and where effective, timely and accurate communication between numerous institutions and actors is critical.

In practical terms, this means that healthcare providers can find in LinkCare technology a supporting tool for those services targeting long-term care. LinkCare facilitates professionals' tasks related to chronic case management, clinical documentation, patient tracking, data analysis, customer relationship management, patient education, professional communication as well as performance evaluation.

The market for information services in chronic care management is truly emerging and LinkCare aims at consolidating a position on it, not only based on the ICT platform but also on the accumulated experience on new models of delivery of care services acquired through the project and through other experiences. This is packed as accompanying consultancy services and/or system integration services to those stakeholders interested in adopting LinkCare in their work practices.

The LinkCare services portfolio includes:

- Electronic case management module and embedded EPR Interface: Capability to support integration and communication with the existing Customer's systems using industry standard protocol (namely, HL7 and XML)
- CRM module—"call-centre" support tools: Core to the LinkCare services is the existence of a single point of access for customers and networked professionals from where the different actions can be decided, ordered and transferred or executed. This requires a call centre supporting advanced CRM (customer relationship management) features. Furthermore, the specificity of the targeted health services means some extra capabilities such as the link to health information resources (corporate HIS, departmental solutions…)
- Professional's mobile support tools: Most of the services that LinkCare will support are based on the mobility of the professionals providing the service. The tools incorporated into LinkCare should fully support these new work practices

allowing the professional to minimise the need to contact the institution. Examples of services should include:

  - Enhanced off-line communication features: possibility of sending messages to pagers, SMS messages/emails/voices messages…
  - Enhanced online communication features: MMS, video-clips, video-conference
  - Automatic tracking of pending tasks with professional agenda update. Warning systems to help in the avoidances of delays

- Patient's mobile support tools: Similar to the previous point, the potential for deploying healthcare services to patients will depend on the level of monitoring capabilities that LinkCare tools could offer. Dealing with more severe patients directly links to the availability of more continuous, long-term monitoring, and the possibility of summarizing data, such as:

  - Services in the area of sleep disorders in COPD patients
  - Distant on-line supervision of rehab sessions.

  Both examples illustrate the need for:

  - extended periods of data collection (1–8 h)
  - significant amount of data transmitted simultaneously
  - need for tools for summarizing the information

- Computer Supported Cooperative Work module including workflow: Support for modelling the clinical processes at customer's site, providing the necessary communication and interaction tools among the involved actors
- On-line education and reference access to current content, docs and research: Emerging models of care provision maximize the importance of patients and care takers as key partners in the management of health conditions. LinkCare should be a comprehensive and trusted repository/broker for information content that could be passively provided (upon user's will, pull paradigm) or actively suggested (push paradigm, in alignment with the program where the patient is currently treated)
- Performance monitoring and evaluation module ("control panel/dashboard" and reporting tools): a set of decision management tools providing essential information about key indicators of the clinical and business processes, to allow timely intervention for corrective interventions and analytical support to improvement actions

11.7.3.2 intLIFE PHR

It is a platform developed internally by Intracom S. A. Telecom Solutions (https://146.124.106.153:8181/intLIFEv1), enhanced and geared to diverse health application through a number of FP7 funded research projects, such as ICT-PSP-NEXES, AAL-PAMAP, FP7-StrokeBack, Artemis-CHIRON, FP7-ARMOR.

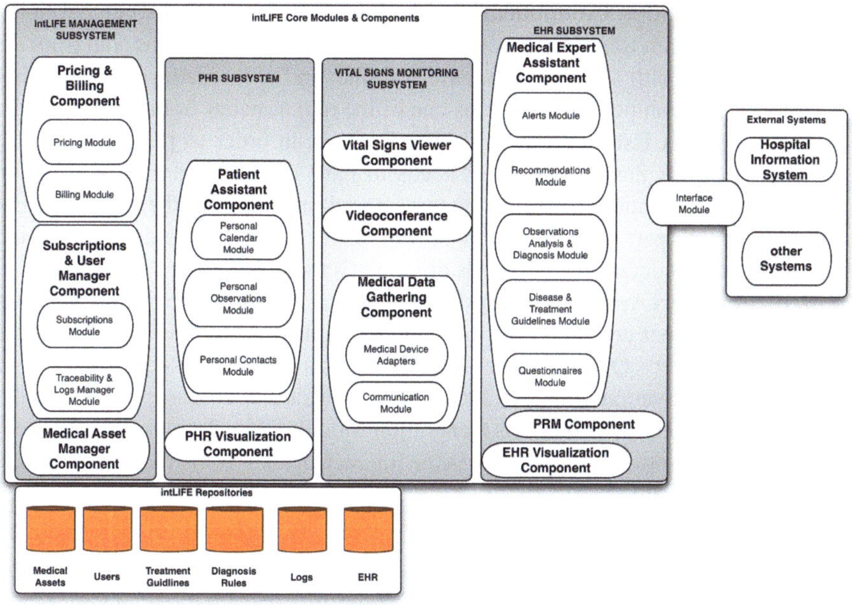

Fig. 11.7 intLIFE core modules and components: electronic health record subsystem

Developed initially with aim to support cardio-vascular services [22], the intLIFE platform has been further enhanced and currently supports the following subsystems (as shown in Fig. 11.7):

- Electronic Health Record Subsystem
- Vital Signs Monitoring Subsystem
- Personal Health Record Subsystem
- intLIFE Management Subsystem

The list features and functionality of the components currently available includes:

- **EHR Subsystem**: the intLIFE EHR application enables clinicians and paramedical personnel to edit/review health related information of the monitored subjects. An Overview tab will provide the user with a quick and printable outline of selected information (e.g. diagnoses, medications, surgeries, specific measurements, etc.). The following information fields are available:
  - Patient's General Health Profile; family health history, habits and social history (e.g. smoking, alcohol consumption), allergies, vaccinations
  - Visits; organ system findings, manual entry of symptoms and measurements
  - Medical Tests; test orders, manual entry of test results, test results overview and graphic representation
  - Diagnosis Management; insert new diagnoses, using the ICD-10 nomenclature, search for past diagnoses
  - Treatment Management; surgeries, medication

- **EHR Visualization Component** is responsible for presenting the content of the Electronic Health Record of a patient. It stores data to and retrieves data from the Electronic Health Record database of the intLIFE platform. The EHR Visualization component comprises one of the major means of interaction of the Medical Expert User with the intLIFE platform. In order to plan the patient's treatment, a Medical Expert User needs to have current data and information concerning the patient's medical history. Based on data from the patient's EHR the Medical Expert User can decide how to proceed, decide whether intervention is required and determine a success of patient's therapy.
- **Medical Expert Assistant Component** is part of the intLIFE EHR and includes a set of tools that are supportive to the utilization of the intLIFE system by the Medical Expert (Questionnaire Module, Alerts Module, Recommendations Module, Observation Analysis and Diagnosis Module, Disease and Treatment Guidelines Module and Reports Generation Module). It is designed in a modular way, so that new tools can be added in the future. Currently, only the Questionnaire Module is available.
- **Vital Signs Monitoring Subsystem**
  - Medical Data Gathering Component; is responsible for gathering measurements from the peripheral medical devices. It is designed in a modular way, so that new devices can be added in the future. It incorporates a Medical Device Adapter Module for each device to communicate with, and a Comms Module to transfer measurements to intLIFE server.
  - Medical Device Adapters Module; They implement the interfaces between the medical devices and the intLIFE platform, translating the vendor- or even device-specific message structures to a common structure, in order for the measurements to be seamlessly integrated to EHR.
  - Communication Module; It securely transfers the measurements collected from the peripheral medical devices. It can use different protocols (HTTP, FTP, etc.) and different encryption algorithms. It provides graphical user interface for configuring the communication parameters.
  - Vital Signs Viewer Component; The Vital Signs Viewer Component provides the Healthcare Professional User with an effective graphical user interface through which he/she may retrieve from the EHR database and visualize measurements from peripheral Medical Devices.
- **Personal Health Record application**: provides an IP-TV interface to the intLIFE EHR. Automatic login, reveals only those EHR tabs that are relevant to the user (e.g. exclude visits, etc.). Moreover, in addition to the typical EHR the PHR application supports personalized messages, personal reminders, personal rehabilitation plans, questionnaires, personal trainer presenting educational material, and videoconference module. This supports also a novel approach to game-based rehabilitation training [23], developed in the frame of an FP7 project StrokeBack.

- **intLIFE Management Subsystem**: the Administrator Web Interface enables the System administrator to have access to a set of administrative tools:
  - Users Administration; this process activates the necessary web user interface controls that enable the administrator to manage intLIFE users. The administrator is able to add (register), update and deactivate or reactivate intLIFE users.
  - Equipment Administration; This process activates the necessary web interface controls that enable the administrator to manage intLIFE equipment, i.e. it is a device manager that associates medical devices, STBs and other terminal equipment to physical or logical entities (patients/healthcare professionals or network nodes, respectively).

11.8 Adoption Problems

Despite the need for centralizing patient information, the adoption of PHR has been very slow. A study [24] made to assess the functionality and utility of online PHRs, identified 19 websites offering different versions of PHRs. Centralized PHRs should help patients relate accurate history during clinical encounters, check for drug interactions, eliminate unnecessary duplication of laboratory tests and diagnostic studies, and serve as an information hub for patients' health management. An analysis of web-based PHR systems has revealed that most websites did provide access to personal medical information. However, each system demonstrated limited capacity in a different way.

From the 19 sites examined, four were applicable only to certain diseases; another four had recurrent technical problems or connections to a specific hospital's information system. The remaining 11 sites did not provide patients with sufficient guidance as to how they should enter personal data. Some of the sites allowed patients to select medical conditions from categorized lists, which did not cover the patients' complete health condition while others allowed free text entry. To formulate medication history, sites that required patients to choose medication from lists requested them to enter a wide range of descriptive information for each medication such as prescribed dose, administration frequency, start date, name of pharmacy that issued the medication and name of provider that prescribed the medication. With respect to laboratory tests, only two allowed patients to import results from outside sources. From these two sites, only one was functional. Not every site allowed patients to enter insurance coverage information. Majority of the sites required patients to enter date and results of diagnostic tests.

Most people do not keep record of minute details of their healthcare experiences and therefore find it difficult to make use of web-based PHRs. Overall, the sites selected for evaluation offered limited functionality to the public. Low adoption of web-based PHRs can be a direct result of limitations in these applications' data entry, validation and information display methods. Hence, the PHR development needs to be guided in the future by ample patient-oriented research.

11.9 Privacy and Ethical Concerns

One of the most controversial issues for PHRs is how the technology could threaten the privacy of patient information. Network computer break-ins are becoming more common, thus storing medical information online can cause fear of the exposure of health information to unauthorized individuals. In addition to height, weight, blood pressure and other quantitative information about a patient's physical body, medical records can reveal very sensitive information, including fertility, surgical procedures, emotional and psychological disorders, and diseases, etc. Various threats exist to patient information confidentiality, example of are:

- Accidental disclosure: during multiple electronic transfers of data to various entities, medical personnel can make innocent mistakes to cause its disclosure.
- Internal leaks: medical personnel may misuse their access to patient information out of curiosity, or leak out personal medical information for spite, profit, revenge, or other purposes.
- Uncontrolled secondary usage: those who are granted access to patient information solely for the purpose of supporting primary care can exploit that permission for reasons not listed in the contract, such as research.
- External intrusion: Former employees, network intruders, hackers, or others may access information, damage systems or disrupt operations

Unlike paper-based records that require manual control, digital health records are secured by technological tools [25] and [26] identifies three general classes of technological interventions that can improve system security:

Deterrents—These depend on the ethical behaviour of people and include controls such as alerts, reminders and education of users. Another useful form of deterrents has been Audit Trails. The system records identity, times and circumstances of users accessing information. If system users are aware of such a record keeping system, it will discourage them from taking ethically inappropriate actions

Technological obstacles—These directly control the ability of a user to access information and ensure that users only access information they need to know according to their job requirements. Examples of technological obstacles include authorization, authentication, encryption, firewalls and more.

System management precautions—This involves proactively examining the information system to ensure that known sources of vulnerability are eliminated. An example of this would be installing antivirus software in the system. The extent of information security concerns surrounding PHRs extends beyond technological issues. Each transfer of information in the treatment process must be authorized by the patient even if it is for patient's benefit. No set of clearly defined architectural requirements and information use policies is available.

11.9.1 Ethical Guidelines Regarding Privacy and Using Medical Data

One of the most controversial issues for PHRs is how the technology could threaten the privacy of patient information. Network computer break-ins are becoming more common, thus storing medical information online can cause fear of the exposure of health information to unauthorized individuals. In addition to height, weight, blood pressure and other quantitative information about a patient's physical body, medical records can reveal very sensitive information, including fertility, surgical procedures, emotional and psychological disorders, and diseases, etc. Various threats exist to patient information confidentiality, example of are:

- *Accidental disclosure*: during multiple electronic transfers of data to various entities, medical personnel can make innocent mistakes to cause its disclosure.
- *Internal leaks*: medical personnel may misuse their access to patient information out of curiosity, or leak out personal medical information for spite, profit, revenge, or other purposes.
- *Uncontrolled secondary usage*: those who are granted access to patient information solely for the purpose of supporting primary care can exploit that permission for reasons not listed in the contract, such as research.
- *External intrusion*: Former employees, network intruders, hackers, or others may access information, damage systems or disrupt operations

Unlike paper-based records that require manual control, digital health records are secured by technological tools. Rindfleisch [25] identifies three general classes of technological interventions that can improve system security:

- *Deterrents*—These depend on the ethical behaviour of people and include controls such as alerts, reminders and education of users. Useful form of deterrents is Audit Trails, recording identity, times and circumstances of users accessing information. Users aware of such a record keeping system, are discouraged from taking ethically inappropriate actions
- *Technological obstacles*—These directly control the ability of a user to access in-formation and ensure that users only access information they need to know ac-cording to their job requirements. Examples of technological obstacles include authorization, authentication, encryption, firewalls and more.
- *System management precautions*—This involves proactively examining the information system to ensure that known sources of vulnerability are eliminated. An example of this would be installing antivirus software in the system

The extent of information security concerns surrounding PHRs extends beyond technological issues. Each transfer of information in treatment process must be authorized by patients even if it is for their benefit. No clearly defined architectural requirements and information use policies are yet available. While the trends and developments of ICT in healthcare have given rise to many positive developments, concerns about the use of ICT in user services mainly concentrate on the difficulty of respecting privacy and confidentiality when third parties may have a strong interest in getting access to personal health data electronically recorded and stored and difficulty in ensuring the security of shared personal data [8]. Therefore the project is dedicated to respecting and protecting the personal data, considered as extremely sensitive since they refer to the identity and private life of the individual. It recognises the intent to create a potential for the circulation of personal data, across local, national and professional borders, giving such data an enhanced European dimension, while respecting the principles of the European Convention of Human Rights, the rules of the Convention of the Council of Europe for the protection of individuals with regard to automatic processing of personal data and especially the European Directive 95/46/EC, for the protection of personal data will be strictly followed when addressing ethical issues.

11.9.2 Involvement of Adult Healthy Volunteers

Potential ethical issues that are addressed in this research will involve end user interviews, questionnaires and trialling of prototype systems during the development and testing. The right to privacy and data protection is a fundamental right and therefore volunteers have the right to remain anonymous and all research will comply with Data Protection legislation regarding ICT research data related to volunteers. During the research in ARMOR only participant who has sufficient cognitive and physical ability to be able to safely participate and clearly give informed consent are asked to participate. Potential ethical issues arise from the fact that participants, especially those who may tire easily or become distressed. Ethical issues may also arise when the system is used to give participant location or wellbeing information to third parties. Here release of this information is subject to informed consent of participants, and subject to the ethical frameworks to restrict knowledge of this information to only those given consent. All participants in the research are volunteers enrolled from the end user groups connected with this research and all ethical criteria are supervised by ethicists. Participants are ensured privacy and technical platform managing private user data is specially geared to enforce ethics.

11.9.3 Tracking the Location of People

Tracking the location of people is tightly linked with services delivered at the location of the user. This requires new look at the new socio-legal issues they raise. In the ARMOR project we only consider laws applicable to protecting privacy of the general population and NOT the laws and regulation specific for the case of the employee tracking and localization.

The European legislation has adopted specific rules requiring that the consent of users or subscribers be obtained before location data are processed, and that the users or subscribers be informed about the terms of such processing. The rule is that the applicable law is that of the Member State where the "controller" is established; and not that of the Member State of which the data subject is a national. If the controller is not established in a Member State, and in that case data protection laws of the 3rd-country should be found adequate by the EU-Commission. Location data collection will be in accordance to some basic principles: finality, transparency, legitimacy, accuracy, proportionality, security and awareness. Access to location data must be restricted to persons who in the course of exercising their duties may legitimately consult them in the light of their purpose. The list of relevant laws includes:

- *Directive 95/46/EC*: Protection of individuals with regard to processing of personal data and free movement of such data
- *Directive 2002/58/EC*: Processing of personal data and the protection of privacy in electronic communications sector
- *Directive 58/2002/EC* of the European Parliament and Council of 12 July 2002

Processing of personal data and the protection of privacy in electronic telecommunications sector is further governed by:

- *Directive 97/66/EC*: Data Protection in the Telecommunications Sector
- *Directive 99/5/EC*: Radio equipment and *telecommunications* terminal equipment and the mutual recognition of their conformity
- *Art. 29—Data Protection Working Party*: Working Document on Privacy on the Internet

11.10 Supporting Medical Research Using PHR

Epilepsy, the propensity for recurrent, unprovoked epileptic seizures, is the most common serious neurological disorder, affecting over 50 million people worldwide. Epileptic seizures manifest with a wide variety of motor, cognitive, affective, and autonomic symptoms and signs and associated changes in the electrical activities of the brain electroencephalography (EEG), heart electrocardiography (ECG), muscle electromyography (EMG), galvanic skin response (GSR), as well as changes in other important measurable biological parameters, such as respiration and blood pressure. Their recognition and full understanding is the basis for their optimal management and treatment, but presently is unsatisfactory in many respects. Epileptic seizures occur unpredictably and typically outside hospital and are often misdiagnosed as other episodic disturbances such as syncope, psychogenic and sleep disorders, with which they may co-exist, blurring the clinical presentation; on the other hand, costs of hospital evaluation are substantial, frequently without the desirable results, due to suboptimal monitoring capabilities.

The consent of users or subscribers shall be obtained before location data needed for supplying a value-added service are processed. Users or subscribers will be informed about the terms of such use. Access to location data must be restricted to persons who in the course of exercising their duties may legitimately consult them in the light of their purpose. All required user profile data are stored upon his/her mobile device and be securely protected. Relevant preferences relate to his/her diet, physical activities, dietary or transport/tourism related preferences, and, in general, simple everyday task preferences will not be stored locally. The user will have the capacity to view/hear, change or delete, as he/she wishes, all stored data by the system (including his/her profile data), with the help of a very simple and multimodal interaction (touch, buttons and voice input supported).

Types of data to be retained under categories identified in Article 4 of Directive 95/58 of 12th July, 2002. Specific safeguards—issues considered by the Article 29 working parties to be addressed with regard to the retention of data processed in connection with the provision of public electronic communication services (21st October 2005 opinion on the same subject directive proposal issued by the EU Commission on 21st September 2005). Reliable diagnosis requires state of the art monitoring and communication technologies providing real-time, accurate and continuous multi-parametric physiological measurements of the brain and the body, suited to the patient's medical condition and normal environment and facing issues of patient and data security, integrity and privacy.

The purpose of the FP7 projects "Advanced multi-parametric monitoring and analysis for diagnosis and optimal management of epilepsy and related brain disorders" (ARMOR) and StrokeBack is to manage and analyse large number of already acquired and new multimodal and advanced medical data from brain and body activities of epileptic patients and controls (MEG, multichannel EEG, video, ECG, GSR, EMG, etc.) aiming to design a holistic, personalized, medically efficient and affordable system for detecting abnormal condition and aid in efficient rehabilitation. New methods and tools have been already developed for multimodal data preprocessing and fusion of information from various sources. Novel approaches for large scale analysis (both real-time and offline) of multi-parametric streaming and archived data have been developed able to discover patterns and associations between external indicators and mental states, detect correlations among parallel observations, and identify vital signs changing significantly. Methods for automatically summarizing results and efficiently managing medical data are also being developed. The project incorporates models derived from data analysis based on already existing communication platform solutions emphasising on security and ethical issues and performing required adaptations to meet specifications.

ARMOR aims to provide flexible monitoring optimized for each patient and will be tested in several case studies and evaluated as a wide use ambulatory monitoring tool for seizures efficient diagnosis and management including possibilities for detecting premonitory signs and feedback to the patient. Therefore, our goal is to develop a personalized system that assists in diagnosis, prognosis and treatment of the disease. Such system should fulfil the following criteria; it should be non-invasive, mobile, continuous and unobtrusive, whereas all possible security and privacy aspects should be taken into account. Since access to large amounts of medical data is required for deriving all necessary models, a special effort is devoted to ensuring data anonymity, protection and restriction of access to private data in whole system.

The core system dealing with patient medical data in e Health related services and applications, like ARMOR, is the Electronic Health Record (EHR), defined as digitally stored health care information about individual's lifetime with the purpose of supporting continuity of care, education and research, and ensuring confidentiality at all times [27]. A patient's healthcare information may be spread out over a number of different institutes that do not interoperate. In order to provide continuity of care, clinicians should be able to capture the complete clinical history of the patient.

The Personal Health Record (PHR) is the electronic part of the health-related in-formation of a person (such as diagnoses, medications, allergies, lab test results, immunization records but also administrative tasks such as appointment or prescription renewals) that can be extracted from multiple sources, but always under the control of the consumer, patient or informal caregiver.

This is the relevant difference between the PHR and the EHR or electronic medical record, which is maintained by the healthcare providers and payers. The EHR standardisation [2] aims to ensure that patient records are used to support shared care among clinicians with different specialisations, while enabling the mobility within and among countries for people who give and receive healthcare. Since EHR systems commonly store sensitive information of patients, Ethical and privacy regulations apply as defined in the ISO/TC 215 Technical Report: "Electronic Health Record Definition, Scope and Context" [3].

11.10.1 Practical Approach to the Use of PHR

The PHR platform developed internally by Intracom S. A. Telecom Solutions, name intLIFE, has been enhanced and geared to diverse health application through a number of FP7 funded research projects, such as ICT-PSP-NEXES, AAL-PAMAP, FP7-StrokeBack, FP7-ARMOR and others (Maharatna, Bonfiglio et al. 2013). Special adaptations made in ARMOR and StrokeBack have been geared to allow safe sharing of patients' clinical data with appropriate measures, as described earlier, for the protection of private data, ensuring controlled access to it while ensuring that any data distributed cannot be traced back to the person from whom the data has been taken from. This way the intLIFE system could be safely applied in ARMOR for the purpose of deriving clinical models build from large amount of data for subsequently allowing more reliable feature based clinical diagnosis on other patients and detection of conditions not possible earlier. In order to provide necessary privacy and security safeguards EHR/PHR, Vital Signs Monitoring and Management subsystems are all connected via secure and encrypted interfaces controlled via authentication, authorisation and anonymizing modules.

The introduction of EHR/PHR systems is the response to the inherent problem of the medical community in dealing with growing amount of papers and printed type of medical records. This becomes also a matter of costs as much time and money is wasted on copying, faxing, and retrieving paper files. Move to electronically stored and managed patient records is both a simplification of the past problems, while adding new ones. Hence, governments demand increasingly secure and standard-compliant health records (http://www.ihtsdo.org/snomed-ct). In today's world, it takes more than a simple document to meet national record keeping guidelines.

Electronic Health Records are an obvious solution to all emerging problems in the medical care, offering simplification of growing patient records, stimulates easier exchange of data among medical professionals, contributes to cutting costs of medical care as a whole. Records are accessible by multiple health providers. Subject to providing sufficient safeguards at every level, a complete security of data may be achieved. Through data encryption, password protection, the electronic health record offers a peace of mind that data is kept away from unauthorised eyes. Nevertheless, although the future of e-Health has never looked so bright, there are still several concerns that needs careful attention.

Growth of e-health systems inherently implies that any patient's data may be stored not in one place, but on several diverse systems implying increased risk of leaking information to unauthorised third parties. Cyber security procedures are also not consistent across various systems, implying that some may be easier to break into and increasing vulnerability of data stored there.

What adds to the problem is lack of seamless interoperability among e-health systems based on electronic records. Since early stages of development of HL7 standard, now one of the base reference standards for e-health, it was considered only as a set of guidelines and not a factual standard to follow. This has resulted in systems being built and deployed that had implemented only a part of the HL7 specification (www.hl7.org) suited to particular needs of a given service provider. Interoperability among such restricted systems is a tiresome process, resulting in exchanging in-complete information.

This deficiency has been recently recognised as critical for future e-health and the HL7 is being evolved to define the base set of interoperability criteria for ensuring smooth collaboration among different health systems. However, this process is still ongoing and requires much more research work, including interoperability at the device level and especially for mobile physiological monitoring. In conclusion we can observe a dramatic changes in the e-health do-main with the introduction of electronic health records, boosting the efficiency of medical services at a lower cost, at the same time offering still a vast range of re-search challenges that we may expect to be pursued and hopefully resolved in the near future.

11.10.2 Seamless Authentication and Authorisation

Authentication and authorisation are two main means for allowing access for the user to a resource. Authentication involves such issues like identifying the user by means of either a simple login/password check to elaborate biometric analysis involving fingerprints, retina scans, and voice and/or face recognition etc. Authorisation then performs checks whether a given user may be granted access to a given resource or not. Such processes have been part of any secure system from the beginning of computing systems. Their complication has increased recently with the rapid growth of the amount of information resources and number of user accounts in each system, platform and/or network. This increases network administrators' work on properly securing their network and databases against un-authorised access at the same time providing users' with uninterrupted access to resources that they should be authorised to access.

However, currently employed methods for performing authentication and authorisation, in most cases, require user authentication every time he moves between resources stored on differently protected sites causing annoyance and loss of time. On the other hand approach to authorising users based on user-resource association requires tedious administrators' job to properly secure access to different resources and gives rise to frequent faults when users are either authorised to access resources that they should not have access to or not being able to access those that they should be able to.

This problem has been identified and addressed in almost all the systems since very long time. Administrators are offered means for specifying access rights per user group, policy definition mechanisms and macros allowing them to simplify management of access right to both existing and new users. Despite the fact that these tools are used the problems of authentication and authorisation still contain loop holes attributed mostly to human errors than to machine security as such. Our proposed seamless authentication and authorisation is aimed to simplify further the process of securing access to multiple-interconnected systems as well as making authorisation less prompt to faults.

Authentication is the process of authenticating a user across multiple account protected resources and platforms, i.e. agreeing between different authentication authorities means of establishing trust relationships and dependability for transferring users authentication status i.e. checking that a user is who they claim to be. The process will include customisation of means of authentication whereby users will not be required to perform authentication while transferring from one trusted party to another using single authentication. In case of moving from a party with lower authentication requirements to one requiring higher level of authentication, user will only be required to perform extra security checks while accessing differently protected resources instead of performing the whole authentication process from the beginning. Approaches like this have been already proposed, most known being Windows Passport where user may move between sites that support this technology. However, such methods do not take into account differences between authentication means required by different sites. This limits applicability of technologies to social WEB sites aiming to keep a record of users accessing their services.

Authorisation in computer terms refers to granting access to a resource to a given user. In most computer systems granting access is related to belonging to a specific user group meant to allowing administrators defining single access rights rules for a group of users. However, this makes it very difficult when it comes to more per-user access granting, especially in systems with large number of user accounts. What we propose is to unify and simplify means of granting user access to given resources. The process will define sets of resource authorisation dependencies whereby access to one resource may be implicitly granted upon prior-assigned access rights to another resource.

This will allow removing the need for the security administrator to provide access rights to every user to every resource, instead concentrating on defining security interdependencies between resources and defining user access right only to key global resources. Note, that this will not remove a possibility to explicitly grant or block access to a given resource to a given user, if required.

11.11 Conclusions and Potential for Future Research

The introduction of EHR/PHR systems is the response to the inherent problem of the medical community in dealing with growing amount of papers and printed type of medical records. This becomes also a matter of costs as much time and money is wasted on copying, faxing, and retrieving paper files. Movement to electronically stored and managed patient records is both a simplification of the past problems, while adding new ones. Reduction of paper records helps in simplifying records and eases exchange of data, though electronic communication implies stronger focus on preventing access to such data. Hence, governments demand increasingly secure and standard-compliant health records. In today's world, it takes more than a simple document to meet national record keeping guidelines. And an increasing number of optional treatments must easily fit into today's recordkeeping systems, complicating matters even further.

Electronic Health Records are an obvious solution to all emerging problems in the medical care, offering simplification of growing patient records, stimulates easier exchange of data among medical professionals, contributes to cutting costs of medical care as a whole. Records are accessible by multiple health providers. They are also fully integrated with other office functions, and interfaces with multiple vendors and diagnostic equipment, allowing for e.g. easy storage or X-ray, MRI, and other images in a standardised way. Subject to providing sufficient safeguards at every level, a complete security of data may be achieved.

Through data encryption, password protection, the electronic health record offers a peace of mind that data is kept away from unauthorised eyes. Nevertheless, although the future of e-Health has never looked so bright, there are still several concerns that needs careful attention. Growth of e-health systems inherently implies that any patient's data may be stored not in one place, but on several diverse systems implying increased risk of leaking information to unauthorised third parties. The cyber security procedures are also not consistent across various systems, implying that some of them may be easier broke into and increasing vulnerability of data stored.

What adds to the problem is lack of seamless interoperability among e-health systems based on electronic records. Since early stages of development of HL7 standard, now one of the base reference standards for e-health, it was considered only as a set of guidelines and not a factual standard to follow. This has resulted in systems being built and deployed that had implemented only a part of the HL7 specification suited to particular needs of a given service provider. Interoperability among such restricted systems is a tiresome process, resulting in exchanging incomplete information. This deficiency has been recently recognised as critical for future e-health and the HL7 is being evolved to define the base set of interoperability criteria for ensuring smooth collaboration among different health systems.

However, this process is still ongoing and requires much more research work, including interoperability at the device level and especially for mobile physiological monitoring. In conclusion we can observe a dramatic changes in the e-health domain with the introduction of electronic health records, boosting the efficiency of medical services at a lower cost, at the same time offering still a vast range of research challenges that we may expect to be pursued and hopefully resolved in the near future.

References

1. Kibbe D, Phillips R, Green L (2004) The continuity of care record. Am Fam Physician 70:1220–1223
2. Kevin H (2008) Standards for PHRs: Are PHRs ready for standards? Are standards ready for PHRs? J AHIMA 79(6):31–36
3. ISO/TC 215 Technical report: electronic health record definition, scope and context, 2nd draft, Aug 2003
4. HL7 (2004) HL7 EHR system functional model: a major development towards consensus on electronic health record system functionality. White Paper. http://www.hl7.org/ehr
5. Detmer D, Bloomrosen M et al (2008) Integrated personal health records: transformative tools for consumer centric care. BMC Med Inform Decis Mak 8:45
6. CEN/TC-251–ISO/TC-215, I. E. Electronic Healthcare Record (EHR) Communication (2010) Parts 1: reference model, part 2: archetype model, part 3: reference archetypes and term lists, part 4: security and part 5: interface specification
7. Muñoz P et al (2011) The ISO/EN 13606 standard for interoperable exchange of electronic health records. J Healthc Eng 2(1):1–24
8. Eichelberg M, Aden T, Riesmeier J, Dogac A, Laleci G (2005) A survey and analysis of electronic healthcare record standards. ACM Comput Surv 37(4):277–315
9. Schloeffel P, Beale T et al (2006) The relationship between CEN 13606, HL7, and OpenEHR. Health Informatics Conference
10. Beale T, Heard S, Kalra D, Lloyd D (2007) The openEHR reference model: EHR information model. openEHR Foundation
11. Ferranti JM, Musser RC, Kawamoto K, Hammond WE (2006) The clinical document architecture and the continuity of care record. J Am Med Inform Assoc 13(3):245
12. Dolin R et al (2006) HL7 clinical document architecture, release 2. J Am Med Inform Assoc 13:30–39
13. Rector A et al (1993) A framework for modelling the electronic medical record. Methods Inf Med 32:109–119
14. Johnson S (1996) Generic data modelling for clinical repositories. J Am Med Inform Assoc 3:328–339
15. Beale T (2002) Archetypes constraint-based domain models for future-proof in-formation systems. In: OOPSLA-2002 workshop on behavioural semantics, Seattle, WA
16. Goossen W, Goossen-Baremans A et al (2010) Detailed clinical models: a review. Healthc Inform Res 16(4):201
17. ASTM International—A century of progress (2005) ASTM international. http://www.astm.org/IMAGES03/Century_of_Progress.pdf
18. McCay C, Kalra D, Worden R (2008) Results of investigating the transformability between HL7 V3, openEHR and EN/ISO 13606. NHS Connecting for Health, Leeds

19. Sundvall E et al (2008) Integration of tools for binding archetypes to SNOMED CT. BMC Med Inform Decis Mak 8(Suppl 1):S7
20. Qamar R, Rector A (2007) Semantic issues in integrating data from different models to achieve data interoperability. Stud Health Technol Inform 129(Pt 1):674–678
21. European Commission (2007) eHealth priorities and strategies in European countries: eHealth ERA report
22. Maharatna K, Bonfiglio S (2013) Next-generation remote healthcare: a practical system design perspective. In: Artur K, Emmanouela V et al (eds) Patient health record (PHR) system. Springer, New York. ISBN 978-1-4614-8842-2
23. Emmanouela V, Krukowski A (2014) Serious games for stroke rehabilitation employing immersive user interfaces in 3D virtual environment. J Health Inform 6:105–113, ISSN 2175-4411
24. Kim M, Johnson K (2002) Personal health records: evaluation of functionality and utility. J Am Med Inform Assoc 9(2):171–180. doi:10.1197/jamia.M0978, PMC 344574. PMID 11861632
25. Rindfleisch TC (2007) Privacy, information technology and health care. Commun ACM 40(8):92–100
26. Ball MJ, Gold J (2006) Banking on health: personal records and information exchange. J Healthc Inf Manag 20(2):71–83, PMID 16669591
27. Krukowski A, Charalambides M, Chouchoulis M (2014) Supporting medical research on chronic diseases using integrated health monitoring platform. 4th International Conference on Wireless Mobile Communication and Healthcare—"Transforming healthcare through innovations in mobile and wireless technologies", special session on "eHealth@Home-Infrastructure and services for remote multi-parametric monitoring, analysis and support", Athens, Greece, 3–5 Nov 2014
28. Beale T, Heard S (2008) openEHR architecture overview. openEHR Foundation

Chapter 12
Offline Analysis Server and Offline Algorithms

Vasileios Megalooikonomou, Dimitrios Triantafyllopoulos, Evangelia I. Zacharaki, and Iosif Mporas

Abstract In this chapter we present algorithmic methodologies and data management system architectures for the analysis of medical data. We focus on offline analysis, in which the results of pattern or motif discovery and association rules extraction are not obtained in real-time. Offline analysis is based on stored data, structured within a database, and usually exploits large amounts of data for statistical processing and analysis.

12.1 Offline Analysis of Medical Data

Monitoring brain and body activities of patients with epilepsy and other brain related disorders is a very important procedure that has proven to be very useful both for clinical and research purposes. Brain and body activity monitoring involves data from a variety of modalities, such as EEG, MEG, ECG, EOG, EMG and fMRI. In the case of epilepsy, EEG is the major modality that is used in the literature for several purposes including diagnosis and seizure detection. In recent years several methodologies for offline analysis of brain activity from epileptic patients were developed. The different methods can be classified depending on the modality used (EEG, ECoG, fMRI, MEG, etc), the objective of the offline analysis (i.e. to detect a seizure related event, to study brain synchronization before and during the event, to localize the seizure origin etc.), or the type of epilepsy (i.e. focal epilepsy, idiopathic generalized epilepsy). The most common objective of methodologies analyzing brain activity of epileptic patients is the automatic detection of epileptic seizure onset.

In this chapter we present offline algorithms and methodologies for the analysis of epileptic data recordings. The algorithms are used for detection of patterns of interest and can be utilized for pattern discovery and for extraction of underlying association rules and motifs. Moreover, we present the database implementation for the analysis of the data.

V. Megalooikonomou (✉) • D. Triantafyllopoulos • E.I. Zacharaki • I. Mporas
Multidimensional Data Analysis and Knowledge Discovery Laboratory, Department of Computer Engineering & Informatics, University of Patras, Patras, Greece
e-mail: vasilis@ceid.upatras.gr

N.S. Voros, C.P. Antonopoulos (eds.), *Cyberphysical Systems for Epilepsy and Related Brain Disorders*, DOI 10.1007/978-3-319-20049-1_12

12.2 Offline Analysis Algorithms

The presented algorithms perform exhaustive analysis of biosignal data for the purpose of detection of patterns and/or events of interest, as well as for the discovery of rules and associations. Since there is no restriction for real-time operation, computationally demanding methodologies can be applied. The events of interest on which the implemented algorithms focus are K-complexes, sleep spindles and seizure onsets.

12.2.1 Detection of K-Complexes

The existence of K-complexes confirms the sleep scoring of stage 2 which makes them key features for the determination of different stages. However, their visual recognition in all-night sleep recordings is time consuming due to the extreme size of data and the presence of noise. Also difficulties arise mainly due to variability of their characteristics across subjects (inter-subject variability) and the different interpretation between scorers due to variation of human perception. From all the above it is apparent that the development of a reliable automatic method for detection of such micro-events is necessary as a preprocessing step for any subsequent analysis of the recordings (mainly EEG). Therefore we have investigated the automated detection of K-complexes and developed a new method utilizing the existing literature.

The method for automatic K-complex detection includes two steps. In the first step, all the K-complex candidate waveforms are extracted based on practices imitating those of manual visual scoring. The rules are based on the values of fundamental features (e.g., peak to peak amplitude, amplitude of the background, duration of the sharp wave, number of appearances in different channels) that a transient event should have in order to be marked as K-complex by a scorer. In the second step, the candidate waveforms are classified into KCs or non KCs in order to reduce the number of false positive detections. Two types of temporal patterns are extracted from the waveforms in order to characterize the detected candidates: (i) the amplitude change over time (signal representation) and the frequency content over time (frequency representation) calculated by summarization of the spectrogram across the frequency axis. The total pipeline of the method is illustrated in Fig. 12.1. The two main steps of our approach are described in more detail next.

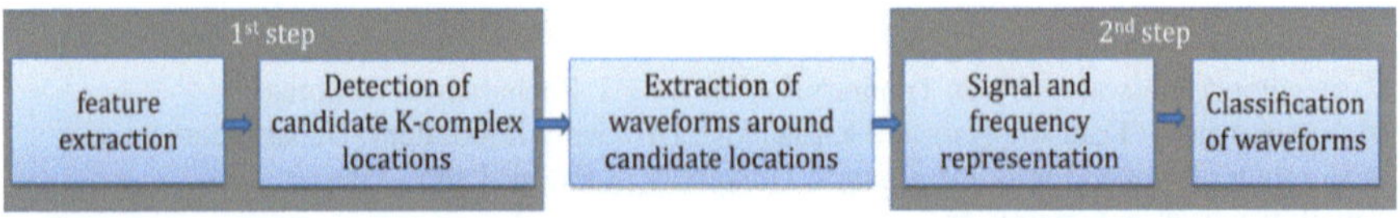

Fig. 12.1 Pipeline of the K-complex detection methodology

First step: In the first step, all the local minima from each available signal are detected and considered as candidate K-complexes. The peaks should have at least a minimum absolute peak height and be at least separated by a distance threshold; thus smaller peaks that may occur in close proximity to a large local peak are ignored. The center of each candidate K-complex is defined as the location of the local minimum. The idea is to limit the number of the candidates using a multi level approach. At each level a new feature related to a rule is calculated. If the value of the feature complies with the rule, the candidate passes on to the next level, otherwise the candidate is rejected.

The first feature is the peak to peak amplitude. The algorithm checks if the local minimum is followed by a positive component and computes the value of their peak to peak amplitude. Then a 10 second epoch around the local minimum is extracted from the EEG signal and a segment of 1 second epoch prior and following the minimum are removed, yielding an epoch indicating the background. The standard deviation of the background EEG amplitude is calculated and the following rules are checked:

- Rule 1: The peak to peak amplitude should be greater than three times the amplitude of the background EEG and between 70 and 560 μV. Also the standard deviation of the wave should be 1.2 times larger than the standard deviation of the background.
- Rule 2: The duration of the negative sharp wave should not be lower than 300 ms and greater than 1 s.
- Rule 3: All the frequencies more than 20 Hz should not have power greater than the 3 % of the total power of the signal. Otherwise it is assumed that muscle noise is contaminating the data.
- The candidates remained from the application of the rules in each channel are fused with the data integration procedure summarized in rule 4: rule 4: The candidate wave should be detected at least in three channels to be a K-complex.

  The majority of false detections even after the application of the above rules in our experiments are delta waves. Delta waves are similar to K-complexes and occur mainly in Slow Wave Sleep. Their characteristic is the fact that they occur repeatedly compared to the single appearances of K-complexes. In order to exclude delta waves from the set of candidates, we apply a simplified estimation of the Slow Wave Sleep epochs and we reject candidates belonging to them. Such a simplified estimation of Slow Wave Sleep is achieved by computing the frequency of appearance of the detected candidates in 10 s epochs. If the frequency of the detected candidates reaches a threshold, all the candidates belonging to the epoch are rejected because it is assumed that the epoch belongs to sleep stage 3 or 4. The threshold is an appropriate percentile of frequency of K-complex appearance estimated by whole night sleep recordings.
- Rule 5: Candidates in very close proximity to each other are rejected.

Second step: The remaining candidates after applying the aforementioned rules are given as input to a classifier. In this step, we implemented two different classification methods; a k-nearest neighbor (kNN) classifier, which uses training samples

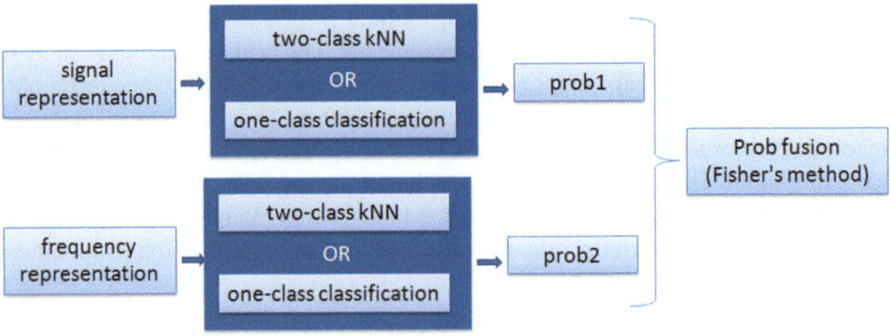

Fig. 12.2 Schematic diagram of K-complex classification (second step of the automatic detection methodology)

from both classes (K-complexes and false detections) and a one-class classifier, which uses training samples only from the positive class (K-complexes). Both classifiers use as input two different representations of the waveforms, the signal and frequency representation and produce a decision function indicating the probability of the input sample to be a K-complex. The two pseudo-probabilities (by each representation) are then fused by Fisher's method and thresholded to obtain the final decision. A schematic diagram of this step is shown above (Fig. 12.2).

KNN classifier uses a training set containing false detections from the first step and true K-complexes, which is balance the by applying k-means clustering in the class with more training samples. For each test sample, the kNN classifier computes the distance to each training sample. The distance is calculated either as Euclidean distance (if fast performance is targeted) or by applying Optimal Subsequence Bijection (OSB) (if accuracy is targeted). OSB is a distance function proposed in [1] and is appropriate for time series. OSB first computes an elastic matching of the two sequences and then calculates the distance using the distances of corresponding elements. Since both the query and the target sequences may be noisy, i.e., contain some outlier elements, the idea is to exclude the outlier elements from matching in order to obtain a robust matching performance.

The test sample's probability of being a K-complex is calculated as the ratio of the number of nearest neighbors being K-complexes divided by the total number of nearest neighbors, k. The two pseudo-probability vectors (for the signal and frequency representation) are fused by Fisher's method [2] which combines p-values from several independent tests into one test statistic that has a chi-squared distribution. Based on the fused vector of probabilities, a test sample is assigned to the K-complex class if its probability is above a threshold. Otherwise, the test sample is considered as non K-complex.

The one-class classifier uses a training set containing only K-complexes as training samples. The spectral clustering algorithm proposed by [3] is applied to the training set, producing a set of clusters. The resulting clusters represent the different patterns of the K-complex. The distribution of the samples within each cluster is learned by calculating all pairwise distances. Given the training clusters, the one-class

Table 12.1 Confusion matrix for kNN and one-class classifier (UoP dataset)

		Expert			
		kNN classifier		One-class classifier	
Algorithm		K-complex	Non K-complex	K-Complex	Non K-complex
	K-complex	214	949	211	1324
	Non K-complex	64	–	67	–

classifier assigns to a test sample a probability of being an outlier (non K-complex) if it deviates from the learned distribution as following. The test sample's distance to each cluster is calculated as the distance to the closest sample in each cluster. The test sample is then assigned to the closest cluster and the significance level (p-value) of the calculated distance for the specific cluster is returned as pseudo-probability of being an outlier.

The method was evaluated on an all night sleep EEG recording collected from a healthy female subject with sampling frequency 2500 Hz. The recording contains 14 excerpts each 30 min long. Before running the detector, each excerpt had been annotated manually by an expert neurophysiologist. In the whole recording 278 K-complexes were manually annotated. The detections of the algorithm were compared to those of the expert. The performance of the detection algorithm was evaluated by calculating the true positive rate (TPR) and the false positive rate (FPR). The True Positive Rate of the kNN classifier used with the OSB distance function was 77 % while False Positive Rate was 29 %. The corresponding confusion matrix, which reports the numbers of K-complexes detected by the algorithm and the expert, is illustrated in Table 12.1 (on the left).

Similar results were obtained by the one class classifier. A True Positive Rate of 76 % was obtained with about 43 % False Positive Rate. Table 12.1 (on the right) illustrates the respective confusion matrix.

Our method was also evaluated in a dataset, which consists of ten excerpts of 30 min EEG recordings, coming from more than one healthy subjects. The dataset is available for use by the University of MONS—TCTS Laboratory and Universite Libre de Bruxelles—CHU de Charleroi Sleep Laboratory. The sampling frequency was 200 Hz. This time, the kNN classifier obtained a 73 % TP rate for 33 % FP rate. The respective statistics for the-one class classifier are 76 % TP rate and 41 % FP rate.

12.2.2 Detection of Sleep Spindles

For the detection of sleep spindles we relied on the combination of discriminative and statistical models. Specifically, the support vector machines and the hidden Markov models (HMMs) were selected due to their advantageous performance in similar signal processing tasks. The sleep spindle detection is performed in two stages. In the first stage the signal is pre-processed, parameterized and processed

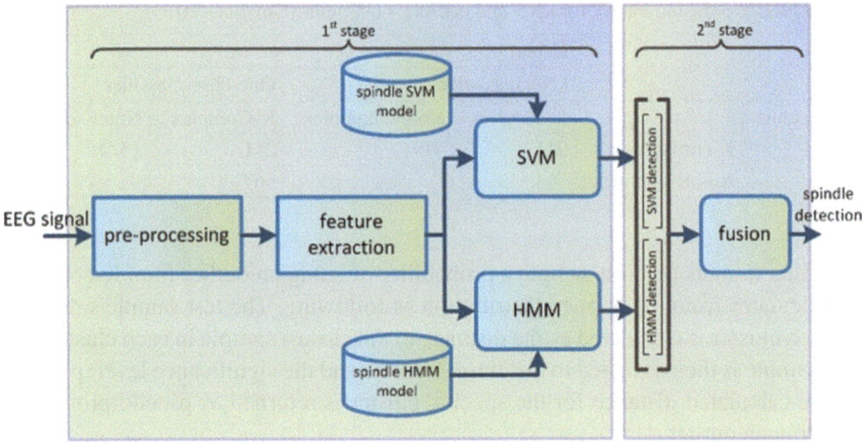

Fig. 12.3 Block diagram of the combined SVM-HMM sleep spindle detection scheme

independently from the discriminative (SVMs) and the statistical (HMMs) models. In the second stage the output recognition results from each model are combined by a fusion method in order to provide the final sleep spindle detection results. The block diagram of the proposed scheme is illustrated in Fig. 12.3.

As shown in Fig. 12.3 an EEG signal is introduced to the system and pre-processed with framing of the signal to blocks of L samples. Each frame is time-shifted by its preceding one by s samples, where $s \leq L$, thus resulting to overlapping frames. At the feature extraction block for each of the computed frames a parametric vector is computed by a signal parameterization algorithm.

The output of the feature extraction block is forwarded in parallel to each of the two models, i.e. the discriminative SVM-based and the statistical HMM-based sleep spindle models. Each of the two models estimates whether each incoming feature vector corresponds to sleep spindle or not, i.e. providing binary classification results with the corresponding recognition score for each of the two classes.

The second stage of the sleep spindle detection scheme exploits the recognition results of the two models, i.e. the SVM-based and the HMM-based, in order to combine them and provide a final decision for each feature vector. Specifically, the recognition results estimated at the first stage by each of the two models, are concatenated in a single vector as shown in Fig. 12.3. A fusion model utilizes the SVM-based and HMM-based predictions, which are included in the single vector, in order to provide the final decision for each frame of the EEG signal. In the present work the fusion model was implemented with the SVM algorithm. For the implementation of the SVM and the HMM models we relied on the WEKA and HTK software toolkits.

The performance of the proposed combined SVM-HMM sleep spindle detection scheme was evaluated on EEG data recorded at the Medical School of the University of Patras. The EEG data were recorded at sampling frequency of 2500 Hz,

Table 12.2 SVM-HMM sleep spindle performance (in percentages)

Recognized as →	Spindle	Non-spindle
Spindle	88.54	11.85
Non-spindle	03.41	96.59

using 64 channels. For the present evaluation we used the CZ electrode recordings from one subject. The duration of the evaluated data is approximately 401 minutes. The sleep recordings were manually annotated by expert sleep technicians of the University of Patras. The evaluated data include 1228 occurrences of sleep spindles and no overlap between training and test data subsets existed. The sleep spindle detection performance is shown in Table 12.2.

12.2.3 Spike Detection

Furthermore, a methodology for the detection of epileptiform discharges (spikes) in EEG that are manifestations of an epileptogenic abnormality of the brain was developed. After preprocessing, the method applies coarse-to-detailed modeling of the spike waveform in a first step and then classifies the transients based on Locality Preserving Projections (LPP) and Support Vector Machines (SVMs) in a second step.

Specifically, in the first step, a peak detection algorithm is applied to detect the primary vertex of the spike in the form of local minima. In order to reduce the number of candidate peaks, only peaks that are at least separated by 100 ms are retained, while small peaks that may occur in close proximity to larger local peaks are ignored. Then, around each detected peak, the EEG signal is extracted within a window (starting 100 ms before the primary vertex and ending 200 ms after it) defining the spike waveform. From each waveform the two half-waves are segmented and four time-domain parameters are calculated: the amplitude difference (A1, A2) and the duration (D1, D2) of each half-wave [4]. These parameters describe the slope around the primary vertex and are calculated as amplitude difference and time interval between the primary vertex (wave minimum) and the two closest local maxima (before and after the minimum). Thresholding of the four parameters is applied to distinguish between candidate spikes and other artifacts. A maximum value on amplitude is used to discard spikes due to noise or movement. It should be noted that the spike amplitudes differ between subjects. In this study we relaxed the threshold constraints to allow detection with high sensitivity. As a consequence, the specificity of this step became especially low; thus a subsequent step is required to reduce false detections using a more elaborate approach.

In the second step of the method, the extracted candidate waveforms are classified into spikes and non-spikes by learning the patterns of interictal discharges based on manually annotated examples. During the training phase of the spike detector a dataset of feature vectors with known class label (spike or non-spike) is used to train a classification model. At the test phase, the existing model is used in order decide for each feature vector the corresponding class. As non-spike examples we used the false detections of the first step instead of random background segments, in order to achieve a more distinctive separation between spikes and spike-like waves.

If the extracted waveform (raw signal around the primary vertex) is used as input to the classifier, classification is deemed to fail due to the high dimensionality of the input pattern. For example, if the sampling frequency is 200 Hz and the extracted time window is 300 ms, the input pattern consists of 60 measurements. When the number of parameters increases, the volume of the space grows so fast that the available data become sparse. In order to obtain statistically sound and reliable results, the sample size should grow exponentially with the dimensionality (effect of dimensionality curse). Large sample size however is difficult to acquire and also increases the computational cost. For all these reasons a common step before classification is feature selection, where only some of the parameters are used for classification, or dimensionality reduction, where the data are mapped into a lower dimensional space.

In this study, we performed Locality Preserving Projections [5, 6] to reduce the dimensionality of the spike-like waveforms. LPP is a linear approximation of the nonlinear Laplacian Eigenmap [7]. It finds a transformation matrix A that maps a set of points $x_i \in \mathcal{R}^d$ $(i=1,\ldots,m)$ into a set of points $y_i \in \mathcal{R}^l$, $y_i = A^T x_i$, such that $l \ll d$. LPP is designed for preserving local structure, thus it is likely that a nearest neighbour search in the low dimensional space will yield similar results to that in the high dimensional space. The mapped (low-dimensional) feature vector is finally used as input to an SVM classifier.

The spike detection tool was evaluated using EEG recordings provided by the Epilepsy Monitoring Unit, St. Luke's Hospital, Thessaloniki, Greece. The data used in this work were acquired during a whole-night sleep EEG of a subject with history of epilepsy. The spikes were visually identified by an experienced neurophysiologist as transients clearly distinguished from background activity with pointed peaks. The markers were manually placed at the peak of the negative phase, but imprecise markings were later corrected by automatically shifting them to the largest negative peak within a predefined neighborhood (±50 ms) around the original marking.

The data consist of 9 h recordings and include 101 marked spikes. The total number of candidate spikes after the first step was 4708 consisting of 99 TP and 4609 FP. Only two spikes were missed. The second step of the method was assessed by tenfold cross validation on the data. The classification of waveforms identified 156 (out of the 4708) as spikes with 98 of them being TP and 58 being FP. Thus the total sensitivity of the method is 0.97 (=98/101) and the number of FP per minute is 0.1. The results of the method are shown in Table 12.3 and are compared against other approaches reviewed by Wilson et al. [8]. Only methods for which both sensitivity and FP/min were reported, are included for comparison. For some methods more than one set of results are reported corresponding to different algorithms or parameters.

Table 12.3 Comparison of our spike detector against other approaches (see [11] for references)

Method	No. patients	Duration (min)	No. spikes	Sensitivity	Fp/min
Our spike detector	1	540	101	0.97	0.1
Davey et al. 1989	1	5.3	23	0.74	0.4
Witte et al. 1991	1	1	50	0.90	4.0
Gabor, Seyal 1992	5	63.8	752	0.97	1.5
Hostetler et al. 1992	5	100	1393	0.76	5.2
				0.87	1.4
Webber et al. 1994	10	40	927	0.73	6.1
Senhadji et al. 1995	1?	10	982	0.86	6.8
Feucht et al. 1997	3	90	1509	0.88	1.8
Dumpelmann, Elger 1999	7 (ECoG)	136	2329	0.32	9.4
				0.23	14.4
				0.28	12.6
				0.41	10.2
Ramabhadran et al. 1999	18	270	982	0.96	0.4
Wilson et al. 1999	50	143	1952	0.47	2.5
				0.15	3.2
				0.70	4.1

Total number of spikes not known. Algorithm detections reviewed after processing

12.2.4 Seizure Detection

The seizure detection module uses as input data segments from electroencephalographic and electrocardiographic recordings. The architecture of the module consists of time and frequency domain feature extraction followed by classification. Four classification algorithms were evaluated on three epileptic subjects in an intra-subject setting based on cross validation. Specifically, we relied on the support vector machines (SVMs), implemented with the sequential minimal optimization method and polynomial kernel function, a two-layered backpropagation multilayer perceptron (MLP) neural network, the k-nearest neighbor (IBK) algorithm and the C4.5 decision tree. We also examined the discriminative ability of the extracted features. For the estimation of the importance of each feature in terms of their binary classification ability, we relied on the ReliefF algorithm. More details can be found in the corresponding publications [9, 10].

The seizure detection architecture was evaluated on polysomnographic recordings of three subjects (denoted here with subject IDs 07-08-09) recruited from the St Thomas' epilepsy clinic and diagnosed with idiopathic generalized epilepsy manifested with absences. Classification was performed with subject-dependent seizure detection models, i.e. the evaluated training and test data subsets consisted of one subject in each experiment. In order to avoid overlap between the training and the test data a tenfold cross validation strategy was followed.

The accuracy of detection of seizure varied from 92.31 % (for subject 07) to 77.78 % (for subject 07), while the detection of clear intervals, i.e. interval recordings without the presence of seizure, was found to be at least 99.24 % for all three subjects. This variation in the accuracy of the seizure detection epochs (approximately 21 %), compared to the more robust detection of non-seizure epochs (with variation of approximately 0.7 %), is mainly owed to the limited amount of training data which does not allow the SVM algorithm to model the seizure characteristics with absolute performance. It is worth mentioning that the duration of each idiopathic generalized seizure occurrence was approximately 2 up to 4 successive epochs, while the available seizure occurrences for subject 08 were significantly fewer than the ones of subject 07, which presented the best seizure detection performance.

12.3 Offline Analysis Server

12.3.1 Overview

Starting from the acquisition phase ARMOR's multimodal data are transmitted through several aggregation points, where they can be accessed by either clinical experts or automated algorithms in order to be processed and/or analyzed. The data acquisition process, illustrated in Fig. 12.4, is described as follows:

Initially, data are collected at the home gateway. Since the uploading process may be delayed or interrupted, the sensor data must be securely stored at the middleware, using the nAssist database.

From the home gateway, the data are uploaded to the central repository, i.e. the Personal Health Record (PHR), where they are permanently stored. The PHR provides the necessary interfaces for up/downloading the data, and serves as the ARMOR's permanent data repository.

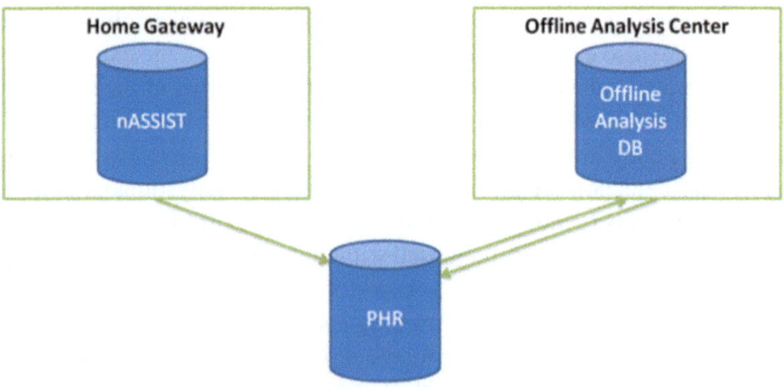

Fig. 12.4 ARMOR's databases

The offline analysis DB is the database that facilitates the offline analysis. The data/metadata stored at the offline analysis DB will be directly queried by the analysis algorithms. Data will be transferred from the PHR to this database in order to be analyzed.

These databases provide the functional, usable and secure environment for ARMOR's data and they correspond to the particular expertise of the involved partners. For example, sensitive information is stored only at the PHR developed by ICOM, since it already provides secure interfaces for up/down loading data. The nAssist database incorporates StreamInsight™, the data stream management system that runs at the middleware, which is within Sensing and Control Systems, S.L. area of expertise. Finally, the offline analysis DB is developed by University of Patras and addresses the requirements of the offline analysis algorithms.

The role of offline analysis DB is to facilitate offline analysis by providing straightforward access methods that will be used by the analysis algorithms. In order to achieve this goal, we identified all the data and metadata that guide the analysis process (and will certainly be queried) and developed a database for their effective storage.

This database is located at the offline analysis center (Fig. 12.5) since it is a key part of the offline analysis. In contrast to the PHR, this database stores all the available data and metadata from the multimodal recordings in a form that can be queried and processed effectively. The schema that we designed for this database (see Sect. 12.3.2) is able to support queries over all this data.

The offline analysis DB doesn't store any information related to the subject (e.g. name) that could be used for the identification of the subject. In contrast, it stores the information that potentially could be used for the analysis. In Sect. 12.3.2 we describe the database schema of the offline analysis DB in detail.

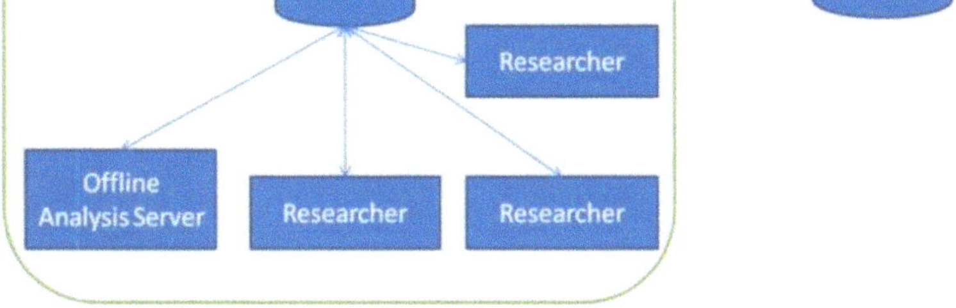

Fig. 12.5 Offline analysis DB

12.3.2 Database Schema for Supporting Offline Analysis

In this section we present the major design decisions for the offline analysis database. Specifically, we show some parts of the entity relationship (ER) diagram that correspond to the most important data to be stored.

The schema presented in Fig. 12.6 was designed to serve both the data that will be recorded with the ARMOR's sensors and the recordings that took place in the past and are part of the offline analysis process (for the initial development of the algorithms). Furthermore, the utilization of this schema gives us the ability to store both clinical and neurophysiological data and metadata and thus, gives us the ability to directly search for and capture clinical—neurophysiological relationships and correlations.

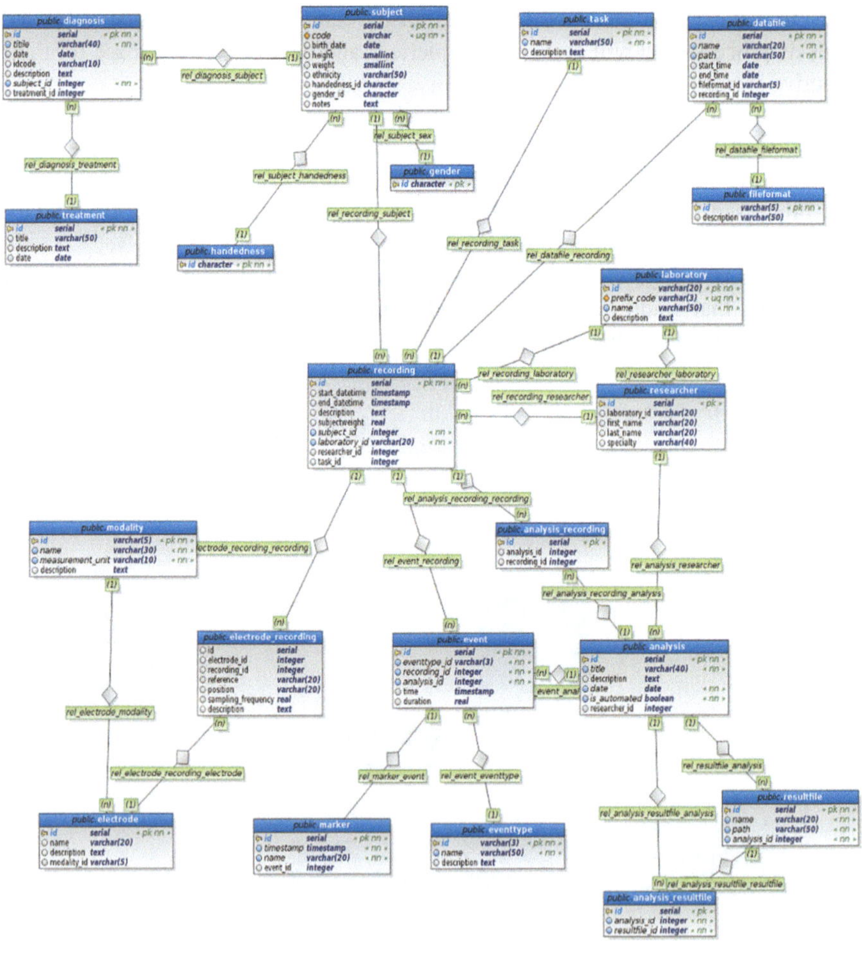

Fig. 12.6 Database schema for offline analysis DB

12.3.2.1 Recording-Related Entities

One of the most important entities of the database is the Recording. By this term we refer to the data acquisition process, consistent in time (i.e. the recording starts when the electrodes start recording the signals and stops when the acquisition is over). Important information that should be stored for every recording includes:

- The subjects' ID.
- The starting and ending time.
- The modalities (e.g. EEG, ECG, GSR).

The actual data (the signals) from the recording are stored in one or multiple files and thus, one recording is associated with a set of files, one for a specific time period. A more detailed description of the recording-related information is shown at the ER diagram in Fig. 12.7.

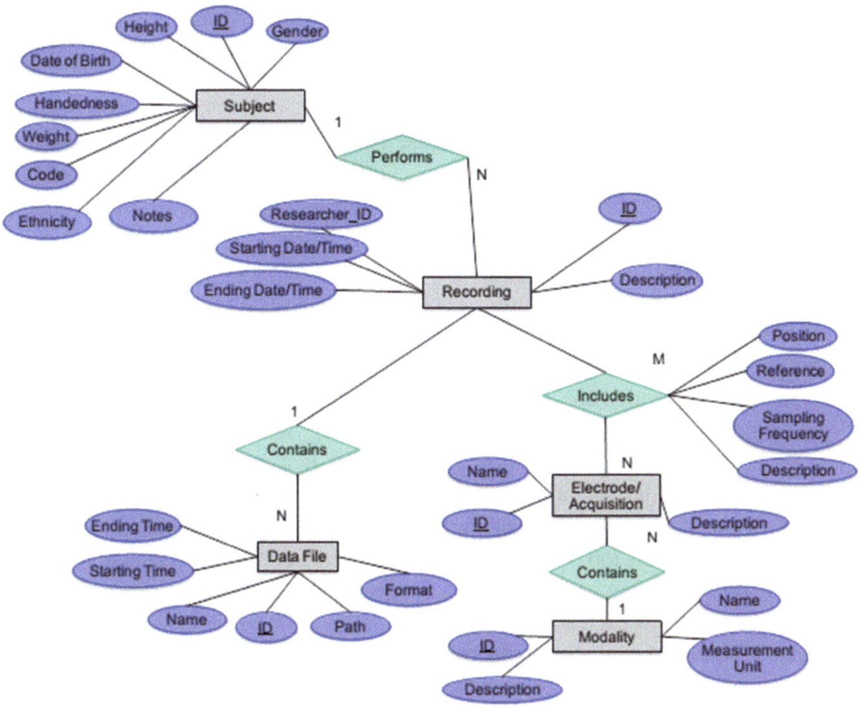

Fig. 12.7 ER diagram for the recording data

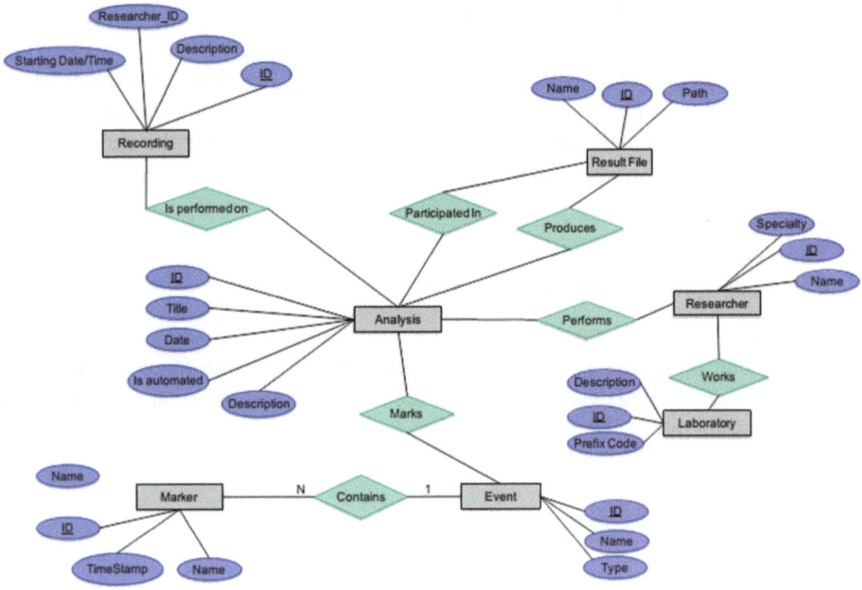

Fig. 12.8 ER diagram for the analysis related data

12.3.2.2 Analysis-Related Entities

Another very important task in the database is to keep track of every analysis that takes place and the corresponding results. For this purpose our database schema includes entities that are able to describe the whole analysis process: the purpose of the analysis, the utilized data and the results (Fig. 12.8).

Anything other than the raw data is considered to be outcome of analysis (either manual or automated). For example, when a sleep technician marks events of interest (e.g. sleep spindles or K-Complexes) he/she performs a manual analysis.

As we mentioned earlier, since marking events of interest is a very common analysis task, we added the Event and the Marker entities for the description of events. One event can be described with a name (e.g. seizure) and a set of markers (e.g. onset, offset, etc). Furthermore, every event is associated with an Analysis process that can either be automated or manual.

This Event/Marker design decision is very flexible and can be used to describe other types of analysis too. For example, sleep scoring can be described in this manner, by using the stages as events (i.e., Stage I,…,Stage IV) and markers for the beginning and end of each stage.

Finally, since the analysis results can be of any type, we included the Resultfiles entity for anything that cannot be described as a set of events/markers. Also, it is very common that the analysis is performed not only on raw data but also on the results of some previous analysis. For this purpose we connect the Analysis with the Resultfile by another relation.

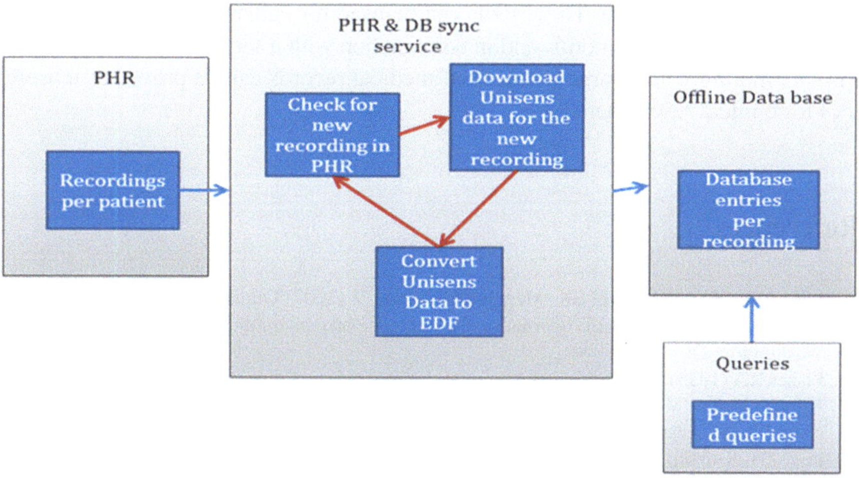

Fig. 12.9 Main components of database synchronization

12.3.2.3 Offline Analysis DB Schema

Offline Analysis Database and PHR Synchronization

Every online recording performed with the ARMOR online platform system is uploaded to the Patient Health Record (PHR) database. From there it is transferred to the offline analysis database. In order for this service to be precise, there are several steps that should be followed. The first one is to check if there is a new recording in PHR. It should be noted that every recording is related to a specific patient, a specific device performing the recording and the data of the recording. The synchronization service checks in specified time intervals for a new recording in the PHR system. If a set of new recordings is detected, then these recordings are requested from the PHR database. The format of the data in each new recording is a Unisens format. Once the whole recording is downloaded and checked for errors (e.g. download errors, missing files etc) the data are transformed in EDF format. Once this step is completed a subroutine will handle the EDF data to extract all the necessary info and store them in the offline analysis database. Then the newly uploaded data can be accessed through prespecified queries.

In the following figure (Fig. 12.9) the structure of the main components of the synchronization of the two databases is presented.

12.4 Conclusions

Offline analysis of medical data using computationally demanding algorithms can be proved useful for the in depth analysis of health-data records. Algorithmic methodologies and data management system architectures for the analysis of medical data were presented here. These tools can be used for pattern discovery, extraction of association rules and motifs, and in combination with a server scheme appropriate for data mining within large volumes of medical records can be proved as a useful tool for clinical researchers.

References

1. Latecki L, Wang Q, Tezel SK, Megalooikonomou V (2007) Optimal subsequence bijection. In: Proceedings of 7th IEEE international conference on data mining. IEEE Computer Society, Washington, DC
2. Fisher RA (1925) Statistical methods for research workers. Oliver and Boyd, Edinburgh
3. Ng AY, Jordan MI, Weiss Y (2001) On spectral clustering: analysis and an algorithm. Adv Neural Inf Process Syst 2:849–856
4. Inan ZH, Kuntalp M (2007) A study on fuzzy C-means clustering-based systems in automatic spike detection. Comput Biol Med 37:1160–1166
5. He X, Niyogi P (2004) Locality preserving projections. In: Advances in neural information processing systems, vol 16. The MIT Press, Cambridge, p 37
6. van der Maaten LJP, Postma EO, van den Herik HJ (2009) Dimensionality reduction: a comparative review. Tilburg University Technical Report, TiCC-TR 2009-005
7. Belkin M, Niyogi P (2002) Laplacian Eigenmaps and spectral techniques for embedding and clustering. In: Advances in neural information processing systems, vol 14. The MIT Press, Cambridge, pp 585–591
8. Wilson SB, Emerson R (2002) Spike detection: a review and comparison of algorithms. Clin Neurophysiol 113:1873–1881
9. Mporas I, Tsirka V, Zacharaki EI, Koutroumanidis M, Megalooikonomou V (2014) Evaluation of time and frequency domain features for seizure detection from combined EEG and ECG signals. In: Proceedings of 7th international conference on PErvasive technologies related to assistive environments (PETRA 2014)
10. Mporas I, Tsirka V, Zacharaki EI, Koutroumanidis M, Megalooikonomou V (2014) Online seizure detection from EEG and ECG signals for monitoring of epileptic patients. In: Proceedings of 8th Hellenic conference on artificial intelligence (SETN 2014)
11. Zacharaki EI, Garganis K, Mporas I, Megalooikonomou V (2014) Spike detection in EEG by LPP and SVM. In: Proceedings of IEEE EMBS international conference on biomedical and health informatics (BHI'2014)

Chapter 13
Personalized Management of Epilepsy Through Smart Use of EEG and Detailed MEG Analysis

Vahe Poghosyan and Andreas A. Ioannides

Abstract ARMOR aspires to use all available information from diverse sources to enhance the management of epilepsy in terms of diagnosis, home monitoring and evaluating the efficacy of treatment. In this chapter we introduce a novel approach aimed at achieving this goal through smart use of available data. The primary focus of the approach is to use magnetoencephalography (MEG) or high-density electroencephalography (hdEEG), when they are available, to improve the effectiveness of routine electroencephalography (EEG) tests. In a nutshell, we acknowledge that with the currently available hardware it is not possible to record MEG or hdEEG on a routine basis, so we want to use one or few such measurements to develop a personalized neurophysiological model of the epileptic condition and then use the model to derive specific biomarkers, which can be measured with simple and more easily available techniques such as few-channel EEG, electrocardiogram (ECG), electrodermal response (EDR) etc. Thus the goal is to use MEG or hdEEG for background reference and a base for development of biomarkers that can address specific clinical questions for a specific patient using simpler devices, which can be easily used in the home environment, such as few EEG electrodes. Although as stated above the primary focus of the approach is the smart use of advanced neurophysiological techniques, such as MEG or hdEEG, the methods developed here can be used with simpler data, such as low-density EEG (e.g. 21-electrode 10–20 system), which is largely available for most epilepsy patients, to significantly improve the EEG setup for further routine measurements.

V. Poghosyan • A.A. Ioannides (✉)
Laboratory for Human Brain Dynamics, AAI Scientific Cultural Services Ltd., Nicosia, Cyprus
e-mail: a.ioannides@aaiscs.com

N.S. Voros, C.P. Antonopoulos (eds.), *Cyberphysical Systems for Epilepsy and Related Brain Disorders*, DOI 10.1007/978-3-319-20049-1_13

13.1 MEG and EEG in Epilepsy

Electroencephalogram (EEG) and magnetoencephalogram (MEG) are the only neuroimaging techniques that offer a time resolution of 1 ms or less, providing unique neurophysiologic data for the investigation of epilepsy. Currently, EEG is the most widely used technique in epilepsy. It is often used in establishing the diagnosis, and, once diagnosed, is the most important tool in defining the type of epilepsy, the prognosis and the initial approach to therapy. Paired with video monitoring, EEG is also commonly used for the localization of epileptogenic regions.

MEG is less accessible technique, because of its high cost and required expertise. Nevertheless, it is very powerful and especially useful when combined with source analysis, because it can produce accurate description of neural activity in the brain with relatively simple head models (e.g. spherical or single-shell realistic head models). When available, usually in big hospitals with large catchment areas, MEG is the technique that must be consulted when traditional EEG studies fail to deliver sufficient information to support clinical decision making.

At present, the most well-established application of MEG in epilepsy is in pre-surgical evaluation, where it is used to localize the irritative and seizure onset zones and to map the adjacent eloquent cortex for the accurate definition of the epileptogenic zone. Successful application of MEG source analysis provides critical information for planning a direct surgical intervention or invasive monitoring. Yet, the predominant method used for this purpose is the equivalent current dipole (ECD) model, which is an old and widely accepted, however limited method (see Sect. 13.1.2). The primary reason for the predominant use of the limited ECD model in clinical practice is the stipulation of authorities overseeing clinical techniques (e.g. only ECD methods have been approved by Food and Drug Administration, FDA, in US).

In general, MEG source analysis during pre-surgical evaluation of patients with epilepsy has clearly demonstrated usefulness, predictive value and positive effect on clinical decision making [1, 2]. Moreover, recent studies have shown that MEG can be a clinically valuable tool also during the primary diagnostic process of epilepsy, and possibly substitute sleep deprived EEG measurements [3, 4], although currently it is not widely used for this purpose. Overall, there is sufficient credible evidence showing the clinical usefulness and added value of MEG in the evaluation of patients with epilepsy [5].

13.1.1 The Relative Advantages and Disadvantages of MEG and EEG

EEG is far more common in hospitals than MEG: there are only a few dozen whole head clinical MEG systems in the world, while EEG is available in every hospital. EEG has two key practical advantages over MEG: it is much cheaper and is wearable, enabling routine laboratory, ambulatory and video-EEG long-term monitoring. However, MEG has other important advantages: in contrast to EEG, MEG signals (magnetic fields) are not attenuated or distorted by the skull and other tissues intervening between the brain and scalp; MEG is reference-free, while EEG relies on a reference, which makes interpretation of the data difficult; MEG measurements are easier to perform, with no electrodes attached to the skin, this may be of special importance when dealing with patients.

Non-invasive source localization is highly recommended during the pre-surgical evaluation of patients with epilepsy, since it is ordinarily much cheaper and safer than invasive studies. Currently, MEG is superior in this area with its better sensitivity and higher source localization accuracy than EEG. However, recent advances in hdEEG recordings combined with advanced modeling techniques slowly overcome some of the source localization challenges of EEG.

MEG is selectively sensitive to tangentially oriented brain sources, while EEG measures both radial and tangential activity, although the radial components dominate the EEG signals at the scalp. A radially oriented current source produces no magnetic field outside a spherically symmetric volume conductor. However, since the human head is not exactly spherically symmetric, the radial orientation is not well defined, and an approximately radial source in the brain is not silent in MEG. Therefore, it is possible that all source orientations generate a magnetic field that is detectable by MEG. Moreover, only very few (<5 %) cortical sources are expected to have the orientation of the lowest sensitivity (close to radial) in MEG [6].

13.1.2 Electromagnetic Source Analysis

Currently, the most common and accepted electromagnetic source localization method in epilepsy is the ECD model, despite its well-documented deficiencies [7]. ECD methods model the brain activity with one or very few point-like generators, which naturally work very well for focal epilepsy with a single restricted epileptogenic cortical zone. However, the recent evidence shows that even *simple linear distributed source models* are still more useful than ECD for the pre-surgical evaluation [8]: they provide more accurate source localization [9–11] and robustly localize onset of spikes even when ECD fails [11]. Moreover, distributed source models can accurately represent the spike propagation through the cortex [11, 12], which can be very valuable in pre-surgical evaluation of epilepsy.

In the current project we rely on an *advanced non-linear distributed source model*, magnetic field tomography (MFT), [13, 14] and on its newly developed adaptation for EEG, EFT, for the accurate source analysis of interictal activity, without any assumptions about the number of the signal generators and minimal a priori constraints about the nature of the source distribution. Currently, MFT is the most robust and accurate method for the MEG source localization [15–20]. However, it has not been widely used in clinical practice, including epilepsy, because it is computationally very intensive—until recently MFT required almost supercomputing resources, and it is quite complex to apply, requiring significant expertise in electromagnetic inverse problem. Furthermore, similar to other non-ECD source models, the lack of approval from relevant authorities (e.g. FDA of US) largely limited the use of MFT to research applications.

In research settings, however, MFT has been extensively used in wide variety of applications, including in ARMOR (see Chap. 4, Chap. 5 and Chap. 13). It was routinely used to map brain activity with almost all multichannel MEG hardware, from the early systems covering only part of the head [21] to most of the whole-head systems used today. MFT provides highly accurate source localization [15, 18], outperforming many other popular methods [20]. Recently we have shown that MFT estimates can be successfully combined with cytoarchitectonic probability maps, effectively mapping brain activity within individual cytoarchitectonic areas [19], and tracing the evolution of stimulus-evoked activity through different cytoarchitectonic areas [18, 22]. Moreover, MFT was successfully used to localize deep subcortical sources in thalamus, amygdala, cerebellum and brainstem [20, 23–25].

13.2 Smart Uses of MEG and EEG in Epilepsy

Clinical MEG is becoming increasingly more popular; however its application in epilepsy and active developments still remain limited mostly to the pre-surgical source localization. We assert that the superior spatial and temporal accuracy of MEG and its fundamental theoretical relationship with EEG, which have been repeatedly demonstrated in controlled experiments, can be successfully exploited in the clinical practice in other ways too. Below we briefly introduce the gist of a novel approach we have developed for ARMOR with a key aim to substantially improve the personalized management of epilepsy and other disorders.

In most cases it is impossible to routinely monitor patients using advanced neurophysiological techniques, such as MEG or hdEEG. The key idea in our approach is to use the advanced techniques for only few well-defined measurements and then utilize the recorded data to plan and improve the effectiveness of the routine measurements with simpler techniques (e.g. only few channel EEG). To be effective, this approach requires high-quality analysis of all available data, both anatomical and neurophysiological. The anatomical data (e.g. high-resolution MRI, CT etc.) are used to build a detailed anatomical (geometrical) and conductivity model of the head (e.g. 5-shell head model comprising the scalp, skull, cerebrospinal fluid, brain and ventricular system). The accuracy of the head anatomical model plays an important role in shaping the neurophysiological activity model derived from MEG or EEG. The neurophysiological data are used to derive the patterns of real time brain activity and EEG signals, based on which the necessary biomarkers will be developed.

Here we describe in some detail one application of this approach in the ARMOR project where a method has been developed for determining the optimal number and position of EEG electrodes for long-term monitoring based on previous MEG or EEG recordings.

13.2.1 Optimal Electrode Selection

The routine EEG measurements for the online monitoring and offline analysis of brain activity should be performed with as few electrodes attached to the patient's head as necessary to achieve the specific clinical requirements. In most cases the exact positions of the few electrodes will significantly affect the amount of useful information that can be extracted. When available, MEG and/or EEG recordings of relevant epileptic events (e.g. ictal or interictal) can be used to determine the optimal electrode positions that allow extraction of maximum useful information. The procedure that we have developed to this end involves two main steps: (1) predicting EEG signals at large number of unmeasured scalp positions (typically 345 positions corresponding to 'EEG 1005' system) from a set of MEG or low-density EEG measurements of relevant epileptic event and (2) selecting the few electrode positions, which in combination provide the most information about the event.

Naturally, when hdEEG data for the specific epileptic event of interest are available, the first step of the procedure can be skipped. However, in most cases, such data are not recorded, because of the cost, expertise and the time and effort it requires. Even in cases when sufficient resources are available to perform hdEEG measurements, often it may be recommended to employ more conventional lower density EEG in the interest of rational utilization of resources and patient comfort.

13.2.1.1 Predicting EEG Signals at Unmeasured Positions from a Set of MEG/EEG Measurements

In this section we describe and test two methods (lead field-based and MFT-based) that can be used to predict EEG signals at unmeasured positions from a set of MEG or EEG measurements. The main purpose of these methods within the ARMOR project is to estimate as closely as possible what would have been measured with hdEEG during the specific epileptic event, when such real measurements are not available. This is the first step of a procedure aimed at determining the case-specific optimal positions for scalp electrodes.

Lead Field-Based Method

This simpler method relies exclusively on lead fields and follows an algorithm similar to the one described in [26]. A lead field describes the sensitivity distribution of a sensor (e.g. electrode, magnetic coil etc.) and is completely determined by the sensor's properties and the conductivity details of the biological medium. Lead fields relate the sources inside the head to the external measurements:

$$d = L \cdot j \tag{13.1}$$

where d represents the MEG/EEG signals measured at the external sensors; it is a N element data vector comprising measurements at N discrete locations in the extra-cranial space, j represents the sources inside the head; it is a $3M$ element current density vector at M locations in the source space (e.g. brain; current density is a vector in 3D space, thus j has $3M$ element for M locations), and L is the linear operator representing the lead field; it is a $N \times 3M$ lead field matrix, which relates current sources to the measured data. Note that $M \gg N$ and L is not invertible.

In practice, the lead fields are obtained by solving the electric or magnetic forward problem for unit dipoles at a set of fixed locations in the conductor. Forward problem is solved using the reciprocity approach [27] for each sensor for three (in 3D space) orthogonal unit dipoles placed at each point of the pre-defined source space within the conductor volume.

In this method the lead field matrix L_{src} for the source sensor array and L_{trg} for the target positions are computed based on the same pre-defined source space and volume conductor model. Singular value decomposition (SVD) is then used to obtain the pseudoinverse of the lead field matrix for the source sensor array:

$$L_{src}^{-1} = V \cdot W^{-1} \cdot U^T$$

where $L_{src} = U \cdot W \cdot V^T$ is the SVD of L_{src} and W^{-1} is computed from W by taking the reciprocal of each non-zero element, after zeroing eigenvalues (diagonal elements of W) smaller than the set tolerance level (e.g. one thousandth of the sum of all eigenvalues). Finally the transfer matrix is computed as:

$$T = L_{trg} \cdot L_{src}^{-1} = L_{trg} \cdot V \cdot W^{-1} \cdot U^T$$

Application of this matrix to the measured source signal provides the estimated signal at unmeasured target positions:

$$d_{unmeasured}^{est} = T \cdot d_{measured}$$

MFT-Based Method

The general idea behind this method for predicting EEG signals is to use the best available MEG/EEG source localization method to estimate sources j^{est} of the measured signals $d_{measured}$ and then use this estimates to predict the signals $d_{unmeasured}{}^{est}$ at the unmeasured target positions (see 13.1):

$$d_{unmeasured}^{est} = L_{trg} \cdot j^{est}$$

Currently, MFT is the most robust and accurate method for the MEG source localization (see the Sect. 12.1.2 above). Thus in this method we use MFT to estimate the sources of the MEG/EEG signals and (13.1) to predict the EEG signals at unmeasured positions over the head.

Test Results

The methods described above were evaluated using simulated as well as real data. The simulated data were used first to test and compare the methods. The methods were then separately evaluated further using real data: the MFT-based method was tested for predicting EEG signals at pre-defined scalp positions from whole head MEG measurements; the lead field-based method was tested for predicting high-resolution EEG signals from low-resolution EEG measurements.

Evaluation Using Simulated Data

MEG and EEG signals (90 channels) were simulated for 57 dipole sources spread throughout the brain using a 4-shell realistic head model (scalp, skull, CSF, brain). Both methods were then used to predict the EEG signals at the same 90 electrode positions from the simulated MEG signals. To evaluate the methods we computed the Pearson correlation coefficients between the predicted and simulated EEG signals. The evaluation results showed that although the MFT-based method for predicting EEG signals is computationally more demanding, it provides significantly better estimates (Fig. 13.1). The considerable deterioration of the predictions for the superficial dipoles can be explained by the higher spatial detail of the map produced by a superficial source.

Evaluation of MFT-Based Method Using Real Data

The MFT-based method was further evaluated using real data recorded in an MEG sleep experiment [24, 28]. The data, at a sampling rate of 625 Hz, included signals from 151 MEG channels, two scalp EEG channels with electrodes placed at approximately C3 and C4 positions and referenced to the contralateral ear, and several auxiliary channels. The MEG signals were processed offline by converting them to third order synthetic gradient, removing DC based on the whole trial, low-pass filtering at 200 Hz, notch filtering at 50 Hz (and its harmonics) and cleaning the eye blink and cardiac artifacts using independent component analysis. For the evaluation purposes we selected sleep segments with prominent signal components (K-Complex, Fig. 13.2).

The method was used to predict EEG signals at 97 positions over the scalp, corresponding to the extended 10–20 system (Fig. 13.3a). The prediction results for the two electrodes for which actually measured EEG signals were available, showed that the MFT-based method can predict real EEG signals with high accuracy (Fig. 13.3b).

The Pearson correlation coefficient was computed between each measured EEG signal (2 channels) and each predicted signal (97 channels). The strongest correlation coefficients were around 0.7 and importantly were found for the corresponding measured and predicted signals (i.e. measured at C3/C4 with predicted for C3/C4).

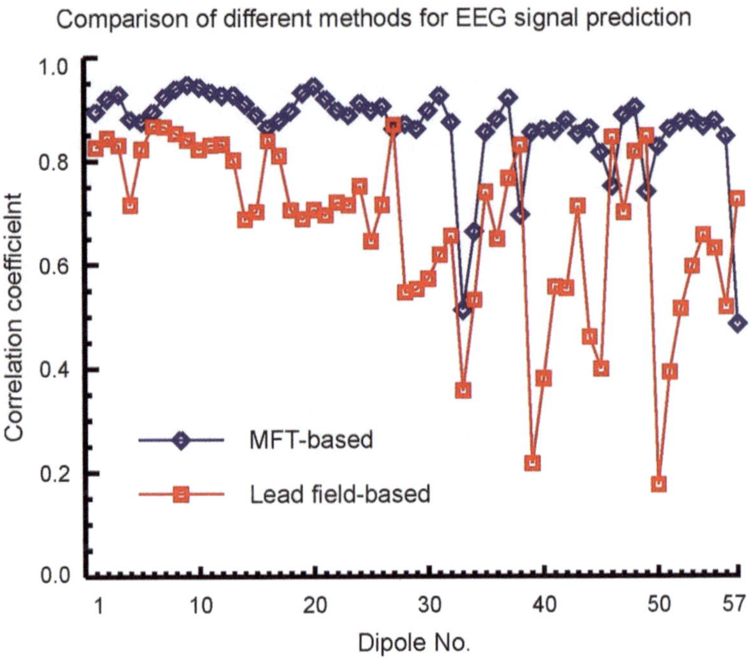

Fig. 13.1 Comparison of methods for predicting EEG signals. EEG signals at 90 positions over the scalp were simulated for 57 dipole sources and predicted from the corresponding MEG simulations. Ordinate shows the Pearson correlation coefficient computed between the 90 predicted and simulated EEG signals; abscissa shows the sequential number of a simulated dipole. The di-poles are arranged roughly in terms of depth, so the early dipoles are deep while the dipoles with higher index are more superficial. The diamonds (*blue*) and squares (*red*) show the correlation coefficients for the MFT- and lead field-based methods respectively. In almost all cases the MFT based method performed better

Evaluation of Lead Field-Based Method Using Real Data

The lead field-based method was further evaluated using real EEG measurements of epileptic events.[1] The data were recorded from a patient using a 64 channel EEG system at a sampling rate of 1000 Hz. The recorded EEG signals were re-referenced to common average reference and were band-pass filtered from 1 to 500 Hz with notches at 50 Hz and its harmonics. For the evaluation purposes we extracted two-second long data epochs around epileptic events (47 events) identified by a specialist (−1 to 1 s with respect to the event onset). Figure 13.4a shows EEG signals in one such typical epoch.

[1]These data were recorded as part of a routine clinical examination at the MINDLab Core Experimental Facility at the Center of Functionally Integrative Neuroscience (CFIN), Aarhus University and Aarhus University Hospitals, Aarhus, Denmark.

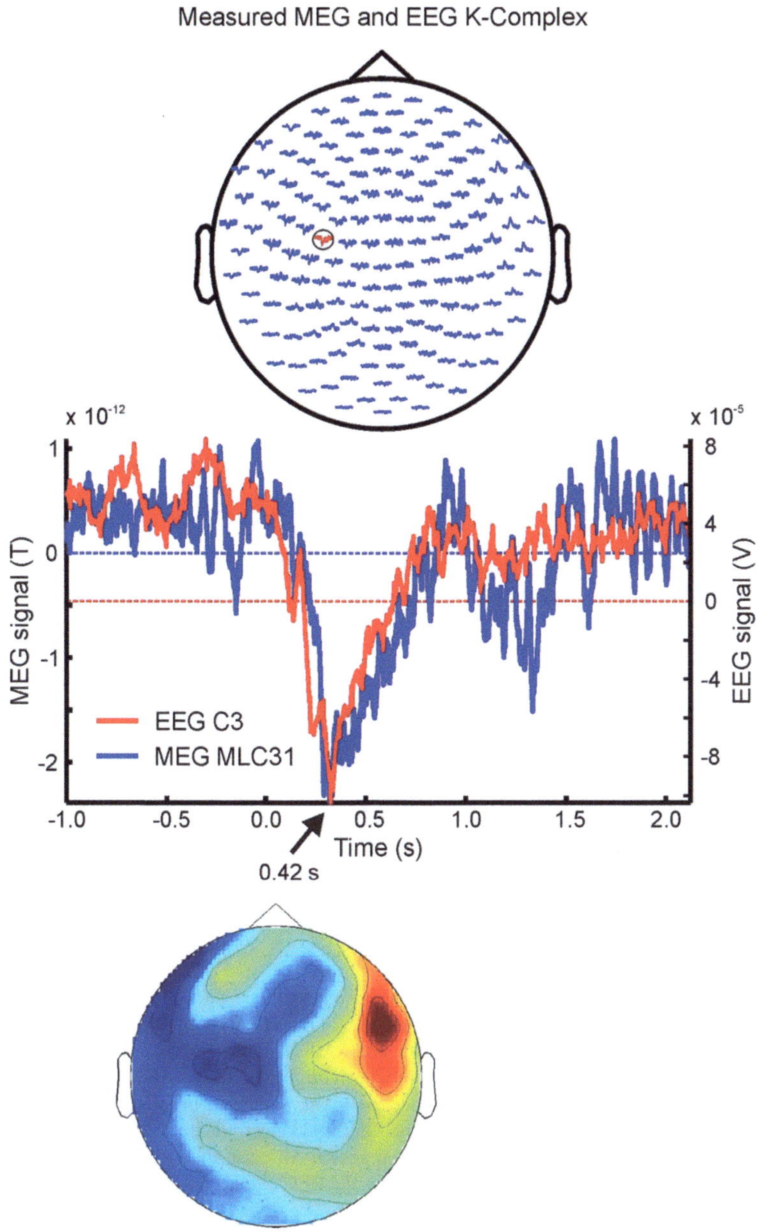

Fig. 13.2 A typical K-Complex recorded during whole-night sleep MEG recording. The first row shows the spatial distribution of MEG signals over the head, viewed from above, with the nose pointing up-wards. Abscissa shows the time with respect to K-Complex onset (from −1 to 2 s); ordinate shows the variation of magnetic field measured by 151 axial gradiometers of the CTF/VSM Omega System. The strongest MEG signal at the peak of the K-Complex (recorded by the MEG channel MLC31) is highlighted in *red*. The second row shows the signals of MEG gradiometer MLC31 (*blue*) and EEG electrode C3 (*red*) during the K-Complex. Abscissa shows the time with respect to K-Complex onset (from −1 to 2 s); ordinate on the *left/right side* shows the MEG/EEG signal magnitude, with the zero level indicated by a *blue/red dashed line*. The third row shows the MEG signal topography over the head at the peak of the K-Complex (0.42 s after the onset)

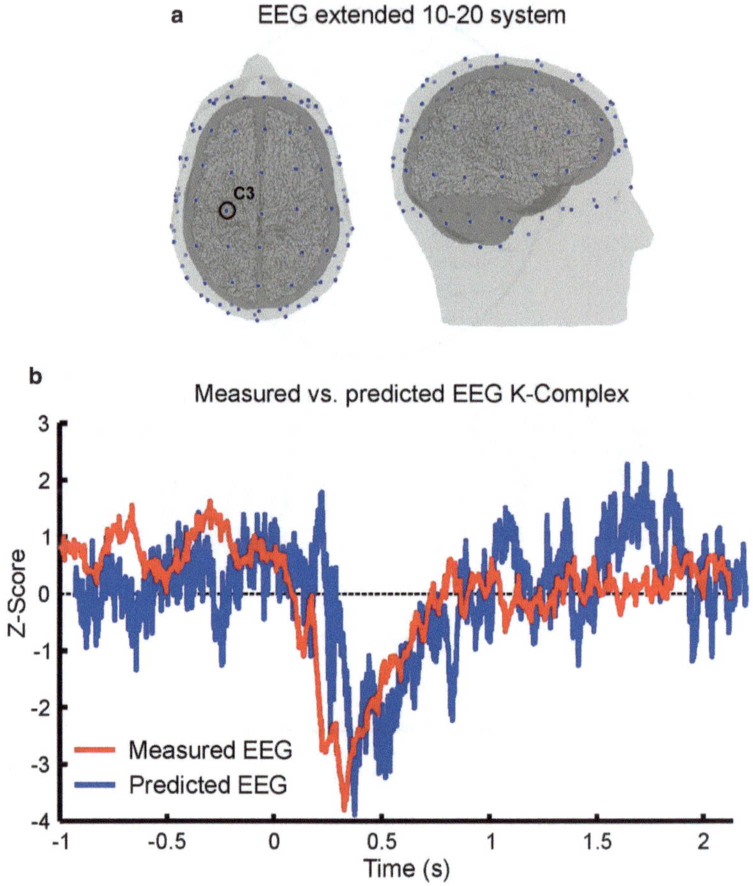

Fig. 13.3 EEG signals predicted from MEG measurements during a K-Complex, using the MFT-based method. (**a**) Positions of the 97 EEG electrodes of the extended 10–20 system over the scalp. EEG signals at these positions were predicted from MEG signals. The C3 electrode, for which actually recorded EEG signal was also available, is highlighted. The electrode positions are shown over the head model: from outside to inside, the scalp, skull and brain layers are shown. (**b**) Measured (*red*) and predicted (*blue*) EEG signals at the C3 electrode during the K-Complex. Both signals were z-transformed to facilitate the comparison. Abscissa shows the time with respect to K-Complex onset (from −1 to 2 s); ordinate shows the EEG signal magnitude in terms of z-score

Fig. 13.4 (continued) 64-channel EEG signals (*red*), measured 21-channel EEG signals (*blue*, subset of 64 channels) and predicted 64-channel EEG signals (*black*, predicted from the 21-channel subset). The time course for each data set was estimated by calculating the difference between the signals with the strongest positive and negative deflections at the peak of the event. Abscissa shows the time with respect to the onset of the event as marked by a specialist (from −1 to 1 s); ordinate shows the magnitude of EEG different signal. (**c**) The EEG scalp topographies at the peak of the event (0.83 s after the marked onset). From *left* to *right*, the topographies are shown for the measured 64-channel EEG (*left*), predicted 64-channel EEG (*middle*, predicted from the 21-channel subset) and measured 21-channel EEG (*right*, subset of 64 channels). On the 64-channel topographies the subset of the used 21 channels is highlighted

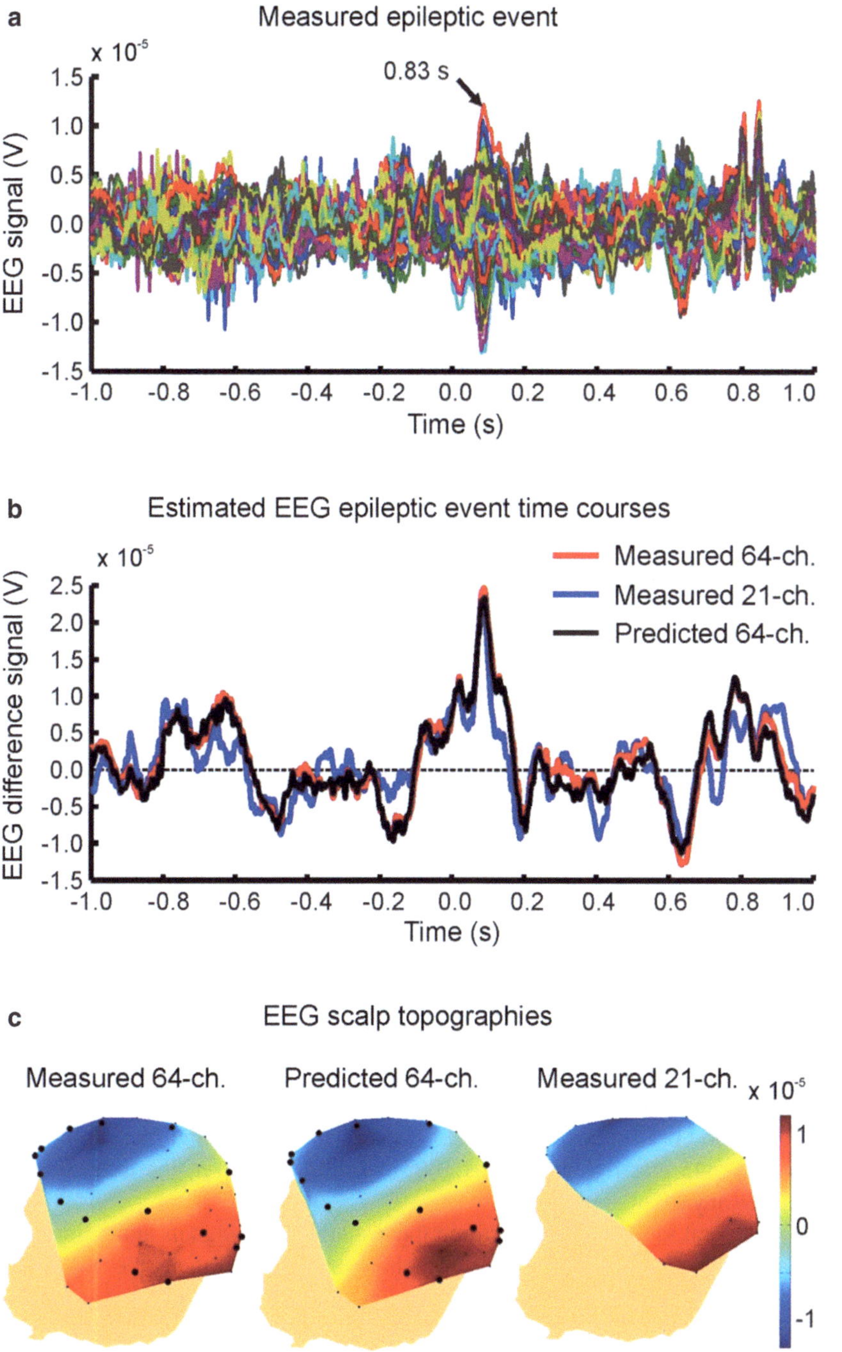

Fig. 13.4 Higher density EEG signals (64-channel) predicted from lower density measurements (21-channel) during an epileptic event, using the lead field-based method. (**a**) The measured epileptic event with signals from all 64 EEG channels overplotted. Abscissa shows the time with respect to the onset of the event as marked by a specialist (from −1 to 1 s); ordinate shows the EEG signal magnitude. (**b**) The time courses of the epileptic event estimated from the measured

For each event/epoch, we then used 21 of the 64 EEG electrodes corresponding to the standard 10–20 system to predict EEG signals at the rest of the 43 electrodes. The method was evaluated by comparing the predicted and actually recorded 64-channel EEG data. We used the recorded 64-channel EEG data as 'gold standard' to assess the time courses and EEG scalp topographies obtained from the predicted 64-channel and recorded 21-channel EEG data. These comparisons demonstrate the added value of using predicted higher density EEG signals instead of simply the recorded low-density signals for the investigation of epileptic events.

To compare the time courses of the epileptic events in the recorded and predicted EEG data, we computed the difference between the signals with the strongest positive and negative deflections at the peak of the event; the latencies of the peaks were determined from the global field power [29] time courses computed for each extracted two-second long epoch. The results showed that the predicted 64-channel EEG data represented the time course of the event very accurately, and substantially better than the 21-channel recorded data (Fig. 13.4b). The scalp topographies of recorded and predicted 64-channel EEG signals at the peaks of the events were also very similar (Fig. 13.4c, correlation coefficient >0.8). For most of the peaks the electrodes with the strongest negative and positive signals were the same in recorded and predicted 64-channel EEG data, while they were usually different (several cm away) in 64- and 21-channel recorded EEG data.

It is obvious that the prediction of unmeasured signals from the measurements within the same modality (e.g. predicting of high- from low-resolution EEG) may be accomplished by many different interpolation methods. However, the methods that we have presented perform significantly better than standard interpolation methods, because they take into account the properties of intracranial currents, and geometrical and conductivity details of the biological medium. In essence, these methods estimate as closely as possible what would have been measured with the target sensors.

13.2.1.2 Selecting the Most Informative Electrode Positions

In this section we describe a method for selecting few electrode positions from a larger set of EEG signals (measured or predicted) that provide maximal information about a specific brain event. This is the second step of a procedure aimed at determining the case-specific optimal positions for scalp electrodes. Typically, the input to this method is a set of predicted EEG signals for an ultra high-density electrode array (e.g. 345 electrodes of 'EEG 1005' system). However, the method can be applied to any set of signals.

The general idea here is to partition the available EEG signals at the time of the relevant event into clusters based on their magnitude and then select from each cluster the electrode with the strongest absolute signal. Specifically, we use K-means clustering algorithm, with Euclidean distance as a similarity measure, to partition the data and the Silhouette method to evaluate the optimal number of clusters (K-means and Silhouette methods are repeatedly applied with increasing number of clusters, and the number with the largest mean Silhouette value across all clustered electrodes is selected as the optimal number of clusters). In principle, other types of clustering algorithms (e.g. hierarchical clustering etc.) or cluster evaluation criteria (e.g. Davies-Bouldin Index etc.) may also be applied.

In practice, we usually have predicted (or measured) hdEEG signals with a number of epileptic events (e.g. interictal spikes) marked by specialists. We then apply the method to each marked event or to a summary of the same marker events, such as their average or pattern-based signal-to-noise ratio (SNR, [30]). In the cases when the method is applied to individual marked events, most commonly identified electrode positions are selected as the optimal set. This application of the method has an added advantage of validating the marked events: if very different sets of optimal electrode positions are identified for some of the same marker events, then the marking must be reviewed and if necessary some of the events must be analysed in more detail, e.g. by way of detailed tomographic analysis. However, typically the method is applied to the averaged signals, which provides one set of optimal electrode positions that can be used to record maximally informative signals about the marked event.

For the brain events with changing EEG topography, possibly indicating activity propagation through the cortex, the method may be applied at several time points, identifying potentially a larger set of optimal electrode positions, which will provide more comprehensive information about the event. In certain cases, additional improvement to the results can be attained by applying the method to the preselected subset of EEG signals that are directly related to the event: a strong signal at the time of the event does not necessarily indicate that it is directly related to the event; the signal at that electrode may be strong also at other times, unrelated to the event, especially when the method is applied to individual events rather than their summary. There is large number of statistical methods that can be used to pre-select the event-related signals. In the simplest case one can identify a baseline window from the recorded EEG data devoid of any large signal components, compute the mean and standard deviation of each electrode's signal across that window, and then pre-select EEG signals that at the time of the event vary at least two (or more) standard deviations from the mean.

The method with and without pre-selecting the electrodes was successfully applied to simulated and real patient data with 21 and 64 EEG electrodes as well as hdEEG (345 electrodes of 'EEG 1005' system) signals predicted from lower density measurements. In the case of the data shown in Fig. 13.4, the method was applied to the *predicted* 64-channel EEG signals at the peak of the event with and without pre-selection. When signals from all 64 electrodes were used (no pre-selecting) the Silhouette method evaluated the optimal number of clusters (and hence electrodes) to be three, with a mean Silhouette value of 0.79 (Fig. 13.5a).

We then applied a simple algorithm to pre-select the electrodes whose signals were most related to the identified epileptic event: we defined the one second period before the onset of the event as baseline and then selected only the signals that at the peak of the event deflected at least two standard deviations from the mean of the baseline. The algorithm selected 35 event-related electrodes and the optimal number of clusters for these electrodes was evaluated to be two, with a mean Silhouette value of 0.95 (Fig. 13.5b). Thus, in this case, two electrodes appear to be sufficient for monitoring the marked epileptic event and can provide useful information with minimal obtrusiveness. To obtain maximum useful information with these two electrodes they must be placed

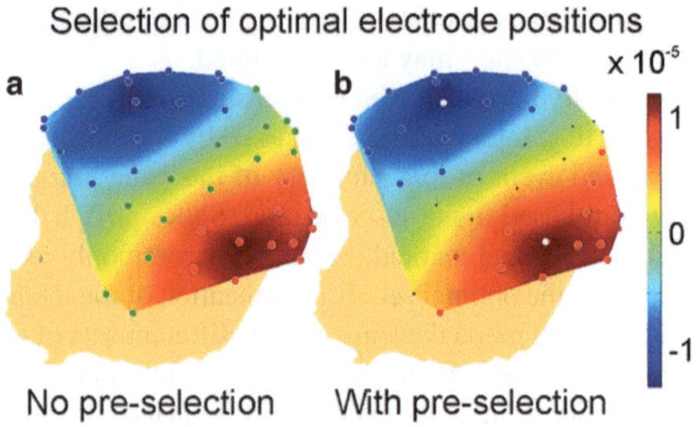

Fig. 13.5 Selection of optimal electrode positions. (**a**) The three clusters of electrodes identified using all 64 signals (no pre-selecting). The electrodes within different cluster are highlighted in *blue*, *red* and *green*. (**b**) The two clusters of electrodes identified after pre-selecting a subset of event-related signals (35 signals were pre-selected). The electrodes within the two clusters are highlighted in *blue* and *red*. The electrodes with the strongest signal (positive or negative) in each cluster are highlighted in *white*; these electrode positions are selected as the most informative for investigating the relevant epileptic event

at the optimal positions indicated by the electrodes with the strongest absolute signal in each identified cluster (highlighted in white in the Fig. 13.5b).

In general, the results obtained from applying this methodology to different kinds of epileptic data showed that in many cases two or three optimally placed electrodes provide sufficient information about an underlying specific brain event.

13.3 Summary

We have briefly introduced the general idea of a novel approach, which we have developed for ARMOR with a key aim to substantially improve the personalized management of epilepsy and other disorder. We have also presented a series of methods, as practical applications of the approach that can be applied offline to MEG or EEG measurements and lead to an optimized set of electrode locations for the routine ARMOR home monitoring.

References

1. Sutherling WW, Mamelak AN, Thyerlei D, Maleeva T, Minazad Y, Philpott L, Lopez N (2008) Influence of magnetic source imaging for planning intracranial EEG in epilepsy. Neurology 71:990–996
2. Knowlton RC, Elgavish RA, Limdi N, Bartolucci A, Ojha B, Blount J, Burneo JG, Ver HL, Paige L, Faught E, Kankirawatana P, Riley K, Kuzniecky R (2008) Functional imaging: I. Relative predictive value of intracranial electroencephalography. Ann Neurol 64:25–34

3. Colon AJ, Ossenblok P, Nieuwenhuis L, Stam KJ, Boon P (2009) Use of routine MEG in the primary diagnostic process of epilepsy. J Clin Neurophysiol 26:326–332
4. Heers M, Rampp S, Kaltenhauser M, Pauli E, Rauch C, Dolken MT, Stefan H (2010) Detection of epileptic spikes by magnetoencephalography and electroencephalography after sleep deprivation. Seizure 19:397–403
5. Bagic A, Funke ME, Ebersole J (2009) American Clinical MEG Society (ACMEGS) position statement: the value of magnetoencephalography (MEG)/magnetic source imaging (MSI) in noninvasive presurgical evaluation of patients with medically intractable localization-related epilepsy. J Clin Neurophysiol 26:290–293
6. Hillebrand A, Barnes GR (2002) A quantitative assessment of the sensitivity of whole-head MEG to activity in the adult human cortex. Neuroimage 16:638–650
7. Ioannides AA (1995) Estimates of 3D brain activity millisecond by millisecond from biomagnetic signals: Method (MFT), results and their significance. In: Eiselt E, Zwiener U, Witte H (eds) Quantitative and Topological EEG and MEG Analysis. Universitatsverlag Druchhaus-Maayer GmbH, Jena, Germany, pp 59–68
8. Tanaka N, Stufflebeam SM (2014) Clinical application of spatiotemporal distributed source analysis in presurgical evaluation of epilepsy. Front Hum Neurosci 8:62
9. Shiraishi H, Ahlfors SP, Stufflebeam SM, Takano K, Okajima M, Knake S, Hatanaka K, Kohsaka S, Saitoh S, Dale AM, Halgren E (2005) Application of magnetoencephalography in epilepsy patients with widespread spike or slow-wave activity. Epilepsia 46:1264–1272
10. Tanaka N, Cole AJ, von Pechmann D, Wakeman DG, Hamalainen MS, Liu H, Madsen JR, Bourgeois BF, Stufflebeam SM (2009) Dynamic statistical parametric mapping for analyzing ictal magnetoencephalographic spikes in patients with intractable frontal lobe epilepsy. Epilepsy Res 85:279–286
11. Kanamori Y, Shigeto H, Hironaga N, Hagiwara K, Uehara T, Chatani H, Sakata A, Hashiguchi K, Morioka T, Tobimatsu S, Kira J (2013) Minimum norm estimates in MEG can delineate the onset of interictal epileptic discharges: a comparison with ECoG findings. Neuroimage Clin 2:663–669
12. Tanaka N, Hamalainen MS, Ahlfors SP, Liu H, Madsen JR, Bourgeois BF, Lee JW, Dworetzky BA, Belliveau JW, Stufflebeam SM (2010) Propagation of epileptic spikes reconstructed from spatiotemporal magnetoencephalographic and electroencephalographic source analysis. Neuroimage 50:217–222
13. Ioannides AA, Bolton JPR, Clarke CJS (1990) Continuous probabilistic solutions to the biomagnetic inverse problem. Inverse Probl 6:523–542
14. Taylor JG, Ioannides AA, Mueller-Gaertner HW (1999) Mathematical analysis of lead field expansions. IEEE Trans Med Imaging 18:151–163
15. Moradi F, Liu LC, Cheng K, Waggoner RA, Tanaka K, Ioannides AA (2003) Consistent and precise localization of brain activity in human primary visual cortex by MEG and fMRI. Neuroimage 18:595–609
16. Ioannides AA, Fenwick PBC, Liu LC (2005) Widely distributed magnetoencephalography spikes related to the planning and execution of human saccades. J Neurosci 25:7950–7967
17. Poghosyan V, Ioannides AA (2008) Attention modulates earliest responses in the primary auditory and visual cortices. Neuron 58:802–813
18. Poghosyan V, Ioannides AA (2007) Precise mapping of early visual responses in space and time. Neuroimage 35:759–770
19. Papadelis C, Eickhoff SB, Zilles K, Ioannides AA (2011) BA3b and BA1 activate in a serial fashion after median nerve stimulation: direct evidence from combining source analysis of evoked fields and cytoarchitectonic probabilistic maps. Neuroimage 54:60–73
20. Papadelis C, Poghosyan V, Fenwick PBC, Ioannides AA (2009) MEG's ability to localise accurately weak transient neural sources. Clin Neurophysiol 120:1958–1970
21. Ribary U, Ioannides AA, Singh KD, Hasson R, Bolton JPR, Lado F, Mogilner A, Llinas R (1991) Magnetic field tomography of coherent thalamocortical 40-Hz oscillations in humans. Proc Natl Acad Sci U S A 88:11037–11041
22. Ioannides AA, Poghosyan V (2012) Spatiotemporal dynamics of early spatial and category-specific attentional modulations. Neuroimage 60:1638–1651

23. Ioannides AA, Poghosyan V, Dammers J, Streit M (2004) Real-time neural activity and connectivity in healthy individuals and schizophrenia patients. Neuroimage 23:473–482
24. Ioannides AA, Corsi-Cabrera M, Fenwick PBC, Del Rio PY, Laskaris NA, Khurshudyan A, Theofilou D, Shibata T, Uchida S, Nakabayashi T, Kostopoulos GK (2004) MEG tomography of human cortex and brainstem activity in waking and REM sleep saccades. Cereb Cortex 14:56–72
25. Ioannides AA, Fenwick PBC (2005) Imaging cerebellum activity in real time with magnetoencephalographic data. Prog Brain Res 148:139–150
26. Knosche TR (2002) Transformation of whole-head MEG recordings between different sensor positions. Biomed Tech (Berl) 47:59–62
27. Malmivuo J, Plonsey R (1995) Bioelectromagnetism—principles and applications of bioelectric and biomagnetic fields. Oxford University Press, New York
28. Ioannides AA, Kostopoulos GK, Liu LC, Fenwick PBC (2009) MEG identifies dorsal medial brain activations during sleep. Neuroimage 44:455–468
29. Lehmann D, Skrandies W (1980) Reference-free identification of components of checkerboard-evoked multichannel potential fields. Electroencephalogr Clin Neurophysiol 48:609–621
30. Laskaris NA, Ioannides AA (2001) Exploratory data analysis of evoked response single trials based on minimal spanning tree. Clin Neurophysiol 112:698–712

Chapter 14
DSMS and Online Algorithms

Vasileios Megalooikonomou, Dimitrios Triantafyllopoulos, Evangelia I. Zacharaki, and Iosif Mporas

Abstract Online (real-time) analysis of medical data is crucial for automatic or semi-automatic monitoring of patients. Technologies involved in online analysis are data streaming, online data management and online data processing. In this chapter we present algorithmic methodologies and implementations for online (real-time) analysis of medical data, i.e. streaming data management and online detection of events of interest.

14.1 Online Analysis of Medical Data

Methods for the online analysis of the above modalities require the identification of features and measures that can be extracted and calculated fast. The detection of seizures at the earliest observable onset of the ictal patterns is one of the main purposes of online analysis and can be used to start more detailed diagnostic procedures during seizures and to differentiate seizures from other related disorders with seizure like symptoms. Several methods have been proposed for this purpose that can be classified depending on the modality of the recorded data (intracranial EEG, scalp EEG and ECG).

In this chapter we present online methodologies for real-time seizure detection from EEG and ECG signals. Moreover we present the implementation of online alpha rhythm detection. Finally, details about the DSMS implementation of the online tools are presented.

V. Megalooikonomou (✉) • D. Triantafyllopoulos • E.I. Zacharaki • I. Mporas
Multidimensional Data Analysis and Knowledge Discovery Laboratory,
Department of Computer Engineering and Informatics,
University of Patras, Patras, Greece
e-mail: vasilis@ceid.upatras.gr

N.S. Voros, C.P. Antonopoulos (eds.), *Cyberphysical Systems for Epilepsy and Related Brain Disorders*, DOI 10.1007/978-3-319-20049-1_14

14.2 Online Analysis Algorithms

In online analysis of medical data the challenge is the development of tools which operate robust enough as well as in real time. Typically, there is a trade-off between performance accuracy and computational complexity of the algorithm. Thus, the aim is the configuration of online methodologies in the correct operational point. In this section we present the algorithm architecture, signal processing and classification setup performed for seizure detection within the ARMOR project, as well as the implementation of it on the Stream-In-Sight (DSMS) framework.

We developed [1] a seizure detector using EEG and ECG signals based on short-time analysis with time-domain and frequency-domain features and classification using support vector machines. We evaluated a large-scale set of time-domain and frequency-domain EEG and ECG features for seizure detection, which are popular in the literature for brain and heart statistical signal processing respectively. Furthermore, we investigated the effect on the detector's performance when using subsets of these features, with respect to a feature ranking evaluation, in order to develop online (real-time) and offline versions of it. Feature ranking investigation and evaluation of the seizure detector using subsets of features showed that the feature vector composed of approximately the 800-best ranked features provides a good trade-off between computational demands and accuracy. The block diagram of the seizure detection architecture adopted in ARMOR is illustrated in Fig. 14.1.

The data captured from the $N+1$ sensors (where N are the EEG electrodes plus one ECG channel) have been synchronized and transmitted as streams of multidimensional signals. Thus, the input to the illustrated in Fig. 14.1 architecture consists of time-synchronous streams of EEG and ECG signal samples. As shown in Fig. 14.1, in a first step the EEG, $x_{EEG} \in \mathbb{R}^N$, and ECG, $x_{ECG} \in \mathbb{R}$, signals are pre-processed. Pre-processing consists of frame blocking of the incoming streams to epochs of constant length w with constant time-shift s. Each epoch is a $(N+1)\times(w)$ matrix, where N is the number of EEG electrodes and $N+1$ is the N-dimensional EEG signal appended by the ECG signal.

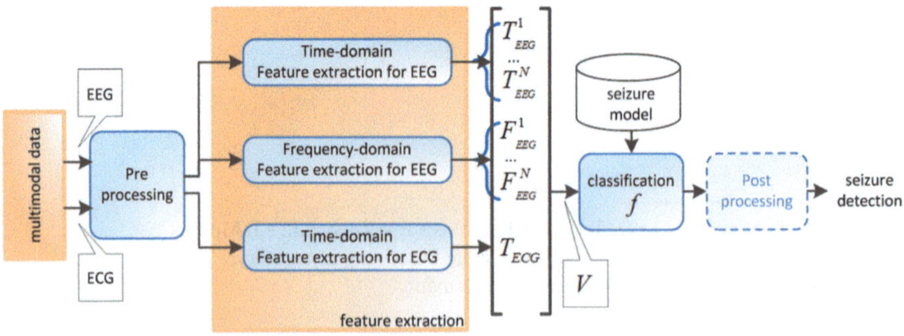

Fig. 14.1 Block diagram of the EEG-ECG based online seizure detection architecture

After pre-processing, the extracted epochs are in parallel processed by time-domain and frequency-domain feature extraction algorithms separately for the N-dimensional EEG and the 1-dimensional ECG signals. In particular, each of the N-dimensions of the EEG signal are processed by time-domain and frequency-domain feature extraction algorithms for EEG, while the ECG signal is processed by time-domain feature extraction algorithms (based on heart rate estimation) dedicated for electrocardiogram, as shown in the block diagram of Fig. 14.1. The extracted time-domain and frequency-domain features for the EEG, $T^i_{EEG} \in \mathbb{R}^{|T_{EEG}|}$ and $F^i_{EEG} \in \mathbb{R}^{|F_{EEG}|}$, with $1 \le i \le N$, and the ECG signal, $T_{ECG} \in \mathbb{R}^{|T_{ECG}|}$, are afterwards concatenated to a single feature vector $V \in \mathbb{R}^{N \cdot (|T_{EEG}|+|F_{EEG}|)+|T_{ECG}|}$ representing each epoch, as shown in Fig. 14.1. The extracted sequences of feature vectors, V, are short-time parametric representations of the EEG and ECG signals representing the time and spectral characteristics of the multimodal signals. This sequence of feature vectors is afterwards used as input to a classification model in order to assign a class label (seizure class or non-seizure class) to each of the vectors, i.e. to the corresponding time-intervals of the vectors.

During the training phase of the seizure detector, a dataset of feature vectors with known class labels (labeled manually by medical experts) is used to train a binary model M (two classes: seizure vs. non-seizure) using a classification algorithm f. At the test phase the existing seizure model, M, is used in order to decide for each epoch's feature vector, V, the corresponding class using the same classification algorithm, f, as in the training phase. Thus, for each epoch i a binary label d_i, i.e. seizure or not, is decided as:

$$d_i = f\left(V_i, M\right) \tag{1}$$

and the sequence of incoming EEG-ECG data is decomposed to time-intervals of seizure or clear (non-seizure) recordings. Post-processing of the automatically detected labels can be performed for improving the performance of the architecture.

During pre-processing the time-synchronized EEG and ECG recordings were frame blocked to epochs of 1-s length, without time-overlap between successive epochs. For each epoch, time-domain and frequency domain features were extracted separately for the each of the 21 EEG channels and the ECG channel.

In particular, each of the EEG channels was parameterized using the following features: time-domain features: minimum value, maximum value, mean, variance, standard deviation, percentiles (25 %, 50 %-median and 75 %), interquartile range, mean absolute deviation, range, skewness, kurtosis, energy, Shannon's entropy, logarithmic energy entropy, number of positive and negative peaks, zero-crossing rate, and frequency-domain features: Sixth order autoregressive-filter (AR) coefficients, power spectral density, frequency with maximum and minimum amplitude, spectral entropy, delta-theta-alpha-beta-gamma band energy, discrete wavelet transform coefficients with mother wavelet function Daubechies 16 and decomposition level equal to eight, thus resulting to a feature vector of dimensionality equal to 55 for each of the 21 EEG channels, i.e. 1155 in total.

The ECG channel was parameterized using the following features: the heart rate absolute value and variability statistics of the heart rate, i.e. minimum value, maximum value, mean, variance, standard deviation, percentiles (25 %, 50 %-median and 75 %), interquartile range, mean absolute deviation, range, thus resulting to a feature vector of dimensionality equal to 12. The heart rate estimation was based on Shannon energy envelope estimation for R-peak detection algorithm, implemented as in [2]. The dimensionality of the overall feature vector V is $1155 + 12 = 1167$.

The computed feature vectors V, one for each EEG-ECG epoch were used to train binary seizure detection models, M. Specifically, we compared the support vector machines (SVMs), implemented with the sequential minimal optimization method and polynomial kernel function, a two-layered back propagation multilayer perceptron (MLP) neural network, the k-nearest neighbour (IBK) algorithm and the C4.5 decision tree. All online seizure detection models were implemented using the WEKA machine learning toolkit software [3].

During the test phase, the EEG and ECG recordings are pre-processed and parameterized as in training. The SVM seizure detection model, M, is used to label each of the incoming EEG-ECG epochs as seizure or clear (non-seizure). In the present evaluation, no post-processing algorithm was applied on the estimated epoch-based results. The seizure detection results with respect to the use of N-best features after feature ranking are illustrated in the following figure (Fig. 14.2).

As can be seen in the evaluation results, for the three evaluated subjects the use of subset of features reduces the precision of the seizure detector. However, the exclusion of the approximately 30 % worst features still offers performance comparable to the best achieved and in combination with the reduction of the computational load of the detection architecture (both in the feature extraction stage and the classification stage) is a valuable solution for the online version.

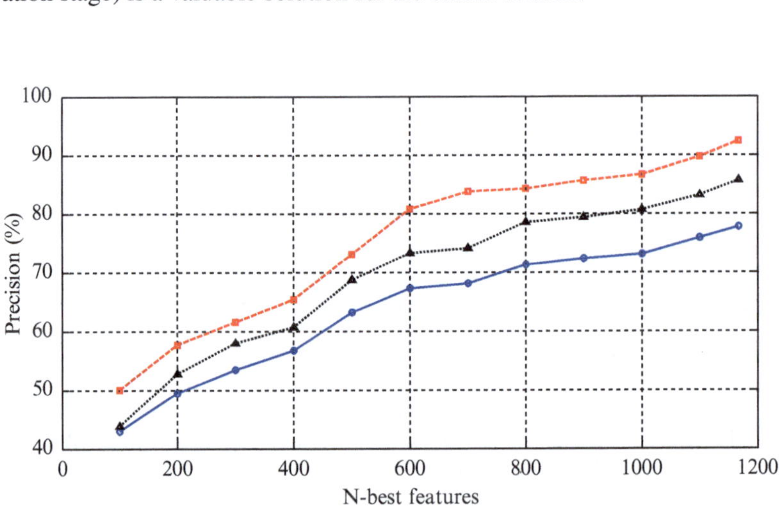

Fig. 14.2 Seizure detection precision (in percentages) for different number of features (N-best) for three subjects

Except online seizure detection, within the ARMOR framework an online alpha rhythm detection tool was developed. The online alpha rhythm detection is a simple algorithmic implementation designed to test the end-to-end system's functionality and show in a relatively easy way it's operation. It is designed in a similar way with the seizure detector, i.e. there is an incoming stream signal which is processed by a signal processing methodology through an online procedure that extracts a specific event (in this case alpha rhythms).

The algorithm estimates if the ratio of the energy in the alpha frequency band to the total signal energy exceeds a previously specified threshold, estimated either from offline analysis or empirically determined. The streamed data are also processed in windowed segments. For each segment, the detector returns a flag, true or false, regarding the existence or not of alpha rhythm. An alarm is sent if three consecutive epochs are detected with big alpha components, in order to reduce the system's alarm load.

14.3 DSMS Implementation

One way of performing an online analysis of sensor data is a multi-layer Data Mining Model. These models are divided into four layers: Data collection, Data management, Event processing layer, Data mining service layer. The ARMOR Online platform consists of a similarly structured system whose components are developed by several partners. The data collection layer is performed from the sensors and using the xAffect tool is streamed to a StreamInsight application [4]. The streamed data are processed with a previously specified and parameterised algorithm, before the detected events are extracted (Fig. 14.3).

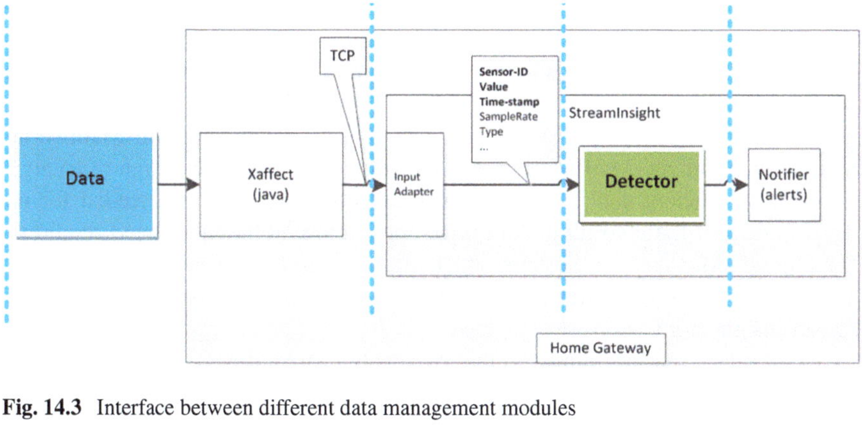

Fig. 14.3 Interface between different data management modules

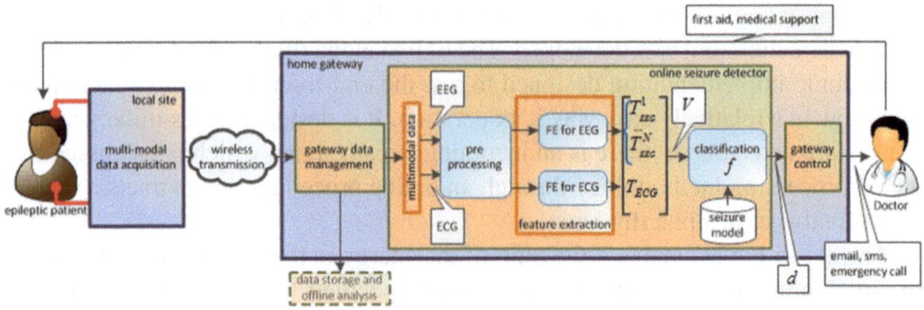

Fig. 14.4 Block diagram of the online seizure detection module

The seizure detection module includes the EEG and ECG recordings, while EMG and EOG, which are mainly used for movement/artifact detection, were not integrated. EMG and EOG were not included in the online seizure detection architecture in order to avoid further increasing the feature vector dimensionality and since the literature review showed limited use of these modalities with no significant advantages in online detection. The block diagram of the seizure detection architecture is illustrated above (Fig. 14.4).

The captured multimodal data (EEG and ECG) are wirelessly transmitted by a wearable solution to a local gateway for online processing. During online seizure detection, the EEG and ECG signals are initially pre-processed.

Real time applications are characterized by a limited amount of time available to process the data stream. The real time processing interval is the time necessary to process each data stream frame which corresponds to 1 s of data. If this interval is bigger than 1 s then the system will eventually reach memory limits and the processing of additional events will be delayed or postponed. Thus for each data block of 1 s the processing time should be less than a second. In our study we tested the performance of the detection algorithm for several numbers of EEG electrodes. The data stream was framed with non overlapping windows of 1 s. In Table 14.1 the average real time processing interval for each set of electrodes is shown. Each experiment was performed using the depicted number of electrodes in addition to the ECG signal.

The core of each of our detection algorithms is implemented in Mathworks' Matlab environment. Matlab offers many features that make the design of an algorithm more efficient and effective as well as shortening the time needed for the development of the algorithm. An online application on the other hand requires a Data Stream Management System, which offers the ability to process online data in a time and memory efficient way. The system used in our applications is StreamInsight from Microsoft.

To introduce our algorithms to the Microsoft's StreamInsight environment, the .Net compiler and API of Mathworks are used. In more detail, Mathworks provide a compiler for .Net packages. The result of this procedure is a set of libraries

Table 14.1 Real time processing for several numbers of EEG electrodes

Number of electrodes	Real time processing
19	0.977
17	0.874
15	0.776
13	0.681
11	0.588

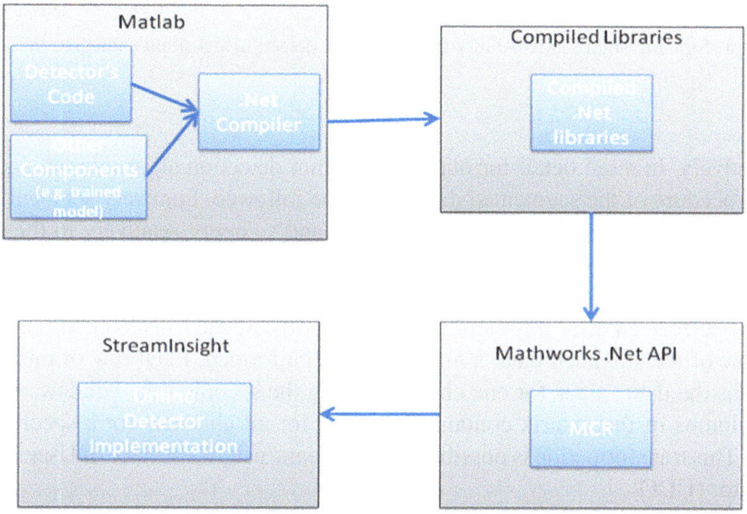

Fig. 14.5 Development scheme for using Matlab implementations of detectors in Microsoft StreamInsight

containing the algorithms and all the necessary components (functions, data such as training models etc.) necessary for each algorithm.

Those compiled libraries can be accessed by a StreamInsight application using the API that Mathworks provides. This API is the Matlab Compiler Runtime, in our case the 8.1 version 32 bit, which provides all the necessary components in order to use the algorithm in a stream application.

The figure below shows the pipeline followed when an online algorithm, designed in Matlab, is introduced in a StreamInsight application (Fig. 14.5).

It should be noted that each algorithm was designed to operate with segments of the data, as is necessary in an online application. In order to pre-process the streamed data, before having them processed with the detection algorithm, a set of StreamInsight tools were used. Depending on the procedure followed by the detection algorithm the streamed data should be aggregated or transformed in matrices whose rows and columns correspond to channels and samples of each segment,

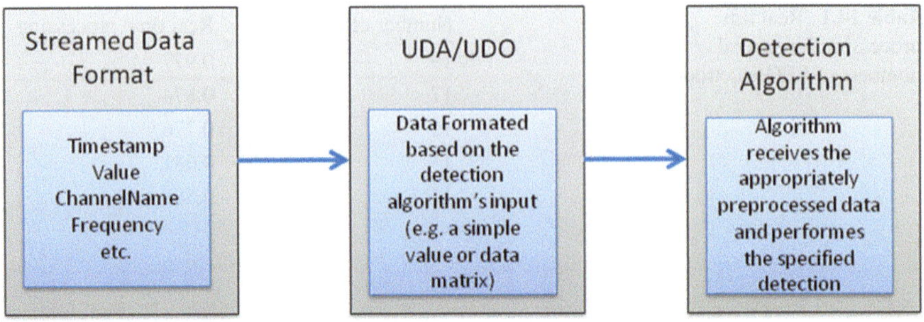

Fig. 14.6 StreamInsight's extensibility framework for detection algorithm's integration

respectively. In more detail for our alpha rhythm detection algorithm, an aggregation procedure of the segmented data should be followed. For every segment of the streamed data, the percentage of the Alpha band's energy relatively to the whole energy of the signal is returned. This is performed by using a User Defined Aggregate (UDA), a service provided by the StreamInsight framework. The detection procedure is performed after this step. In the case of the seizure detector for each time window of the streamed data, a matrix should be formed. Each row of this matrix contains the data values for one channel during the specific time window, whereas the columns of this matrix contain the values for all channels for a specific time point. This transformation is possible due to StreamInsight's service of User Defined Operator (UDO).

The figure above shows the procedure followed in order to adapt the streamed data to the input format of a detection algorithm (Fig. 14.6).

14.4 Conclusions

We presented algorithmic methodologies for online analysis of medical data. These methodologies serve as tools for online analysis of streaming data operated by a DSMS system. Both the online (real-time) processing of the medical data and the management of them are crucial when monitoring patients, and especially for the case of chronic diseases such as epilepsy. In contrast to offline analysis where performance is the key, when designing and implementing online tools there is a trade-off between tool's performance and computational complexity. Thus, fine-tuning of the tools in order to meet real-time processing demands with acceptable performance, i.e. accuracy in online detection of events of interest, is essential.

References

1. Mporas I, Tsirka V, Zacharaki EI, Koutroumanidis M, Megalooikonomou V (2014) Online seizure detection from EEG and ECG signals for monitoring of epileptic patients. In: 8th Hellenic conference on artificial intelligence (SETN 2014), Ioannina
2. Sabarimalai MM, Soman KP (2012) A novel method for detecting R-peaks in electrocardiogram (ECG) signal. Biomed Signal Process Control 7(2):118–128
3. Witten HI, Frank E (2005) Data mining: practical machine learning tools and techniques. Morgan Kaufmann, San Francisco
4. Ali M et al (2009) Microsoft CEP server and online behavioral targeting. In: VLDB 2009 (demonstration)

MIX
Papier aus verantwortungsvollen Quellen
Paper from responsible sources
FSC® C105338

If you have any concerns about our products,
you can contact us on
ProductSafety@springernature.com

In case Publisher is established outside the EU,
the EU authorized representative is:
Springer Nature Customer Service Center GmbH
Europaplatz 3, 69115 Heidelberg, Germany

Printed by Libri Plureos GmbH
in Hamburg, Germany